Radar Technology for Coastal Areas and Open Sea Monitoring

Radar Technology for Coastal Areas and Open Sea Monitoring

Editors

Giovanni Ludeno
Marco Uttieri

MDPI • Basel • Beijing • Wuhan • Barcelona • Belgrade • Manchester • Tokyo • Cluj • Tianjin

Editors
Giovanni Ludeno
Institute for Electromagnetic Sensing of the Environment (IREA)—National Research Council of Italy (CNR)
Italy

Marco Uttieri
Department of Integrative Marine Ecology, Stazione Zoologica Anton Dohrn
Italy

Editorial Office
MDPI
St. Alban-Anlage 66
4052 Basel, Switzerland

This is a reprint of articles from the Special Issue published online in the open access journal *Journal of Marine Science and Engineering* (ISSN 2077-1312) (available at: https://www.mdpi.com/journal/jmse/special_issues/radar_technology).

For citation purposes, cite each article independently as indicated on the article page online and as indicated below:

LastName, A.A.; LastName, B.B.; LastName, C.C. Article Title. *Journal Name* **Year**, *Article Number*, Page Range.

ISBN 978-3-03936-972-0 (Hbk)
ISBN 978-3-03936-973-7 (PDF)

Cover image courtesy of Giovanni Ludeno.

Contents

About the Editors

Giovanni Ludeno (Ph.D.) is a researcher at the Institute for Electromagnetic Sensing of the Environment (IREA), National Research Council of Italy (CNR) of Naples, Italy. His research interests concern the field of applied electromagnetism, with special regard to remote and in situ sensing, and include electromagnetic diagnostics, inverse scattering and ground penetrating radar (GPR) surveys. Specifically, his research activities deal with: a) the development and assessment of innovative strategies for the estimation of sea state parameters, such as surface currents and bathymetry, from high-resolution nautical X-band radar data; b) design and validation of strategies for target imaging from data collected by means of radar system collected by different observation platform, and material characterization by using THz systems. He is participating/has participated in European and Italian research projects. Giovanni Ludeno has authored and co-authored more than 60 papers, mainly in scientific journals, proceedings of international conferences and books. He is a reviewer for several international journals.

Marco Uttieri (Ph.D.) is a researcher at the Department of Integrative Marine Ecology—Stazione Zoologica Anton Dohrn (Naples, Italy). His research interests include the study of surface circulation and wave motion through high frequency (HF) coastal radars, in situ and remote sensing tools, and numerical models. In addition, his research focuses on the biology and ecology of zooplanktonic organisms from both marine and freshwater environments. He is the author of >40 contributions indexed in Scopus, and serves as an Editorial Board Member and Reviewer for numerous journals.

Preface to "Radar Technology for Coastal Areas and Open Sea Monitoring"

Monitoring oceans and coastal areas has a fundamental social impact, and this scenario is made still more challenging with the present and future issues related to climate change. In this context, radar systems have gained increasing interest, since they are remote sensing devices capable of providing information about sea waves, currents, tides, bathymetry, and wind. Moreover, radar systems can be designed to perform both large-scale and small-scale monitoring, with different spatial and temporal resolutions, and can be installed on different observation platforms (ship-based, ground-based, airborne, satellite or drones). In this regard, this book aims at engendering a virtual forum for ocean radar researchers, where state-of-the-art methodologies and applications concerning ocean monitoring by means of radar technologies are reviewed and discussed.

Giovanni Ludeno , Marco Uttieri
Editors

Journal of
Marine Science and Engineering

MDPI

Editorial

Editorial for Special Issue "Radar Technology for Coastal Areas and Open Sea Monitoring"

Giovanni Ludeno [1,*] and Marco Uttieri [2,3,*]

1 Institute for Electromagnetic Sensing of the Environment (IREA), National Research Council of Italy (CNR), 80124 Naples, Italy
2 Department of Integrative Marine Ecology, Stazione Zoologica Anton Dohrn, 80121 Naples, Italy
3 CoNISMa, ULR Parthenope, Piazzale Flaminio 9, 00196 Rome, Italy
* Correspondence: ludeno.g@irea.cnr.it (G.L.); marco.uttieri@szn.it (M.U.)

Received: 19 July 2020; Accepted: 21 July 2020; Published: 25 July 2020

Keywords: remote sensing; HF radar; X-band radar; synthetic aperture radar; radar altimeter; sea surface current; significant wave height; sea state monitoring

---

1. Overview

The sea has always played a fundamental role in the social and economic development, as well as in the shaping and functioning of natural ecosystems and services. This has led to growing attention towards the sea and its behaviour by the scientific community. In order to characterise coastal and open-ocean dynamics at the proper spatial and temporal scales, over the past few years the development of marine observatories has been spurred and promoted (e.g., [1,2]). These observation infrastructures can be composed of different platforms, each providing specific information on one or more parameters (either physical, hydrological, biological or chemical), which can then be used to reconstruct the status of the system.

Within this framework, radar systems represent a useful, non-invasive technology for retrieving sea state information. The radar systems employed for the remote sensing of the ocean surface refer to three main typologies: microwave radar, microwave radiometers and high-frequency (HF, or decametric) radar [3]. In recent decades, radar oceanography has fostered the comprehension of sea surface dynamics, including the measurement of ocean currents, waves and wind, but also target detection and operational activities (e.g., oil spill monitoring, search and rescue operations). Reviews on the different types of radars and their applications are available in the literature, providing specific details and insight [4–6].

In this regard, this Special Issue (SI) on "Radar Technology for Coastal Areas and Open Sea Monitoring", published in the *Journal of Marine Science and Engineering*, aims at engendering a virtual forum for ocean radar researchers, where state-of-the-art methodologies and applications concerning ocean monitoring by means of radar technologies are reviewed and discussed. In particular, for this SI we invited papers considering various topics, including ocean dynamics, open sea and coastal area monitoring, sea safety and protection, signal processing, and physical–biological interactions. After a rigorous peer review process, we accepted 13 papers, which cover a wide range of topics related to sea state monitoring in both coastal areas and the open sea.

A general classification of the contributions included in the present SI may be provided on the basis of the adopted remote sensing monitoring systems. Specifically, some of the presented studies deal with analyses of data acquired from radar systems installed in coastal areas, such as HF radar [7–14] and X-band nautical radar systems [15,16], while others were mounted on satellite platforms, such as Synthetic Aperture Radar (SAR) [17,18] and Radar Altimeters (RAs) [19]. An overview of the contributions published is provided in the following section, showing the wide range of applications and methodologies covered in the SI.

2. Contributions

A great proportion of the contributions included in the SI refer to HF radar applications. Since the 1970s, HF radars have increasingly demonstrated their ability to efficiently resolve surface currents and wave fields over extended coastal basins. Three lines of investigation can be identified in the presented SI papers: characterisation of surface current fields; HF-derived surface wave measurements; baseline research on signal processing and analysis.

Surface current dynamics are investigated in three papers. The contribution by [7] reports on the long-term monitoring of the Strait of Gibraltar and the associated Atlantic inflow in the Mediterranean Sea. The authors show seasonal patterns in surface structures, but also reversal episodes, unveiling relationships with atmospheric and oceanic forcings, as well as with tides. The authors in [8] provide a new application for soft-computing techniques (random forest, RF) to forecast surface current fields in Galway Bay (Ireland). In this study, HF radar-derived fields, coupled with numerical outputs, are used as inputs for the RF model. The forecasted fields show a high consistency with the HF ones, pointing at promising implementations of these techniques for long-term simulations. The surface circulation along the southwest coast of Australia is investigated by [9]. Data merged from two systems (SeaSonde and WERA) identify a seasonal signal in the Leeuwin Current, with zonal migrations and yearly variations in magnitude. The study also highlights a modest tidal contribution, and the occurrence of persistent sub-mesoscale eddies.

The majority of the HF radar literature focuses on applications to the surface current field. HF radars, however, have the potential to measure gravity waves as well, and increasing attention has been paid to this specific topic in recent years. The validation of HF-derived waves against measurements retrieved by a wave buoy and two ADCPs is the focus of the work by [10]. Radar and in situ data show a good consistency, pointing to the crucial selection of the integration time to calculate radar directional spectra. These results indicate that the inversion method used by the authors is appropriate, leading to the absence of commonly observed errors (i.e., overestimation of wave heights and noise in short-term measurements). A wave climatology for the Gulf of Naples (Italy) is built by [11] by integrating HF radar and in situ (ADCP) measurements. The patterns reconstructed by the two platforms are consistent both at seasonal and interannual scales, and reveal specificities for the different sub-basins of the study area. The work supports the use of HF radars to properly characterise coastal processes.

The acquisition and elaboration of signals is central to the functioning of any radar system, and the quality of the data retrieved by these systems is deeply dependent upon validation and inversion procedures. The accuracy of HF measurements is strongly dependent on radiation patterns, and numerous quality control (QC) methods have been proposed in the literature. In their contribution to the SI, the authors in [12] present a new QC method for radial current velocities based on internal consistency checks, and implement an extension of the least squares direction-finding algorithm. The automation of the procedure and its application to wave measurements are proposed as next steps. A neural network is employed by [13] to invert the HF radar Doppler spectrum, and derive directional ocean spectra. Data on wave direction, peak wave direction, peak period, period energy and significant wave height outputted by the neural network are compared with measurements retrieved through a wave buoy and a pre-existing inversion method. Good correlations confirm the robustness of the method and open the way to further implementations. The author in [14] reports on the application of a listen-before-talk (LBT) operation mode to a SeaSonde system, by which a radar scans its radio environment before transmitting. The outcomes of the investigation indicate that SeaSonde radars can perform an LBT radio scan with their existing hardware and software, although the system may lose efficiency with variable radial range resolution. Recommendations are thus made to optimise the functioning of LBT in these systems.

Remaining in the context of coastal basins, two papers related to the use of the X-band radar system for sea state and shoreline monitoring are published in the SI. The authors in [15] present an automatic modified Temporal Waterline Method (mTWM) to extract, from time stack X-band radar images, a time series of shoreline positions and intertidal foreshore slopes in a sandy, micro-tidal beach site at Hasaki Oceanographical Research Station (HORS) in Hasaki, Japan. The comparison with the survey observations demonstrates both the accuracy and efficiency, as well as the robustness, of the proposed method. Moreover, the authors believe that the proposed tool will be useful to help authorities understand coastal changes, facilitating coastal protection and sustainable development in coastal zones. Concerning sea state monitoring, the estimation of the significant wave height (Hs) from X-band radar images is one of the most interesting and challenging tasks. The authors in [16] propose a novel and alternative strategy for estimating Hs, which allows avoiding the calibration of the wave spectrum by means of an external sensor reference during the installation of a wave radar system. The validation is performed by considering simulated wave fields generated under various sea state conditions. The encouraging results obtained from the analysis of the simulated data mean that the work provides a proof of concept. Therefore, the application of the proposed approach on real-world data and its testing at various sites is expected.

Remote sensing satellites are equipped with sensors that allow the observation of the whole globe, or an assigned part of it within a defined time period. In the context of this SI, the sensors deployed on satellite platforms are Synthetic Aperture Radar (SAR) systems and Radar Altimeters (RAs). The authors in [17] present a feasibility study to investigate the potentialities and limitations in the extraction, from SAR data, of information about the sea surface currents in the coastal area of the Gulf of Naples (South Italy) in the Mediterranean Sea. The study shows that, generally, wind plays a direct and significant role in the observed Doppler surface current. Moreover, through the availability of an oceanographic numerical model for one of the analysed cases, the authors are able to interpret the effect of the typical thermohaline circulation pattern on the Doppler anomaly. The authors in [18], instead, present a *ready-to-use* approach, based on the level 2 (L2) ocean product of Sentinel-1 (S1) images, for detecting cyanobacteria blooms in the Curonian Lagoon. The use of L2 S1 products improves the spatio–temporal detection of algal blooms. Moreover, this approach is easy and self-sustainable, and is able to provide observations independent of the presence of atmospheric haze or cloud cover. Finally, the authors in [19] investigate the R-effect, i.e., the Doppler shift deriving from the orbital velocity of sea wave particles, which can have an effect on the response of Delay Doppler Altimeters (DDAs). The results show that, when the wavelength of sea waves is of the same order of magnitude as the altimeter resolution, the waveform might be significantly influenced by the R-effect. This phenomenon is particularly important for the monitoring of long swells, often taking place in open oceans.

3. Conclusions

The broad coverage of the research topics addressed in this SI demonstrates that radar technology is an active and expanding field of investigation, and the papers published confirm the flexibility of these approaches to advance our current knowledge of coastal and open sea dynamics. The scientific community is actively engaged in these lines of investigation, and the contributions included in the present SI will herald further developments and applications in the near future.

Author Contributions: Conceptualization, G.L. and M.U.; writing—review and editing, G.L. and M.U. All authors have read and agreed to the published version of the manuscript.

Funding: This research received no external funding.

Acknowledgments: We would like to thank all contributing authors, whose interest in this collection has made it possible to realise a SI that we are confident will have an impact on the scientific community. As SI Editors, we also owe a debt of gratitude to the reviewers, who had a key role in selecting the most appropriate papers, as well as providing useful insights and comments.

Conflicts of Interest: The authors declare no conflict of interest.

References

1. Bastos, L.; Bio, A.; Iglesias, I. The importance of marine observatories and of RAIA in particular. *Front. Mar. Sci.* **2016**, *3*, 140. [CrossRef]

2. Crise, A.; Ribera d'Alcalà, M.; Mariani, P.; Petihakis, G.; Robidart, J.; Iudicone, D.; Bachmayer, R.; Malfatti, F. A conceptual framework for developing the next generation of Marine OBservatories (MOBs) for science and society. *Front. Mar. Sci.* **2018**, *5*, 318. [CrossRef]
3. Shearman, E.D.R. Radio science and oceanography. *Radio Sci.* **1983**, *18*, 299–320. [CrossRef]
4. Paduan, J.D.; Washburn, L. High-Frequency radar observations of ocean surface currents. *Ann. Rev. Mar. Sci.* **2013**, *5*, 115–136. [CrossRef] [PubMed]
5. Huang, W.; Wu, X.; Lund, B.; El-Darymli, K. Advances in coastal HF and Microwave (S- or X-band) radars. *Int. J. Antennas Propag.* **2017**, *2017*, 3089046. [CrossRef]
6. Yang, X.; Li, X.; Nunziata, F.; Mouche, A. (Eds.) *Ocean Remote Sensing with Synthetic Aperture Radar*; MDPI: Basel, Switzerland, 2018; p. 352. [CrossRef]
7. Lorente, P.; Piedracoba, S.; Sotillo, M.G.; Álvarez-Fanjul, E. Long-term monitoring of the Atlantic Jet through the Strait of Gibraltar with HF radar observations. *J. Mar. Sci. Eng.* **2019**, *7*, 3. [CrossRef]
8. Ren, L.; Miao, J.; Li, Y.; Luo, X.; Li, J.; Hartnett, M. Estimation of coastal currents using a soft computing method: A case study in Galway Bay, Ireland. *J. Mar. Sci. Eng.* **2019**, *7*, 157. [CrossRef]
9. Cosoli, S.; Pattiaratchi, C.; Hetzel, Y. High-Frequency radar observations of surface circulation features along the South-Western Australian coast. *J. Mar. Sci. Eng.* **2020**, *8*, 97. [CrossRef]
10. Lopez, G.; Conley, D.C. Comparison of HF radar fields of directional wave spectra against in situ measurements at multiple locations. *J. Mar. Sci. Eng.* **2019**, *7*, 271. [CrossRef]
11. Saviano, S.; Cianelli, D.; Zambianchi, E.; Conversano, F.; Uttieri, M. An integrated reconstruction of the multiannual wave pattern in the Gulf of Naples (South-Eastern Tyrrhenian Sea, Western Mediterranean Sea). *J. Mar. Sci. Eng.* **2020**, *8*, 372. [CrossRef]
12. Lipa, B.; Barrick, D.; Whelan, C. A quality control method for broad-beam HF radar current velocity measurements. *J. Mar. Sci. Eng.* **2019**, *7*, 112. [CrossRef]
13. Hardman, R.L.; Wyatt, L.R. Inversion of HF radar Doppler spectra using a neural network. *J. Mar. Sci. Eng.* **2019**, *7*, 255. [CrossRef]
14. Cosoli, S. Implementation of the Listen-Before-Talk mode for SeaSonde High-Frequency ocean radars. *J. Mar. Sci. Eng.* **2020**, *8*, 57. [CrossRef]
15. Kumar, D.; Takewaka, S. Automatic shoreline position and intertidal foreshore slope detection from X-band radar images using modified Temporal Waterline Method with corrected wave run-up. *J. Mar. Sci. Eng.* **2019**, *7*, 45. [CrossRef]
16. Ludeno, G.; Serafino, F. Estimation of the significant wave height from marine radar images without external reference. *J. Mar. Sci. Eng.* **2019**, *7*, 432. [CrossRef]
17. Zamparelli, V.; De Santi, F.; Cucco, A.; Zecchetto, S.; De Carolis, G.; Fornaro, G. Surface currents derived from SAR Doppler processing: An analysis over the Naples coastal region in South Italy. *J. Mar. Sci. Eng.* **2020**, *8*, 203. [CrossRef]
18. De Santi, F.; Luciani, G.; Bresciani, M.; Giardino, C.; Lovergine, F.P.; Pasquariello, G.; Vaiciute, D.; De Carolis, G. Synergistic use of Synthetic Aperture Radar and optical imagery to monitor surface accumulation of cyanobacteria in the Curonian Lagoon. *J. Mar. Sci. Eng.* **2019**, *7*, 461. [CrossRef]
19. Reale, F.; Pugliese Carratelli, E.; Di Leo, A.; Dentale, F. Wave orbital velocity effects on radar Doppler altimeter for sea monitoring. *J. Mar. Sci. Eng.* **2020**, *8*, 447. [CrossRef]

Journal of
Marine Science and Engineering

MDPI

Article

Long-Term Monitoring of the Atlantic Jet through the Strait of Gibraltar with HF Radar Observations

Pablo Lorente [1,*], Silvia Piedracoba [2], Marcos G. Sotillo [1] and Enrique Álvarez-Fanjul [1]

[1] Puertos del Estado, 28042 Madrid, Spain; marcos@puertos.es (M.G.S.); enrique@puertos.es (E.Á.-F.)
[2] Centro Tecnológico del Mar (CETMAR), 36208 Vigo, Spain; spiedracoba@cetmar.org
[*] Correspondence: plorente@puertos.es; Tel.: +34-917-229-816

Received: 19 November 2018; Accepted: 20 December 2018; Published: 2 January 2019

Abstract: The present work focuses on the long-term coastal monitoring of the Atlantic surface inflow into the Mediterranean basin through the Strait of Gibraltar. Hourly current maps provided during 2016–2017 by a High Frequency radar (HFR) system were used to characterize the Atlantic Jet (AJ) since changes in its speed and direction modulate the upper-layer circulation of the Western Alboran Gyre (WAG). The AJ pattern was observed to follow a marked seasonal cycle. A stronger AJ flowed north-eastwards during autumn and winter, while a weaker AJ was directed more southwardly during the middle of the year, reaching a minimum of intensity during summertime. A strong relationship between AJ speeds and angles was evidenced: the AJ appeared to be frequently locked at an angle around 63°, measured clockwise from the North. The AJ speed usually fluctuated between 50 $cm \cdot s^{-1}$ and 170 $cm \cdot s^{-1}$, with occasional drops below 50 $cm \cdot s^{-1}$ which were coincident with abrupt modifications in AJ orientation. Peaks of current speed clearly reached values up to 250 $cm \cdot s^{-1}$, regardless of the season. A number of persistent full reversal episodes of the surface inflow were analyzed in terms of triggering synoptic conditions and the related wind-driven circulation patterns. High sea level pressures and intense (above 10 $m \cdot s^{-1}$), permanent and spatially-uniform easterlies prevailed over the study domain during the AJ collapse events analyzed. By contrast, tides seemed to play a secondary role by partially speeding up or slowing down the westward currents, depending on the phase of the tide. A detailed characterization of this unusual phenomenon in the Strait of Gibraltar is relevant from diverse aspects, encompassing search and rescue operations, the management of accidental marine pollution episodes or efficient ship routing.

Keywords: HF radar; monitoring; circulation; Atlantic Jet; flow reversal; Gibraltar; Alboran Sea

1. Introduction

The Strait of Gibraltar (SoG hereinafter), the unique connection between the semi-enclosed Mediterranean basin and the open Atlantic Ocean, is characterized by a two-layer baroclinic exchange which is hydraulically controlled at Camarinal Sill (Figure 1a). Whilst saltier Mediterranean water flows out at depth, an eastward surface jet of relatively fresh Atlantic water (AJ) flows into the Alboran Sea by surrounding the quasi-permanent Western Anticyclonic Gyre (WAG) and the more elusive Eastern Anticyclonic Gyre (EAG) in a wavelike path. As the WAG owes its existence to the input of new Atlantic waters provided by the AJ, both structures are widely considered to be coupled and usually referred to as the AJ-WAG system [1]. A significant number of analytical, field and modeling studies have previously attempted to disentangle the AJ-WAG system and properly explain the underlying physical processes [1–3].

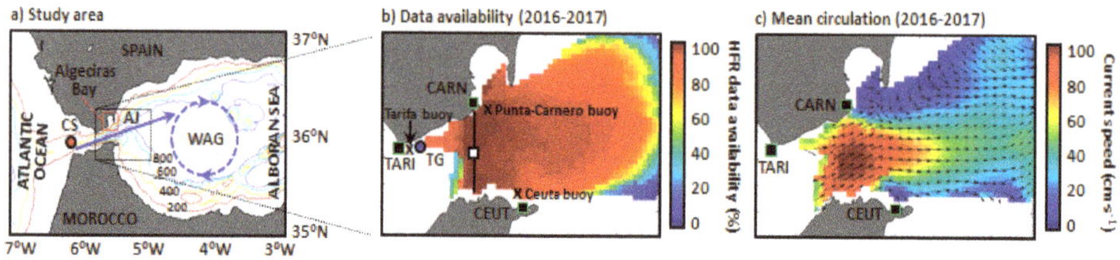

Figure 1. (**a**) Study area: surface Atlantic Jet (AJ) flowing through the Strait of Gibraltar into the Alboran Sea, feeding the Western Alboran Gyre (WAG); isobath depths are labeled every 200 m. Red dot indicates a topographic feature: Camarinal Sill (CS). (**b**) HFR hourly data availability for the period 2016–2017: solid black squares represent radar sites, black crosses indicate buoys locations, blue dot denotes Tarifa tide-gauge (TG) situation. Black line and the related white square indicate the selected transect and its midpoint, respectively; (**c**) HFR-derived mean surface circulation pattern for 2016–2017.

The position, intensity and direction of the AJ fluctuate in a broad range of temporal scales, driving the upper-layer circulation of the Alboran Sea with subsequent physical and biological implications [4–6]. For instance, the presence of an intense AJ close to the northern shore of the Alboran Sea reinforces the coastal upwelling and therefore increases both the nearshore chlorophyll concentration and the spawning of fish in this region [5,7]. By contrast, meteorologically-induced inflow interruptions can trigger the weakening and even the decoupling of the AJ-WAG system, the subsequent eastward migration of the WAG and the genesis of a new gyre that coexists with the other two, giving rise to a three-anticyclonic-gyre situation [8].

Within this context, the AJ pattern has been traditionally described to oscillate between two main circulation modes at seasonal scale [9,10]: (i) a stronger AJ flows northeastwards during the first half of the year; (ii) a weaker AJ flows more southwardly towards the end of the year. Sea level pressure (SLP) variations over the Western Mediterranean basin and local zonal wind (U) fluctuations in the Alboran Sea have been usually considered to be the main factors controlling and modulating the AJ [11,12]. At higher time scales, diurnal and semidiurnal variations of the mean flow through the SoG are mainly due to interaction of tidal currents with the intricate topography [13]. In particular, local wind has been largely invoked as the primary driving agent to explain both the intensification of the surface inflow during prevalent westerlies and also extreme AJ collapse events recorded when intense easterlies are predominant [14]. The zonal wind intensity has been reported to follow an annual cycle with more westerly (easterly) winds during winter (summer) months [15]. The seasonal variability and occasional interruptions of the Atlantic inflow due to meteorological forcing have been previously investigated with in situ data from fixed moorings [10,11]. More recently, a considerable number of satellite tracked drifters were released on both sides of the SoG within the framework of MEDESS-4MS project, thus providing a complete Lagrangian view of the Atlantic waters inflow into the Alboran Sea [16,17].

The main goal of this work is twofold: firstly, to build up upon previous works and characterize the AJ dynamic during a 2-year period (2016–2017) by using novel remote-sensed high-resolution current estimations. Secondly, to provide further insight into a singular event: the quasi-permanent full reversal of the AJ surface inflow through the entire selected transect (Figure 1b) during, at least, 48 h when intense easterlies episodes were prevalent. The triggering met-ocean factors (i.e., the atmospheric and tidal forcing) were explored along with the physical implications in terms of sharp variations in the sea surface temperature during summertime. A variety of full reversal episodes was detected and the related wind-induced circulation patterns were interpreted.

To this end, a High-Frequency radar (HFR) system was used, since it provides quality-controlled hourly maps of the surface currents in this first-order geostrategic spot (Figure 1b). As shown, data availability was significantly high (almost 100%) in the selected transect for the study period, decreasing in the easternmost sectors of the coverage domain. The accuracy of HFR observations, which are often affected by intrinsic uncertainties (radio frequency interferences, antenna pattern distortions, environmental noise, etc.), was previously assessed by comparing them against in situ data provided by a current meter [18]. Skill metrics improved when calibrated antenna patterns were used to process radial data. Recent research with this HFR network successfully investigated the water exchange between Algeciras Bay (Figure 1a) and the SoG [19], the impact of the atmospheric pressure fluctuations on the mesoscale water dynamics of the SoG and the Alboran Sea [20], or the dominant modes of spatio-temporal variability of the surface circulation [21].

As a consequence, this HFR network can be considered to be an appropriate device to effectively monitor the AJ dynamics in near real-time, with current pulses often exceeding 200 cm·s^{-1} and time-averaged north-eastward speeds around 100 cm·s^{-1} in the narrowest section of the SoG (Figure 1c). In the same line, this observing platform is also assumed to be capable of reliably portraying the AJ variability and the aforementioned episodic full reversals of the surface flow thanks to its significant spatio-temporal resolution, hence constituting an additional asset for wise decision-making of Algeciras Bay Harbor operators [22]. A detailed characterization of this unusual phenomenon in the SoG is relevant from diverse aspects, encompassing search and rescue operations (to adequately expand westwards the search area), the management of accidental marine pollution episodes (to establish alternative contingency plans), or safe ship routing (to maximize fuel efficiency).

The paper is organized as follows. Section 2 outlines the specific instrumentation used in this study. Section 3 describes the main results obtained. Finally, a detailed discussion of the results, along with future plans, is presented in Section 4.

2. Materials and Methods

The study domain includes an array of three coastal buoys and a tide-gauge operated by Puertos del Estado (Figure 1b), providing quality-controlled hourly-averaged observations of sea surface temperature (SST) and sea surface height (SSH), respectively. To ensure the continuity of the data record, occasional gaps detected in time series (not larger than 6 h) were linearly interpolated. Basic features of each in-situ instrument are described in Table 1.

Table 1. Description of in situ instruments operated by Puertos del Estado in the Strait of Gibraltar.

Name	Punta Carnero	Tarifa	Ceuta	Tarifa
Instrument	Buoy	Buoy	Buoy	Tide-gauge
Type	Directional	Directional	Directional	Radar
Manufacturer	WatchKeeper	Triaxys	Triaxys	Miros
Year of deployment	2010	2009	1985	2009
Longitude	5.42° W	5.59° W	5.33° W	5.60° W
Latitude	36.07° N	36.00° N	35.90° N	36.01° N
Depth	40 m	33 m	21 m	——
Frequency sampling	60 min	60 min	60 min	1 min

The HFR system employed in the present study consists of a three-site shore-based CODAR Seasonde network, installed within the framework of TRADE (Trans-regional Radars for Environmental applications) project and supported by European FEDER funding. The system covers the easternmost area of the SoG, including the mouth of the Algeciras Bay and part of the western Alboran Sea (Figure 1b,c). The sites, currently owned and operated by Puertos del Estado, were deployed in two different stages. The first two sites started data collection on May 2011: Ceuta (39.90° N, 5.31° W) and Punta Carnero (36.08° N, 5.43° W). Afterwards, the third site Tarifa (36.00° N, 5.61° W) was installed in October 2012 in order to properly resolve the baseline between Ceuta and Punta Carnero and also to gain accuracy over the entire radar coverage. Hereafter the sites will be referred to by their four-letter site codes: CEUT, CARN, and TARI, respectively (Figure 1b).

Each site operates at a central frequency of 26.8 MHz with a 150 KHz bandwidth, providing hourly radial measurements with a cut-off filter of 250 cm·s^{-1}, a threshold defined according to the historical values observed in the area. The maximum horizontal range and angular resolution are

40 km and 5°, respectively. The estimated current velocities are representative of the upper 0.5 m of the water column. Only calibrated antenna patterns are used to process radial data due to their positive impact in HFR accuracy, as previously proved by Reference [18]. Radial current measurements from the three sites are geometrically combined with an averaging radius set to 3 km, in order to estimate hourly total current vectors on a Cartesian regular grid of 1 × 1 km horizontal resolution.

A source of error to be considered in the computation of the total vectors is the so-called Geometrical Dilution of Precision (GDOP). The GDOP is defined as a dimensionless coefficient of uncertainty that characterizes how radar system geometry may impact on the measurements accuracy and position determination errors, owing to the angle at which radial vectors intersect [23]. Maps of east and north GDOP for an HFR system generally follow a pattern where their values increase with the distance from the radar sites and along the baselines (lines connecting two HFR sites) due to poor intersecting beam geometry, as the combining radial vectors are increasingly parallel and the orthogonal component tends to zero.

Figure 2a,b illustrates the zonal and meridional components of the GDOP over the HFR domain, estimated according to the formulation of Reference [23]. The values of GDOP in the baselines are low, less than 1.5 for both components. The spatial distribution of the sites allows resolving the central region of the SoG, minimizing the GDOP of the system in this sensitive area. The baseline between CEUT and CARN is resolved using radial vectors from TARI, while the baseline between TARI and CEUT is resolved using radial vectors from CARN. On the other hand, the GDOP for the meridional component increases in the eastern region of the domain, reaching values of 2.5 in the boundary (Figure 2b). In this area, only radial vectors from CARN and CEUT are used to compute the total velocities and the direction of their radial vectors is almost parallel. In this context, the transect here used to examine the AJ surface inflow was readily selected as the associated total GDOP (Figure 2c) was reduced (below 1.3) and the spatial and temporal data coverage were optimal during 2016–2017 (Figure 1b). A similar approach was previously adopted in Reference [24] to evaluate the water renewal mechanism and the related inshore-offshore exchanges in the Gulf of Naples. From an oceanographic perspective, the election of such transect was also convenient to better characterize both the intensity and direction of the AJ, since its midpoint covers the area where the highest peak of current speed is usually detected and also where the inflow orientation is not influenced yet by the water exchange between Algeciras Bay and the Strait of Gibraltar, as shown in Figure 1c.

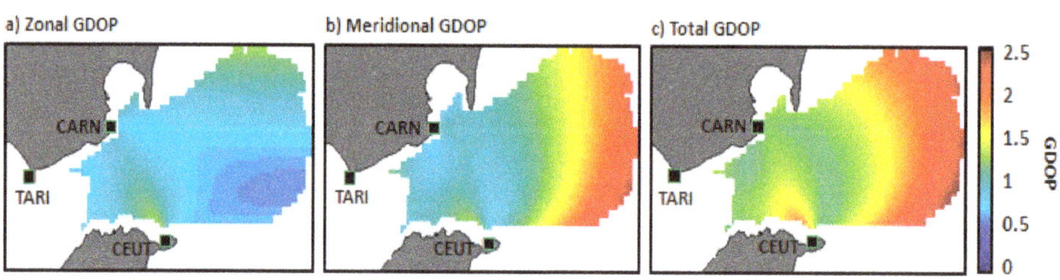

Figure 2. Geometric Dilution of Precision (GDOP) for the geometry of Gibraltar HFR system: (**a**) Zonal GDOP; (**b**) Meridional GDOP; (**c**) Total GDOP.

Finally, the prevailing synoptic atmospheric conditions were inferred from three-hourly predictions of sea level pressure (SLP) and zonal wind at 10 m height (U-10), provided by the European Center of Medium Weather Forecast (ECMWF).

3. Results

The scatter plot of HFR-derived eastward current speed against direction at the midpoint of the selected transect revealed that currents stronger than 50 cm·s^{-1} flowed predominantly towards the North-East, forming, on average, an angle of 63° measured clockwise from the North (Figure 3a).

The mean eastward current speed was 109 cm·s^{-1} for the analyzed 2-year period and speed peaks clearly reached values up to 250 cm·s^{-1}. Velocities below 50 cm·s^{-1} were registered along the entire range of eastward directions (0–180°). Westwards currents, albeit minority (1256 hourly observations during 2016–2017, 8% of the time), were also observed and tended to predominantly form an angle of 274° with averaged speeds about 35 cm·s^{-1} (Figure 3b). Higher velocities (>50 cm·s^{-1}) were scarcely observed for directions ranging from 240° to 290°. Since the AJ flowed predominantly north-eastwards (15,487 observations, 88% of the time), a scatter plot of zonal versus meridional current speed was also computed to infer the relationship between both components in terms of strength of the flow (Figure 3c). According to the results derived from the best linear fit, they were highly and positively correlated (0.74). Moreover, the zonal velocity component emerged to be, on average, almost three times stronger than the meridional current speed (as reflected by a slope of 0.37), thus highlighting the relevance of the along-shore transport in the water exchange between the Atlantic Ocean and the Mediterranean Basin, in agreement with [4].

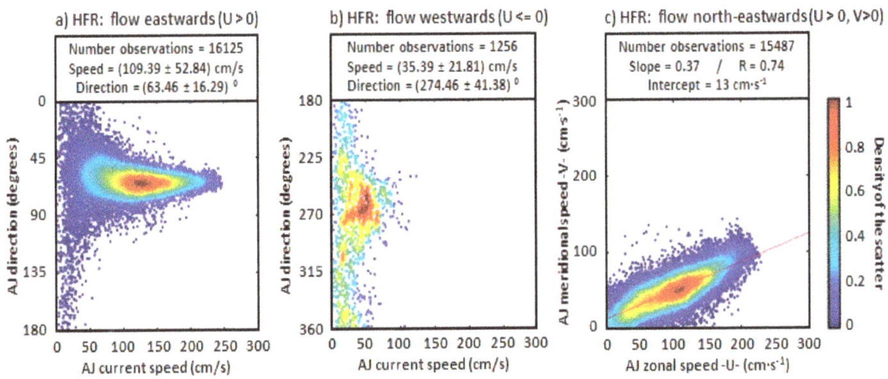

Figure 3. 2-year scatter plot of hourly Atlantic Jet (AJ) current direction (angle measured clockwise from the North) versus speed estimations at the midpoint of the selected transect: (**a**) AJ flowing eastwards (positive zonal speed, U > 0); (**b**) AJ flowing westwards (U < 0). Mean and standard deviation values of both AJ speed and direction computed for 2016–2017; (**c**) Scatter plot of meridional versus zonal current speed when the AJ flows north-eastwards. Results of the best linear fit are provided, where R is the correlation coefficient.

Scatter plots were also computed on a quarterly basis to infer potential differences for each season (Figure 4). During winter (defined as January-February-March –JFM-), the AJ reached its highest quarterly-averaged speed (123 cm·s^{-1}) with a related mean orientation of 63° (Figure 4a). Most of the directional estimations were comprised in the first quadrant (0–90°), whereas only isolated values lay in the range of 90–180° (i.e., flowing towards the South-East). Likewise, weaker westward currents were residually observed (7% of the wintertime), mainly flowing to the NW with a mean angle around 283° (Figure 4b).

During spring (defined as April-May-June –AMJ-), both AJ mean speed and angle were lower, 113 cm·s^{-1} and 60°, respectively (Figure 4c). Again, the AJ flowed mostly towards the North-East, with scarce observations of surface currents flowing to other quadrants (4% of the time, Figure 4d). In summer (July-August-September -JAS-) some noticeable differences were observed: the eastward AJ speed (angle) quarterly-averaged value reached a minimum (maximum) of 99 cm·s^{-1} (68°), with the AJ directed more southwardly. Higher scatter was detected along the directional axis, with surface currents often flowing towards the South-East (Figure 4e), the South-West and the North-West (Figure 4f). By the end of the year (October-November-December -OND-), the AJ mean speed partially recovered (102 cm·s^{-1}) and the currents were again directed predominantly to the North-East with an averaged angle of 63° (Figure 4g), thereby completing the annual cycle. Likewise, flow reversals were sporadically observed (9% of the autumn time), with a prevailing direction towards the west, forming an angle of 277° (Figure 4h).

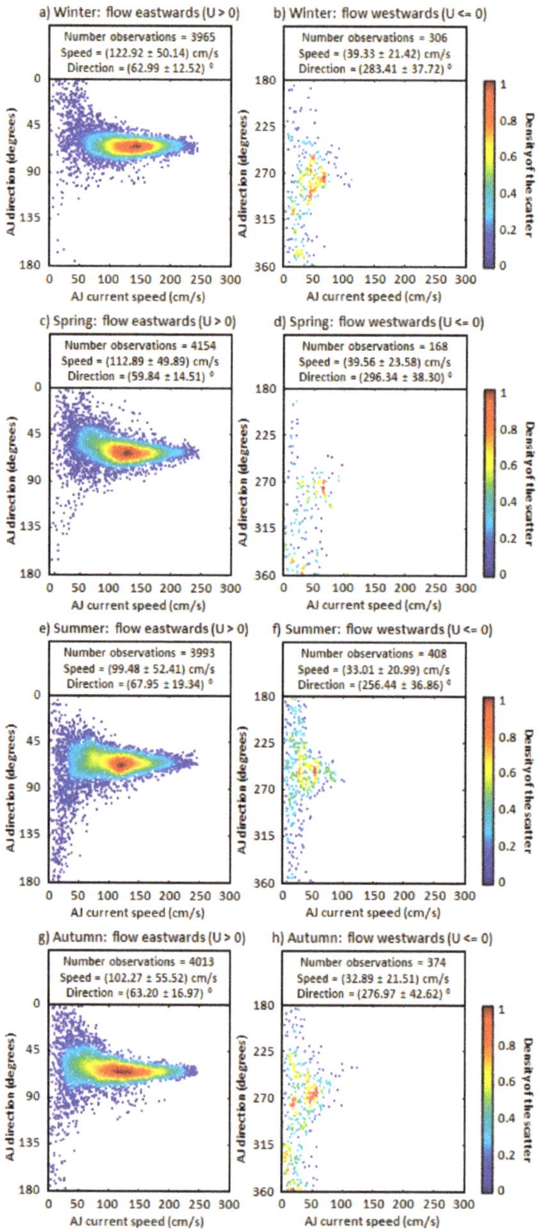

Figure 4. Seasonal scatter plots of hourly Atlantic Jet (AJ) current direction (angle measured clockwise from the North) versus speed estimations at the midpoint of the selected transect, with AJ flowing eastwards (left) and westwards (right): (**a**,**b**) Winter (JFM); (**c**,**d**) Spring (AMJ); (**e**,**f**) Summer (JAS); (**g**,**h**) Autumn (OND). Quarterly mean and standard deviation values of both AJ speed and direction are provided. Analysis performed for a 2-year period (2016–2017).

The long-term coastal monitoring of AJ speed and direction at the midpoint of the selected transect is presented in Figure 5. A strong relationship between remotely observed AJ speeds (red line) and angles (blue line) was revealed: with current velocities above 50 cm·s^{-1}, the AJ appeared to be locked at an angle around 63°. It is also noticeable that AJ hourly speed usually fluctuated between 50 cm·s^{-1} and 170 cm·s^{-1}, with peaks of 250 cm·s^{-1} in the form of eastward pulses. Occasional drops below 50 cm·s^{-1} were coincident with abrupt changes in AJ orientation. While some of those sharp variations in directions were of short duration (i.e., a few hours) in response to the joint action of local wind forcing (moderate and/or time-variable easterlies) and tidal currents during the outflow phase [25], others were persistent enough (lasting more than 48 h) to give rise to full surface inflow reversals.

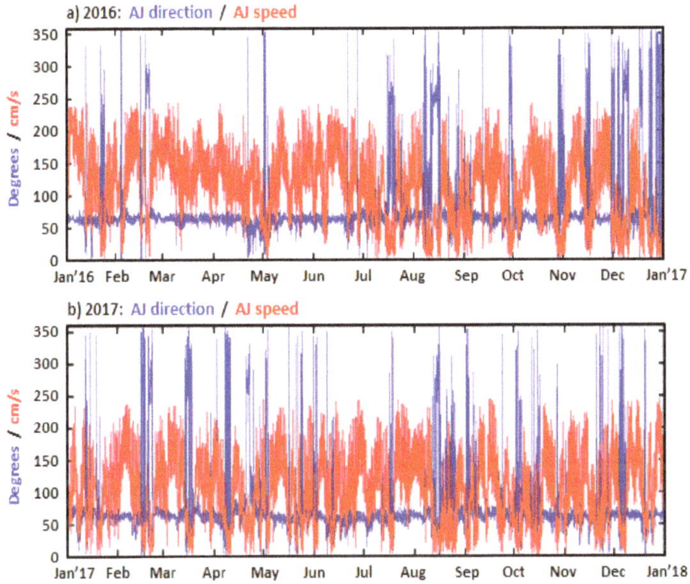

Figure 5. Annual evolution of the AJ orientation (blue line) and current speed (red line) for (**a**) 2016 and (**b**) 2017, as derived from hourly HFR estimations at the midpoint of the selected transect in the Strait of Gibraltar (Figure 1b).

Complete collapse of the AJ and quasi-permanent inversion of the surface inflow during prevalent intense easterlies is a singular phenomenon that deserved detailed exploration. Under this temporal premise, monthly Hovmöller diagrams were computed for HFR-derived zonal currents at the selected transect to easily detect and categorize six 2-day full reversal episodes, represented by black boxes in Figure 6. In December 2016 (Figure 6b) and February 2017 (Figure 6c), brief inversions (related to less intense easterlies) preceded the full reversal of the surface flow. The event detected in March 2017 consisted of an abrupt interruption of the eastward inflow and complete reversal of the surface stream (Figure 6d). By contrast, in August 2016 (Figure 6a), August 2017 (Figure 6e) and December 2017 (Figure 6f), the classical AJ inflow into the Mediterranean Sea (red color) was only observed in the southern part of the transect, whereas a weaker coastal counter current (CCC) was detected flowing westwards and bordering the northern Spanish shoreline.

The prevailing atmospheric synoptic conditions were inferred from ECMWF predictions of SLP and U-10, as shown in Figure 7. A significant latitudinal gradient of SLP was observed in four episodes that took place, respectively, in August 2016, February 2017, March 2017, and December 2017, with high pressures over the Gulf of Biscay and isobars closely spaced in the SoG (Figure 7a,c,d,f), leading to extremely intense easterlies (above 10 m·s^{-1}), channeled through the Strait due to its specific geometric configuration (Figure 7g,i,j,l). In December 2016, high pressures dominated the entire Western Mediterranean (Figure 7b), equally yielding strong along-shore westward winds along both sides of the SoG (Figure 7h). In August 2017, the typical summer weather type was observed with Azores High pressures governing the Mid-Atlantic Area (Figure 7e) with moderate but persistent easterly winds blowing through the entire Western Mediterranean (Figure 7k).

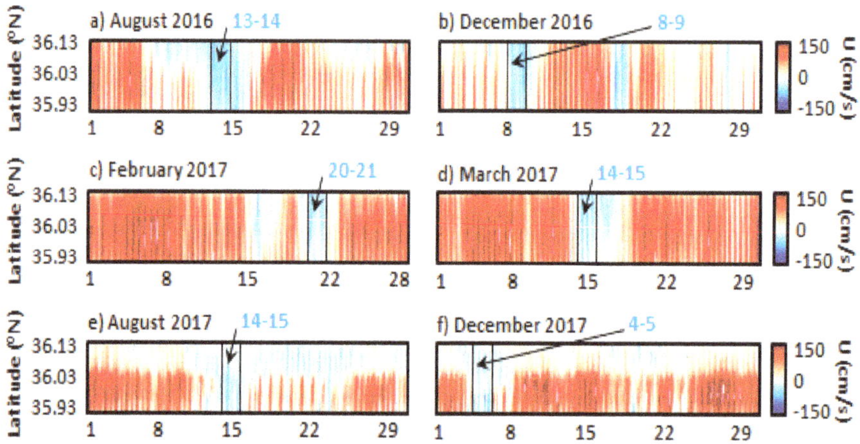

Figure 6. Monthly Hovmöller diagrams of High-Frequency Radar (HFR)-derived zonal current speed at the selected transect in the Strait of Gibraltar (Figure 1b). Red (blue) colors represent eastward (westward) surface flow. 2-day episodes of permanent flow reversal are marked with black boxes.

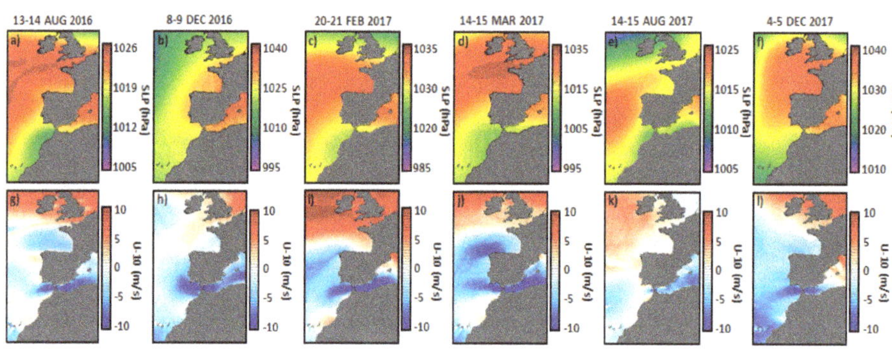

Figure 7. Synoptic patterns related to the six Atlantic Jet (AJ) inflow reversal events analyzed during the period 2016–2017: 2-day averaged maps of sea level pressure (SLP: **a–f**) and zonal wind at 10 m height (U-10: **g–l**), as provided by the European Center of Medium Weather Forecast (ECMWF).

2-day averaged maps of modeled wind field at 10 m height are presented in Figure 8 for each of the six events analyzed. As shown, intense easterlies prevailed over the entire study domain, with uniform velocities usually above 10 m·s^{-1}. Only the fifth episode, corresponding to August 2017 (Figure 8e), exhibited lower wind speed (in accordance with Figure 7k) and more spatially-variable wind conditions.

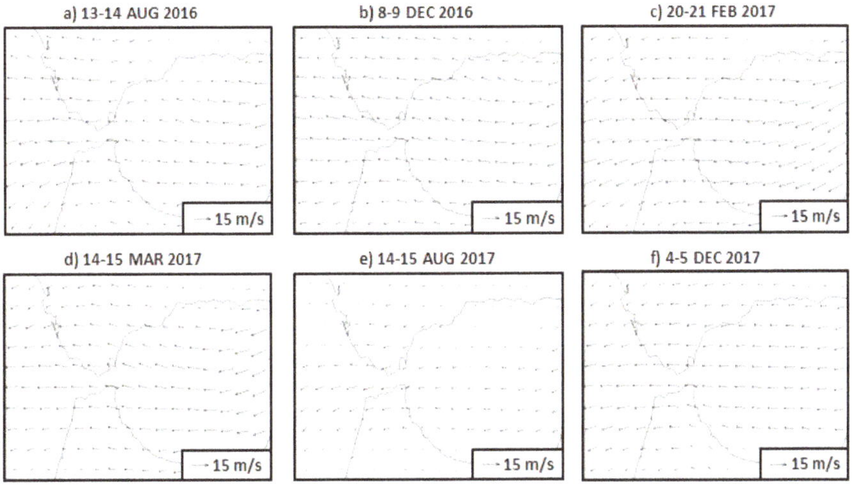

Figure 8. 2-day averaged maps of wind at 10 m height (provided by the European Center of Medium Weather Forecast (ECMWF)) related to the six Atlantic Jet (AJ) inflow reversal events analyzed during the period 2016–2017.

Short-lived reversals of the surface inflow have been previously reported to occur almost every tidal cycle in Camarinal Sill (western end of SoG: Figure 1a) mainly due to the contribution of the semidiurnal tidal component M_2 [25–28]. Since the mean inflow of Atlantic water is modulated by barotropic tidal currents, hourly-averaged SSH observations provided by Tarifa tide-gauge were used to elucidate if the six 2-day inflow reversal events in the eastern end of the Strait could have been mostly influenced by spring-neap tidal cycle fluctuations (Figure 9). Although the fortnightly variability was clearly observable in a monthly time series of SSH, no cause-effect relationship could be visually inferred from the inspection of zonal velocities at the selected transect (Figure 6). Apparently, evidence of preference for a specific tidal cycle was not observed as the six flow reversal episodes took place during different tidal conditions, ranging from neap tides (Figure 9a,c) to spring tides (Figure 9d,f).

To support the previous statement, hourly maps of HFR-derived surface circulation were depicted for eight different test-cases (Figure 10), defined according to the prevailing met-ocean conditions (Table 2). The first two analyzed events took place in February 2017 under strong westerlies (Figure 10a) during spring tides (Figure 10c), when the tide was at flood (event 1) and later at ebb (event 2). In both cases, the wind-driven flow was directed eastwards (Figure 10g,i), although more vigorously in the case of event 1. By contrast, events 3 and 4 occurred under extremely intense easterlies (above 15 m·s^{-1}) during neap tides. Regardless of the tide height, the surface circulation patterns were similar in both cases (Figure 10k,m), with a manifest reversal of the main flow towards the west, as reflected in the monthly Hovmöller diagram (Figure 10e). In December 2017, events 5 and 6 were registered again under strong easterlies (Figure 10b), but this time during spring tides (Figure 10d). According to the hourly HFR-derived maps, the westward inversion of the flow was evidenced independently of the tidal phase (Figure 10h,j). Finally, events 7 and 8 took place during neap tides, when persistent and very intense westerlies were blowing. The associated surface circulation maps exhibited an acceleration of the Atlantic inflow into the Mediterranean (Figure 10l,n), reaching velocities above 160 cm·s^{-1} at the selected transect (Figure 10f).

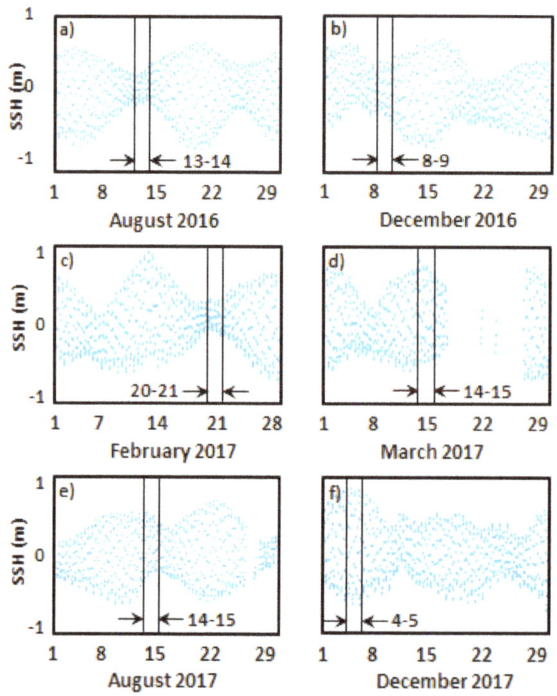

Figure 9. Monthly time series of hourly-averaged sea surface height (SSH) provided by Tarifa tide-gauge (Figure 1b). 2-day episodes of permanent inflow reversal are marked with black boxes.

The 2-day averaged HFR-derived circulation patterns associated with the six events here studied are depicted in Figure 11. Some common peculiarities were exposed, such us the westward outflow through the narrowest section of the SoG or the prevalent anticyclonic inflow into the Algeciras Bay (with a direct impact on larger residence times and reduced water renewal). Three case studies revealed a predominant circulation towards the West/South-West, together with a marked acceleration of the flow in the vicinity of Algeciras Bay, reaching speeds above 50 $cm{\cdot}s^{-1}$ (Figure 11a,c,e). On the contrary, two episodes illustrated how the circulation in the easternmost region of the study domain followed a clockwise rotation, likely feeding the WAG, which was out of the picture (Figure 11d,f). In the first event, corresponding to March 2017, the anticyclonic gyre was accelerated up to 80 $cm{\cdot}s^{-1}$ (Figure 11d). The second case (December 2017) was substantially less energetic and exhibited besides a counter-clockwise recirculation structure in the entrance to the strait. (Figure 11f). The event identified in December 2016 represented an overall cyclonic recirculation cell with a marked intensification of the flow reversal nearby the Algeciras Bay when westward currents were channeled through the narrow SoG (Figure 11b). Finally, two small-scale counter-clockwise eddies were observed detached from the shore and enclosed in the general westward flow (Figure 11c,e).

Among the physical implications of the surface inflow reversal, abrupt increases in the SST field were revealed, especially during summertime when warmer Mediterranean waters outflowed into the mild Atlantic Ocean (Figure 12). During August 2016, the aforementioned CCC showed up the 8th and lasted 10 days (Figure 12a). During this period (delimited by green dotted lines in Figure 12a–d), the SST field experienced a net increase of 6 °C, from 18 °C to 24 °C, according to the in situ estimations provided by Tarifa and Punta-Carnero buoys (Figure 12b,c). Soon afterwards, a sharp decrease to 17 °C was recorded in few hours (steeper in the case of Tarifa buoy), as a consequence of the reestablishment of the intense (above 130 $cm{\cdot}s^{-1}$) AJ flowing eastwards (Figure 12a). In the southern region, the SST remained rather constant (~16.5 °C) and obviously unaffected by the northern CCC (Figure 12d). Notwithstanding, as easterly winds progressively dominated the study area and persisted enough, the CCC broadened and the complete inflow reversal transported warmer Mediterranean waters to the west through the entire transect, as reflected by the pronounced SST maximum (~22.5 °C) recorded in Ceuta buoy but surprisingly delayed by several hours. A secondary peak of SST was monitored only in the northern shoreline by the end of the month (delimited by purple dotted lines in Figure 9b,c). The temperature increased from 18 to 22 °C, approximately, whereas in the southern area of the SoG no relevant changes in SST were noticed (Figure 12d).

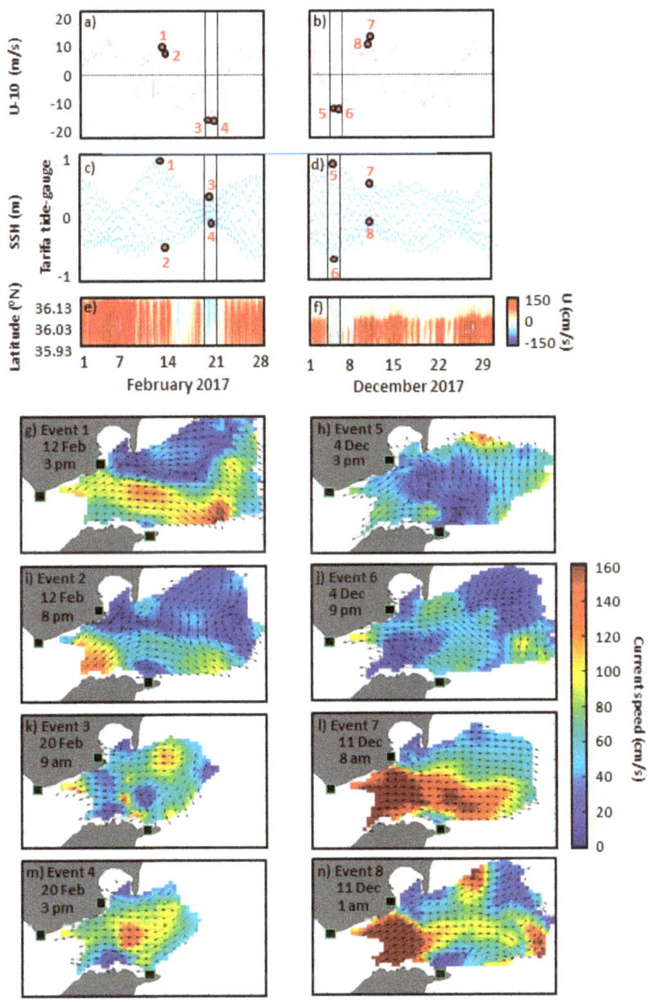

Figure 10. (**a**,**b**) Monthly time series of European Center of Medium Weather Forecast (ECMWF) 3-h zonal wind at 10 m height (U-10), spatially-averaged over the Strait of Gibraltar, for February and December 2017, respectively. Red dots denote specific selected events, defined in Table 2; (**c**,**d**) Monthly time series of hourly sea surface height. (**e**,**f**) Monthly Hovmöller diagrams of High-Frequency Radar (HFR)-derived zonal current speed at the selected transect in the Strait of Gibraltar. 2-day episodes of permanent flow reversal are marked with black boxes; (**g**–**n**) Hourly maps of HFR-derived surface circulation, corresponding to the eight selected events, labeled in the upper panels.

Table 2. Description of eight events categorized according to different met-ocean conditions.

	Westerlies (U-10 > 0)		Easterlies (U-10 < 0)	
Tidal Cycle	**SSH (m)**	**Event (date, hour)**	**SSH (m)**	**Event (date, hour)**
Spring tide	0.77	1 (12 Feb, 3 pm)	0.84	5 (4 Dec, 2 pm)
Spring tide	−0.38	2 (12 Feb, 8 pm)	−0.50	6 (4 Dec, 9 pm)
Neap tide	0.52	7 (11 Dec, 8 am)	0.30	3 (20 Feb, 9 am)
Neap tide	−0.20	8 (11 Dec, 1 am)	−0.07	4 (20 Feb, 3 pm)

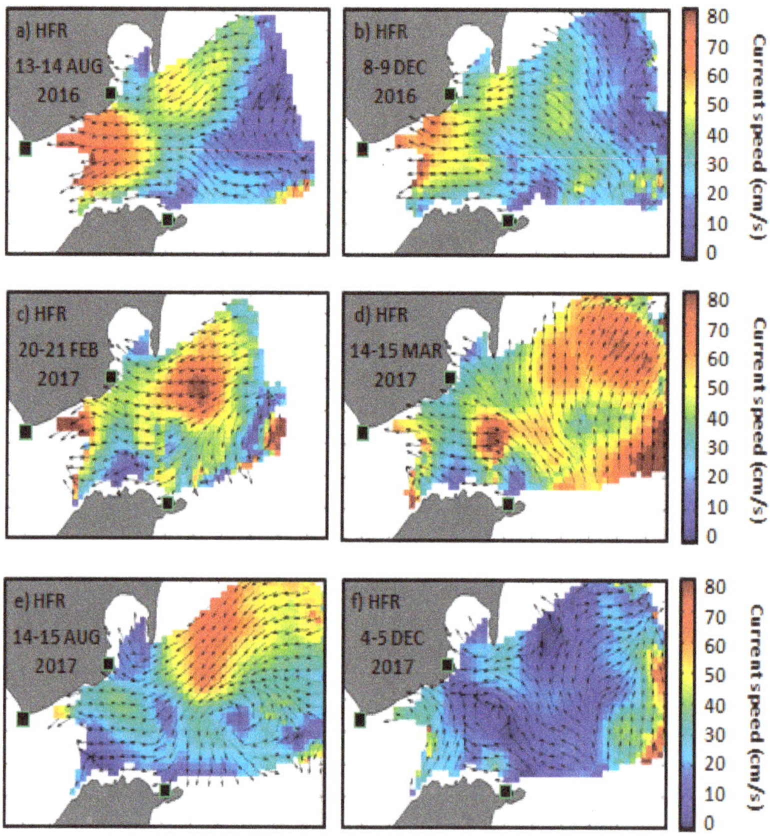

Figure 11. 2-day averaged maps of High-Frequency Radar (HFR)-derived surface circulation for each of the six Atlantic Jet (AJ) inflow reversal events analyzed during the period 2016–2017.

Equally, a similar situation was encountered from 11th to 25th of August 2017 (Figure 12e), with a progressive warming of 7.5 °C at the upper ocean layer of the northern shoreline (Figure 12g). The westward CCC was revealed from the 11th towards the end of the month, and confined at higher latitudes except for the already analyzed 2-day event of 14th–15th, when a full flow reversal was documented (Figure 12e) and an abrupt increase of 4 °C was observed in Ceuta buoy (Figure 12h).

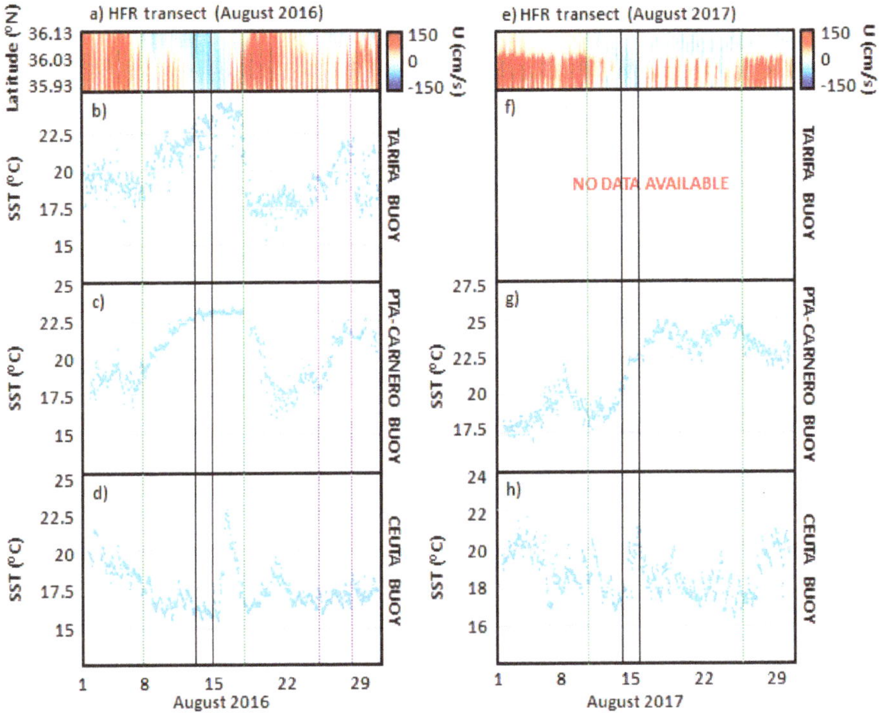

Figure 12. (**a–d**) Monthly Hovmöller diagram of High-Frequency Radar (HFR)-derived zonal current speed at the selected transect in the Strait of Gibraltar for August 2016 along with monthly timeseries of in situ sea surface temperature (SST) recorded by three coastal buoys deployed in the study area: Tarifa, Punta Carnero and Ceuta. (**e–h**) Idem, for August 2017. Two full Atlantic Jet (AJ) reversal episodes marked with black boxes for the 13th–14th August 2016 (left) and 14th–15th August 2017 (right). Green and purple dotted lines denote the beginning and end of SST increase as a result of the partial inversion of the flow in the northern coastal area of the Strait, and the subsequent intrusion of warmer and saltier Mediterranean waters into the Strait.

4. Discussion

Motivated by the critical relevance of AJ over the upper-layer circulation in the Western Alboran Sea, the ability of the HFR system to properly portray the AJ intensity and direction at the midpoint of the selected transect (Figure 1b) was evaluated. A strong relationship between remotely observed AJ speeds and angles was evidenced: with current velocities above 50 cm·s^{-1}, the AJ appeared to be locked at an angle around 63° (Figures 3a and 5). This finding is in close agreement with previous results reported in Reference [11], where an averaged angle of 60° was obtained by using satellite-derived observations and hydrodynamic model simulations. The role of the underlying topography (Figure 1a) on the prevailing current direction is also noteworthy in steeper regions like the SoG since it acts as physical constraint, enclosing the directional variability of the flow.

The AJ speed usually fluctuated between 50 cm·s^{-1} and 170 cm·s^{-1} (Figure 3a), with periodic tidal-induced pulses reaching peaks above 200 cm·s^{-1} as a result of increases in current speed as they encountered the constrictions of the local bathymetry [28]. Occasional drops below 50 cm·s^{-1} were coincident with abrupt changes in AJ orientation (Figure 5). A significant direct relationship between the zonal and meridional velocity components was showcased, with the former almost three times stronger than the latter (Figure 3c). Such dependence was confirmed by a high and positive correlation coefficient of 0.74, thereby explaining the recurrent propagation direction of the surface stream when it was directed towards the North-East.

The AJ pattern was observed to follow a marked seasonal cycle (Figure 4), in high accordance with the sequence documented in the literature [9,14]. A stronger AJ flowed towards the NE during winter, whereas a steady weakening in mean AJ speed was noticed along springtime. A minimum was reached by summer when a weaker AJ was directed more southwardly. During autumn, the AJ intensity partially recovered and the currents flowed again predominantly to the NE. Notwithstanding, the main discrepancy with earlier works lies in the higher AJ intensity here reported: whereas HFR-derived mean speed was about 110 $cm \cdot s^{-1}$ (reaching peaks of 200–250 $cm \cdot s^{-1}$ regardless of the season), previous modeling works substantially underestimated monthly-averaged AJ speeds, always reporting values below 50 $cm \cdot s^{-1}$ [14].

Six full reversal episodes of the surface inflow through the SoG, persistent during at least 48 h, were detected by means of Hövmoller diagrams of HFR-derived zonal velocity component (Figure 6) and subsequently analyzed in terms of prevailing synoptic conditions (Figures 7 and 8) and wind-driven circulation patterns (Figure 11). High pressures and intense (above 10 $m \cdot s^{-1}$), permanent, and spatially-uniform easterlies prevailed over the entire study domain, inducing a westward outflow through the SoG as revealed by the 2-day averaged HFR circulation maps. Local wind forcing at this scale seemed to play a primary role in explaining such AJ collapse and the related inflow reversals. It has been previously well-documented how the interface fluctuates at the northeastern entrance of the Strait in response to local wind forcing [25]. Under strong westerlies, the interface depth rises (reducing the upper layer thickness) and the eastward HFR-derived current speed increases, being fully representative of the entire upper layer flow. Conversely, under prevailing intense easterlies, the interface becomes deeper (increasing the upper layer thickness) and HFR estimations are thereby illustrative of only the shallowest part of such a layer. Future efforts should address the synergistic integration of HFR estimations and outputs from a 3D high-resolution coastal model to provide further insight into the entire water column and comprehensively characterize the circulation in the SoG and the water exchanges at different depth levels.

According to both Table 2 and Figure 10, tides appeared to play a secondary role in the full flow inversion as the six episodes were observed during strong easterlies, regardless of the tidal phase. The westward inflow reversal was moderately accelerated (decelerated) during the tidal outflow (inflow). On the contrary, the Atlantic inflow into the Mediterranean was favored and intensified under westerlies: the tidal flow only partially sped up or slowed down the eastern flowing current, depending on the phase of the tide. These results are in accordance with previous modeling studies [26] where the contribution of the semidiurnal tidal component to the transport was proved to be relevant over the Camarinal Sill, (incrementing the mean transport by about 30%, both for the inflow and the outflow), whereas it was almost negligible at the eastern end of the Strait. Equally, Reference [27] described the structure of the barotropic tide (sea level oscillation) and showed that the amplitude of the prevailing semidiurnal constituents diminished more than 50% from the western to the eastern end of the Strait and had little cross-strait structure.

One of the direct physical consequences of persistent inflow inversion was the sudden and abrupt rise of SST (above 5°) in the SoG during summertime when warmer Mediterranean waters were injected into the mild Atlantic Ocean (Figure 12). Some of those SST increases were confined to the northern shoreline of the Strait due to the presence of the CCC. Such coastal flow inversion has been previously described and is the subject of further analysis in Reference [25], where the CCC was demonstrated to exist regardless of the wind direction, being substantially intensified and broadened during predominant strong easterlies.

Ongoing efforts are devoted to conduct a multi-model inter-comparison exercise, ranging from global ocean models to regional and high-resolution coastal models in order to elucidate: (i) the accuracy of each system to characterize the AJ dynamic; (ii) their ability to adequately reproduce the singular and sporadic small-scale coastal ocean process here described, the quasi-permanent full reversal of the AJ surface flow; and (iii) quantify the added-value of dynamical downscaling approaches in coastal areas.

As an essential part of the coastal ocean observing system [29], HFR only provide information of surface currents. To better study three-dimensional coastal ocean circulation variability, it would be useful to adopt a multi-platform approach by combining HFR estimations with other observational data sets, such as in situ observation from moored ADCPs and satellite remote sensing products [30,31].

Finally, it is worth mentioning that the present three-site HFR network is planned to be extended, with the imminent integration of a new site (denominated Camarinal and managed by the University of Cádiz, Cádiz, Spain) to better monitor the coastal ocean dynamics in the westernmost section of the SoG. Within this framework, the top priority issue is not only the maintenance of continued financial support to preserve the infrastructure core service already implemented, but also the pursuit of funding to extend the network for an overall coverage.

Author Contributions: conceptualization, P.L. and S.P.; methodology, P.L.; software, P.L.; investigation, P.L.; writing—original draft preparation, P.L. and S.P.; writing—review and editing, M.G.S.; supervision, E.Á.-F.

Funding: This research received no external funding.

Acknowledgments: The authors gratefully acknowledge Qualitas Remos Company (Madrid, Spain) (partner of CODAR) for their useful technical suggestions.

Conflicts of Interest: The authors declare no conflict of interest.

References

1. Sánchez-Garrido, J.C.; García-Lafuente, J.; Álvarez-Fanjul, E.; Sotillo, M.G.; De los Santos, F.J. What does cause the collapse of the Western Alboran Gyre? Results of an operational ocean circulation system. *Prog. Oceanogrogr.* **2013**, *116*, 142–153. [CrossRef]
2. Macías, D.; Martín, A.P.; García-Lafuente, J.; García, C.M.; Yool, A.; Bruno, M.; Vázquez-Escobar, A.; Izquierdo, A.; Sein, D.V.; Echevarría, F. Analysis of mixing and biogeochemical effects induced by tides on the Atlantic-Mediterranean flow in the Strait of Gibraltar through a physical-biological coupled model. *Prog. Oceanogr.* **2007**, *74*, 252–272. [CrossRef]
3. Viúdez, A. An explanation for the curvature of the Atlantic Jet past the Strait of Gibraltar. *J. Phys. Oceanogr.* **1997**, *27*, 1804–1810. [CrossRef]
4. Solé, J.; Ballabrera-Poy, J.; Macías, D.; Catalán, I.A. The role of ocean velocity in chlorophyll variability. A modelling study in the Alboran Sea. *Stientia Marina* **2016**, *80*, 249–256. [CrossRef]
5. Ruiz, J.; Macías, D.; Rincón, M.; Pascual, A.; Catalan, I.A.; Navarro, G. Recruiting at the edge: Kinetic energy inhibits anchovy populations in the Western Mediterranean. *PLoS ONE* **2013**, *8*, e55523. [CrossRef] [PubMed]
6. Navarro, G.; Vázquez, A.; Macías, D.; Bruno, M.; Ruiz, J. Understanding the patterns of biological response to physical forcing in the Alboran Sea (western Mediterranean). *Geophys. Res. Lett.* **2011**, *38*, L23606. [CrossRef]
7. Macías, D.; Bruno, M.; Echevarría, F.; Vázquez, A.; Garcia, C.M. Meteorologically-induced mesoscale variability of the north-western Alboran sea (southern Spain) and related biological patterns. *Estuar. Coast. Shelf Sci.* **2008**, *78*, 250–266. [CrossRef]
8. Viúdez, A.; Pinot, J.M.; Harney, R.L. On the upper layer circulation in the Alboran Sea. *J. Geophys. Res.* **1998**, *103*, 21653–21666. [CrossRef]
9. Vargas-Yáñez, M.; Plaza, F.; García-Lafuente, J.; Sarhan, T.; Vargas, J.M.; Vélez-Belchí, P. About the seasonal variability of the Alboran Sea circulation. *J. Mar. Syst.* **2002**, *35*, 229–248. [CrossRef]
10. Bormans, M.; Garret, C.; Thompson, K.R. Seasonal variability of the surface inflow through the Strait of Gibraltar. *Oceanol. Acta* **1986**, *9*, 403–414.
11. García-Lafuente, J.; Delgado, J.; Criado, F. Inflow interruption by meteorological forcing in the Strait of Gibraltar. *Geophys. Res. Lett.* **2002**, *29*, 1914. [CrossRef]
12. Macías, D.; Navarro, G.; Echevarría, F.; Garcia, C.M.; Cueto, J.L. Phytoplankton pigment distribution in the north-western Alboran Sea and meteorological forcing: A remote sensing study. *J. Mar. Res* **2007**, *65*, 523–543. [CrossRef]

13. Sánchez-Román, A.; García-Lafuente, J.; Delgado, J.; Sánchez-Garrido, J.C.; Naranjo, C. Spatial and temporal variability of tidal flow at the Strait of Gibraltar. *J. Mar. Syst.* **2012**, *98–99*, 9–17. [CrossRef]
14. Macías, D.; García-Gorriz, E.; Stips, A. The seasonal cycle of the Atlantic Jet dynamics in the Alboran sea: Direct atmospheric forcing versus Mediterranean thermohaline circulation. *Ocean Dyn.* **2016**, *66*, 137–151. [CrossRef]
15. Dorman, C.E.; Beardsley, R.C.; Limeburner, R. Winds in the Strait of Gibraltar. *Q. J. Meteorol. Soc.* **1995**, *121*, 1903–1921. [CrossRef]
16. Sotillo, M.G.; Amo-Baladrón, A.; Padorno, E.; García-Ladona, E.; Orfila, A.; Rodríguez-Rubio, P.; Conte, D.; Jiménez, J.A.; De los Santos, F.J.; Álvarez-Fanjul, E. How is the surface Atlantic water inflow through the Strait of Gibraltar forecasted? A lagrangian validation of operational oceanographic services in the Alboran Sea and Western Mediterranean. *Deep Sea Res.* **2016**, *133*, 100–117. [CrossRef]
17. Sotillo, M.G.; García-Ladona, E.; Orfila, A.; Rodríguez-Rubio, P.; Maraver, J.C.; Conti, D.; Padorno, E.; Jiménez, J.A.; Capó, E.; Pérez, F.; et al. The MEDESS-GIB database: Tracking the Atlantic water inflow. *Earth Syst. Sci. Data* **2016**, *8*, 141–149. [CrossRef]
18. Lorente, P.; Piedracoba, S.; Soto-Navarro, J.; Álvarez-Fanjul, E. Accuracy assessment of high frequency radar current measurements in the Strait of Gibraltar. *J. Oper. Oceanogr.* **2014**, *7*, 59–73. [CrossRef]
19. Chioua, J.; Dastis, C.; Gonzalez, C.J.; Reyes, E.; Mañanes, R.; Ruiz, M.I.; Álvarez-Fanjul, E.; Yanguas, F.; Romero, J.; Álvarez, O.; et al. Water exchange between Algeciras Bay and the Strait of Gibraltar: A study based on HF coastal radar. *Estuar. Coast. Shelf Sci.* **2017**, *196*, 109–122. [CrossRef]
20. Dastis, C.; Izquierdo, A.; Bruno, M.; Reyes, E.; Sofina, E.V.; Plink, N.L. Influence of the atmospheric pressure fluctuations over the Mediterranean Sea on the mesoscale water dynamics of the Strait of Gibraltar and the Alboran Sea. *Fundamentalnaya Prikladnaya Gidrofisika* **2018**, *11*, 28–29. [CrossRef]
21. Soto-Navarro, J.; Lorente, P.; Álvarez-Fanjul, E.; Sánchez-Garrido, J.C.; García-Lafuente, J. Surface circulation at the Strait of Gibraltar: A combined HF radar and high resolution model study. *J. Geophys. Res. Oceans* **2016**, *121*. [CrossRef]
22. Fanjul, E.A.; Sotillo, M.G.; Gómez, B.P.; Valdecasas, J.M.G.; Rubio, S.P.; Lorente, P.; Dapena, Á.R.; Marco, I.M.; Luna, Y.; Padorno, E.; et al. Operational Oceanography at the Service of the Ports. In *New Frontiers in Operational Oceanography*; Chassignet, E., Pascual, A., Tintoré, J., Verron, J., Eds.; GODAE OceanView: Exeter, UK, 2018; pp. 729–736. [CrossRef]
23. Barrick, D. *Geometrical Dilution of Statistical Accuracy (GDOSA) in Multi-Static HF Radar Networks*; CODAR Ocean Sensors Ltd.: Mountain View, California, USA, 2006.
24. Cianelli, D.; Falco, P.; Iermano, I.; Mozzillo, P.; Uttieri, M.; Buonocore, B.; Zambardino, G.; Zambianchi, E. Inshore/Offshore wáter Exchange in the Gulf of Naples. *J. Mar. Syst.* **2015**, *145*, 37–52. [CrossRef]
25. Reyes, E. A High-Resolution Modeling Study of the Ocean Response to Wind Forcing within the Strait of Gibraltar. Ph.D. Dissertation, University of Cadiz, Cadiz, Spain, 2015.
26. Sannino, G.; Bargagli, A.; Artale, V. Numerical modeling of the semidiurnal tidal exchange through the Strait of Gibraltar. *J. Geophys. Res.* **2004**, *109*, C05011. [CrossRef]
27. García Lafuente, J.; Almazán, J.L.; Fernández, F.; Khribeche, A.; Hakimi, A. Sea level in the Strait of Gibraltar: Tides. *Int. Hydrogr. Rev.* **1990**, *LXVII*, 111–130.
28. La Violette, P.E.; Lacombe, H. Tidal-induced pulses in the flow through the strait of Gibraltar. In *Oceanol. Acta, Special Issue 0399-1784*. Available online: https://archimer.ifremer.fr/doc/00267/37794/ (accessed on 28 December 2018).
29. Liu, Y.; Kerkering, H.; Weisberg, R.H. (Eds.) *Coastal Ocean Observing Systems*; Elsevier: London, UK, 2015; 461p.
30. Liu, Y.; Weisberg, R.H.; Shay, L.K. Current patterns on the West Florida Shelf from joint self-organizing map analyses of HF radar and ADCP data. *J. Atmos. Ocean. Technol.* **2007**, *24*, 702–712. [CrossRef]
31. Manso-Narvarte, I.; Caballero, A.; Rubio, A.; Dufau, C.; Birol, F. Joint analysis of coastal altimetry and high-frequency (HF) radar data: Observability of seasonal and mesoscale ocean dynamics in the Bay of Biscay. *Ocean Sci.* **2018**, *14*, 1265–1281. [CrossRef]

Journal of
Marine Science and Engineering

MDPI

Article

Automatic Shoreline Position and Intertidal Foreshore Slope Detection from X-Band Radar Images Using Modified Temporal Waterline Method with Corrected Wave Run-up

Dipankar Kumar [1,2,*] and Satoshi Takewaka [3]

1 Graduate School of Systems and Information Engineering, University of Tsukuba, Tsukuba 305-8573, Ibaraki, Japan

2 Department of Mathematics, Bangabandhu Sheikh Mujibur Rahman Science and Technology University, Gopalganj 8100, Bangladesh

3 Faculty of Engineering, Information and Systems, University of Tsukuba, Tsukuba 305-8573, Ibaraki, Japan; takewaka@kz.tsukuba.ac.jp

* Correspondence: dks.bsmrstu@gmail.com; Tel.: +81-80-5456-2220

Received: 22 December 2018; Accepted: 8 February 2019; Published: 12 February 2019

Abstract: Automatic and accurate shoreline position and intertidal foreshore slope detection are challenging and significantly important for coastal dynamics. In the present study, a time series shoreline position and intertidal foreshore slope have been automatically detected using modified Temporal Waterline Method (mTWM) from time-averaged X-band radar images captured throughout the course of two-week tidal cycle variation over an area spanning 5.6 km on the Hasaki coast between 12 April 2005 and 31 December 2008. The methodology is based on the correlation map between the pixel intensity variation of the time-averaged X-band radar images and the binary signal of the tide level ranging from −0.8 m to 0.8 m. In order to ensure the binary signal represented each of the water levels in the intertidal shore profile, determining the water level direction-wise bottom elevation is considered as the modification. Random gaps were detected in the captured images owing to the unclear or oversaturation of the waterline signal. A horizontal shift in the detected shoreline positions was observed compared to the survey data previously collected at Hasaki Oceanographical Research Station (HORS). This horizontal shift can be attributed to wave breaking and high wave conditions. Wave set-up and run-up are the effects of wave breaking and high wave conditions, respectively. The correction of the wave set-up and run-up is considered to allow the upward shift of the water level position, as well as shoreline position, to the landward direction. The findings indicate that the shoreline positions derived by mTWM with the corrected wave run-up reasonably agree with the survey data. The mean absolute bias (MAB) between the survey data and the shoreline positions detected using mTWM with the corrected wave run-up is approximately 5.9 m, which is theoretically smaller than the spatial resolution of the radar measurements. The random gaps in the mTWM-derived shoreline positions are filled by Garcia's data filling algorithm which is a Penalized Least Squares regression method by means of the Discrete Cosine Transform (PLS-DCT). The MAB between survey data and the gap filled shoreline positions detected using TWM with corrected wave run-up is approximately 5.9 m. The obtained results indicate the accuracy of the mTWM with corrected wave run-up integrated with Garcia's method compared to the survey observations. The executed approach in this study is considered as an efficient and robust tool to automatically detect shoreline positions and intertidal foreshore slopes extracted from X-band radar images with the consideration of wave run-up correction.

Keywords: X-band radar; tidal variation; modified temporal waterline method; shoreline position; intertidal foreshore slope; wave run-up correction

1. Introduction

Monitoring and managing shorelines have received great attention owing to their social and economic significance for coastal regions around the world. In this context, coastal scientists and engineers are continuously seeking better tools to determine the accurate position of shorelines and to analyze the variations in shoreline positions. Shoreline variability and trends must be analyzed considering both their temporal and spatial changes. Shoreline is a well-known term for coastal scientists and engineers, which is ideally defined as the physical interface of the coastal land and the mean water level position [1,2].

Several data sources, such as aerial photographs, beach profile surveys, LiDAR (Light detection and ranging) surveys, video camera analysis, satellite imagery, and X-band radar images, have been utilized to derive shoreline positions and to investigate their variability [3–11].

Aerial images typically have broad spatial coverage; however, their temporal coverage is limited by acquisition time. LiDAR systems can cover large areas over short time periods and provide both accurate and more effective high-resolution digital elevation models (DEM). However, the data source is usually limited due to the high processing cost.

Video imaging systems are becoming more popular for continuously monitoring the coastal morphological behaviors worldwide. It can also be employed to track the sea surface features with high temporal resolution, as usual contributing several images per second in addition to trace stretches of intertidal sandy beaches at some sites around the world. Several video monitoring techniques have been used for the long-term monitoring of shoreline positions and nearshore beach processes [12–20].

Video cameras can generate color images, which allow the identification of wave breaking, suspension of foams, and sediment concentrations etc. and to seek out their temporal and spatial variations. It is often mounted on towers standing along the shoreline to provide slanted views, and the rectified montages of video images from different cameras allow for the measurement of waves and current fields, sediment transport, and morphological characteristics [21]. However, the application of video monitoring is only limited to daylight hours and fair-weather conditions. The invention of thermal infrared cameras allows the use of the data collected by video monitoring both during the day and at night to procure intertidal DEM [22]. In addition, thermal infrared cameras operate in low-light conditions to observe hydrodynamics in the nearshore zone. Such monitoring systems operate well in low visibility, have excellent temporal and spatial coverage, and can provide a data quality identical to that produced by a camera at slightly lower resolution, but they are limited to a remarkably higher range regardless of light conditions [23].

X-band radar is an integral part of nearshore the remote sensing infrastructure system, which can be used to overcome some of the limitations of video imaging [24]. It has been employed to trace the significant movement of wave crests over areas spanning several kilometers to detect coastal features. The most attractive feature of X-band radar systems applications is their ability to continuously and remotely collect data that allow the proper understanding of the nearshore coastal processes along the coastal areas under different weather conditions. The most significant advantage of shoreline monitoring with X-band radar is that it can provide real-time and uninterrupted observations even in bad weather conditions. Since the last two decades, land-based remote sensing monitoring systems such as X-band radar have become popular in coastal studies [6,25–29].

Many effective attempts to extract waterline positions from different types of remote sensing data such as synthetic aperture radar (SAR) images, satellite images, and X-band radar images have been reported. The waterline position is one type of shoreline indicator, which is used to describe the instantaneous position of land and the water interface from remotely sensed data. Due to the rhythmic rise and fall of tide levels with time, the instantaneous waterline position shifts horizontally landward or seaward. Generally, the shoreline position is defined at the mean water level (MWL). After an induced tide correction, the waterline position is known as the shoreline position.

In the past, several researchers [6,30–35] applied the waterline approach to data collected by Synthetic Aperture Radar (SAR) images, satellite images, and X-band radar images. For instance,

Koopmans and Wang [30] applied the waterline approach to SAR data, identified the waterlines of the intertidal areas of Wadden Sea, and used a tidal model to assign those contours to water elevations. Mason et al. [31] adopted the same approach using SAR images of the extensive intertidal areas of Morecambe Bay in the United Kingdom. Heygster et al. [32] also used SAR images and applied the waterline approach to generate the tidal flats topography along the German coast of the Wadden Sea during the period 1996 to 1999. Ryu et al. [33] extracted the waterline of a tidal flat in Gomso Bay, Korea from Landsat Thematic Mapper (TM) data, and Zhao et al. [34] applied an identical approach to multi-temporal satellite images of the Yangtze Delta, China. Recently, Xu et al. [35] have conducted a study to estimate the temporal and seasonal topographic changes associated with two major tidal flats in Gomso and Hampyeong Bay in the southern part of the West Sea of South Korea based on the waterline approach using 18 scenes of Landsat TM and Enhanced Thematic Mapper Plus (ETM+) data considering the corresponding tidal gauge observation data, covering the period 2003–2004. Takewaka [6] employed X-band marine radar to detect the shoreline positions and intertidal foreshore slopes by imaging the waterlines in the spatial domain and describing beach contour lines using time-averaged images and tidal records. The findings revealed an acceptable agreement with the survey data, concluding that the radar estimation can be considered as a mighty tool to track the morphology of sandy coasts. Furthermore, the shoreline positions and intertidal foreshore slopes were manually digitized using the tidal records and waterlines extracted from time-averaged X-band radar images.

The waterline extraction approach, which is based on visual interpretation, is considered an effective and straightforward procedure that can be extensively applied to remote sensing data. This method relies on the human eye's capability of precisely detecting the boundaries between the coastal land and water from aerial photographs, SAR images, optical satellite images, and X-band radar images. However, the manual digitization method involves very tedious, time-consuming, and labor-intensive practices to measure the boundaries between coastal land and water. Hence, developing an automated process to measure the shoreline positions was urgently needed. In this regard, numerous automatic digitization approaches [28,36–39] have been developed and implemented to detect expected objects from different types of remotely sensed data. In the present study, we focus on an automatic digitization approach to detect shoreline positions and intertidal foreshore slopes from time-averaged X-band radar images.

Recently, Bell et al. [28] developed the Temporal Waterline Method (TWM) as a robust way to automatically detect intertidal shore profiles using pixel intensity variations in time-averaged X-band radar images and binary signals of tidal records. For validation, Bell et al. [28] adopted the method to develop a morphological map of the target intertidal shore profile using radar dataset received from Hilbre Island at the mouth of Dee Estuary, UK. The resulting elevation maps presented the intertidal region with a radial range of 4 km of that area. The accuracy of the results obtained by the TWM was verified in comparison to airborne LIDAR data surveyed throughout the same area and during the radar survey period. The vertical elevation bias between the compared results was approximately ± 0.5 m, indicating that a relatively stable macrotidal environment was utilized as the test case. Furthermore, Bird et al. [29] employed the TWM to monitor inter- and intra-annual intertidal morphological changes and described the seasonal variations in the morphology of Hilbre Island at the mouth of Dee Estuary, UK. The heterogeneous study area examined by Bird et al. [29] found sandy, sandbank, intertidal sand flats, and saltmarsh beaches, along with several rocky outcroppings.

In the current study, an entirely sandy, highly varied, and micro-tidal area is examined. In such beach conditions, the original TWM does not perform correctly to detect the intertidal shore profile and is unable to determine the time-series shoreline positions and intertidal foreshore slopes from radar images. Hence, the modified Temporal Waterline Method (mTWM) is introduced and applied to extract the time-series of shoreline positions and intertidal foreshore slopes in a sandy, micro-tidal beach site at Hasaki Oceanographical Research Station (HORS) in Hasaki, Japan. The limitations of the TWM and advantages of the mTWM will be discussed.

The mTWM exploits the excellent temporal update rates provided by land-based remote sensing, while it can uniformly be used for a wide range of remote sensing data with an adequate number of images. Accordingly, in this study, 31,888 X-band radar images captured during the period 12 April 2005 to 31 December 2008 were used to measure the shoreline positions and intertidal foreshore slopes.

The main objective of this study is to focus on application of the mTWM to a highly movable micro-tidal sandy beach exposed to energetic waves of the south Pacific Ocean with a comparison of the shoreline position and intertidal foreshore slope results with survey results along the pier in the four years from 2005 to 2008. More precisely, the study is intended to validate the accuracy of the mTWM extracted shoreline positions and intertidal foreshore slopes compared to survey data. The objective of the study involves the following:

(i) to implement an automated mTWM to detect the time series of shoreline positions and intertidal foreshore slopes extracted from time stack X-band radar images considering tidal variations in the abovementioned entirely sandy and highly varied study site;
(ii) to validate the derived temporal updates of shoreline positions and intertidal foreshore slopes at the research pier in comparison to the previously collected survey data;
(iii) to compare the temporal updates of the extracted shoreline positions with corrected wave set-up and run-up with survey data.

2. Radar Observation and Data Processing

2.1. Study Site

The Kashima Coast, which is located in the Ibaraki Prefecture of Japan facing the South Pacific Ocean, is almost straight with a sandy beach that spread over an area spanning approximately 17 km. It is bounded on the south end by the Choshi Fishery Port and the north end by the Kashima Port. The study area is marked with the yellow rectangle box shown in Figure 1. The 400 m long field observation research pier is located in the center of HORS with the research building located on the backshore approximately 110 m backward from the mean shoreline position. The median sediment diameter is 0.18 mm and occasionally increases to 1.00 mm around troughs after severe storms [40]. Based on previous studies [41,42] at the single-bar Hasaki coast, the mean beach slope was 1/50, ranging from −60 to 200 m seaward and 1/200 in the deeper area.

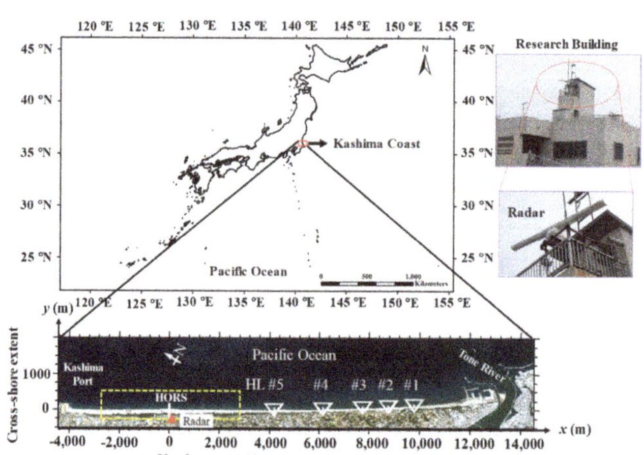

Figure 1. A location map of the study area: Kashima Coast, Japan. The red triangle indicates the radar location. The origin of the coordinate system is located at the base of HORS. The yellow rectangular box indicates the area covered by the radar observation, and the white triangles indicate the positions of the artificial headlands. Photos of the research building and radar are on the right.

2.2. Tide and Wave Data

The Japanese Meteorological Agency collects hourly tidal elevation data at Choshi Fishery Port, and the interannual variation is shown in Figure 2a. The tidal environment of this area is micro-tidal with a tidal variability range of approximately 1.5 m. Figure 2b–d show the variations in the significant wave height, period, and incidence angle, respectively. These data are measured every two hours at Kashima Port Station, where the mean water depth is approximately 24 m, as part of the Nationwide Ocean Wave Information Network for Ports and Harbors (NOWPHAS: http://nowphas.mlit.go.jp/eng.html).

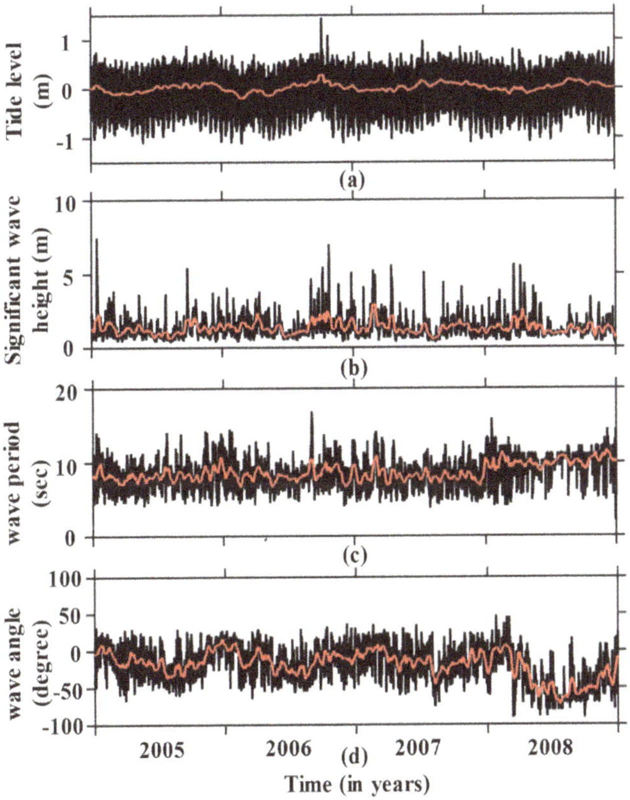

Figure 2. The temporal variations of (**a**) the tide level, (**b**) the significant wave height, (**c**) the wave period, and (**d**) the wave direction during 2005–2008. The red line indicates the 14-day moving average of the respective data.

2.3. X-Band Radar System and Time-Averaged Images

A conventional X-band radar measurement was conducted at HORS research pier of the Port and Airport Research Institute (PARI) in Hasaki, Japan. The 2.8 m radar antenna was mounted on the roof of the research building 17 m above mean sea level. The antenna rotated with a period of approximately 2.5 s and transmitted with a beam width of 0.8° in the horizontal direction and 25.0° in the vertical direction. The echo signals from the sea surface were grabbed with a specially designed A/D board installed on a computer, with sampling rate of 20 Mhz [6,43].

Individual echo images sampled every 2 s were averaged yielding a "time-averaged image" or so-called "time exposure" [6]. Ensembles of 512 individual echo images over 17 min (512 × 2 s = 1024 s) were processed to time-averaged radar images for every hour. Figure 3a shows an example of a time-averaged X-band radar image and coordinate system. The x-axis corresponds to the longshore extent, and the y-axis is taken in the cross-shore direction, positive toward the offshore. The echo signals are converted into a rectangular image size of 1024 pixels horizontal (longshore extent) and 512 pixels vertical (cross-shore extent). The pixel size of the rectangular image is approximately 5.42 m, whereas the theoretical spatial resolution of the radar system is approximately 7.5 m. Thus, the image covers an area spanning approximately 5556 m (1024 pixels) in the alongshore direction and 2778 m (512 pixels) in the cross-shore direction. The pixel intensities of the time-averaged gray images are belonging 0 (no backscatter) to 255 (saturation). Owing to reflections, individual waves disappear in the time-averaged radar images. A horizontal and bright edge extending in the longshore corresponds to the shoreline, which is marked in Figure 3a; the radar position is located at the center of the bottom edge in the figure, indicated by a black oval. The time-averaged images enable the identification of the intertidal bathymetry, breaker zone, rip current, bar crest locations, mega-cusp migrations, wave run-up, and other features [6,26,27,43–45]. Previously, Takewaka [6] examined the accuracy of the intertidal morphological feature using time-averaged images compared to field survey results. Hourly processed time-averaged images have been collected since 2005; however, some data gaps were reported due to mechanical trouble.

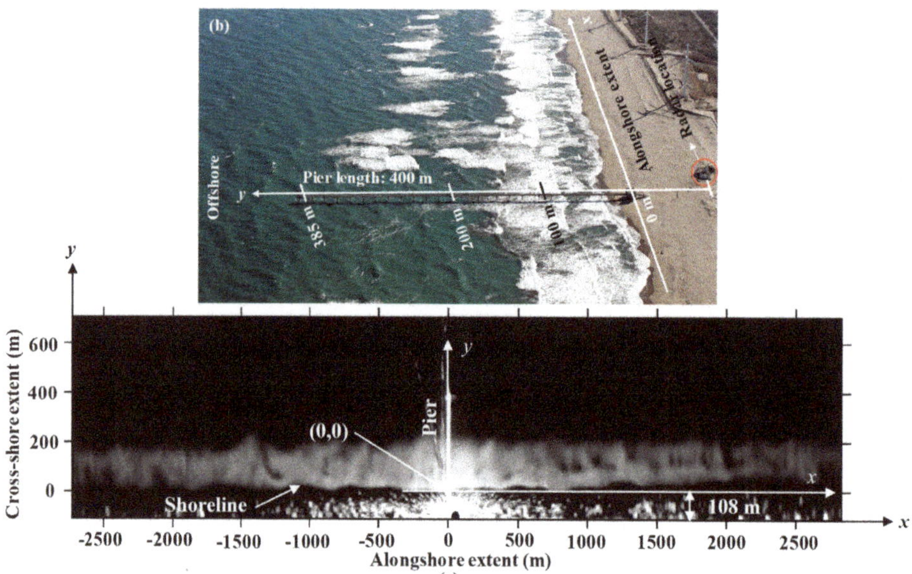

Figure 3. (**a**) The time-averaged X-band radar echo image and coordinate system (16 April 2005 at 12:00) and (**b**) the aerial view of the research pier. Survey data are measured along the pier from radar −110 m to 385 m.

3. Beach Profile along the Pier

Beach profiles along the research pier were measured at 5 m intervals every day from −110 m to 385 m, except on weekends and national holidays. Figure 3b shows an aerial photo of the research pier. Figure 4a illustrates an example of a surveyed beach profile, and Figure 4b shows the variation in beach profile. Figure 4c displays the mean, maximum, and minimum beach profiles, whereas Figure 4d depicts their standard deviation. Small standard deviation values and narrow envelopes indicate stable regions, while large standard deviations and wide envelopes are associated with regions of high variability.

The shoreline position is defined where the bottom elevation is equal to 0.0 m (Tokyo Peil, T.P). The intertidal foreshore slope is defined as the linear slope of the beach profile ranging between the height of −0.8 m and 0.8 m. Figure 5a,b show variations in the shoreline positions and intertidal foreshore slopes during 2005–2008.

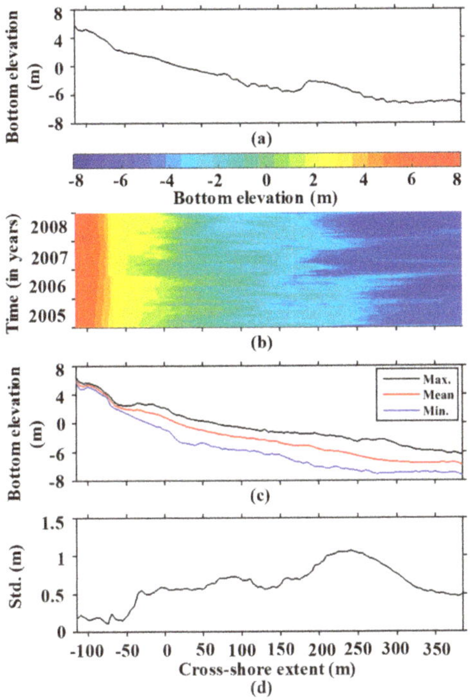

Figure 4. (**a**) An example of the beach profile variation along the pier (18 April 2005 at 7:00 am); (**b**) the beach profile variation during 2005–2008; (**c**) the mean, minimum, and maximum range of the beach profile; and (**d**) the standard deviation.

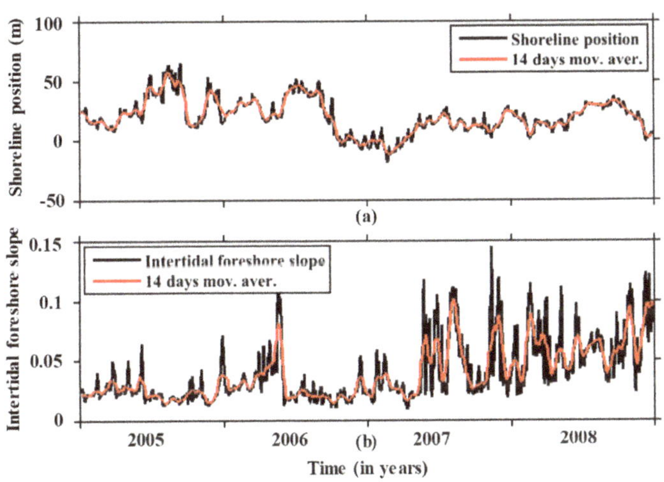

Figure 5. The temporal variations of (**a**) the shoreline positions and (**b**) the intertidal foreshore slopes at the research pier. The red line indicates the 14-day moving average shoreline positions and slopes.

4. Shoreline Position and Intertidal Foreshore Slope Estimation

4.1. Intertidal Beach Profile and Shoreline Estimation Using TWM and mTWM

The TWM was developed by Bell et al. [28] to extract an intertidal beach profile from time-averaged X-band radar images considering the tidal variation, and the mTWM is the modified form of Bell et al.'s [28] approach. First, the basic concept of TWM and then our modification are described in this subsection.

Time-series of individual radar pixel intensities are gathered from hourly time-averaged images across the selected timescale of two weeks, including a full spring-neap cycle (as an example, June 17–30, 2005). Figure 6a displays the cross-shore time stack images within the range between $y = 5$ m and $y = 103$ m. The red line marked in Figure 6b indicates the variation of the instantaneous waterline position digitized manually by visual inspection, and Figure 6c is the concurrent tidal records. In general, the instantaneous waterline position is the boundary between the coastal land and water at one instant in every measurement. It can be clearly seen that the instantaneous waterline position moves seaward when the tide ebbs and vice versa (see Figure 6b,c).

At a longshore point x, the pixel intensity from the intertidal zone at y is extracted from time-averaged images and repeated for two weeks, which yields $P(y, n)$ where n is time. Figure 7a–d illustrate the variations in pixel intensities over a two weeks period, showing the periodic episode. High pixel intensities are acquired from time-averaged images when the waterline clearly appears, and vice versa.

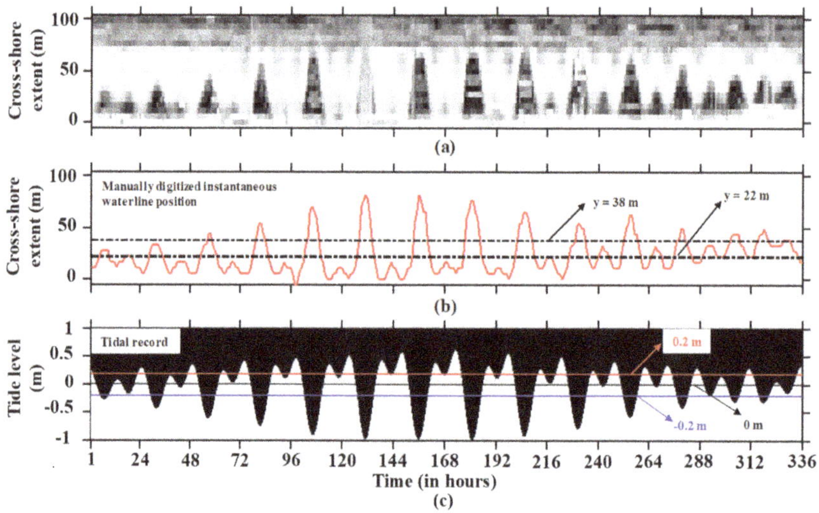

Figure 6. (**a**) A cross-shore time stack image between the cross-shore extent, $y = -5 - 103$ m at $x = -49$ m; (**b**) the manually digitized waterline positions from the cross-shore time stack image; and (**c**) the tidal record during 17–30 June 2005. The instantaneous waterline position shifts seawards when the tide falls and vice versa.

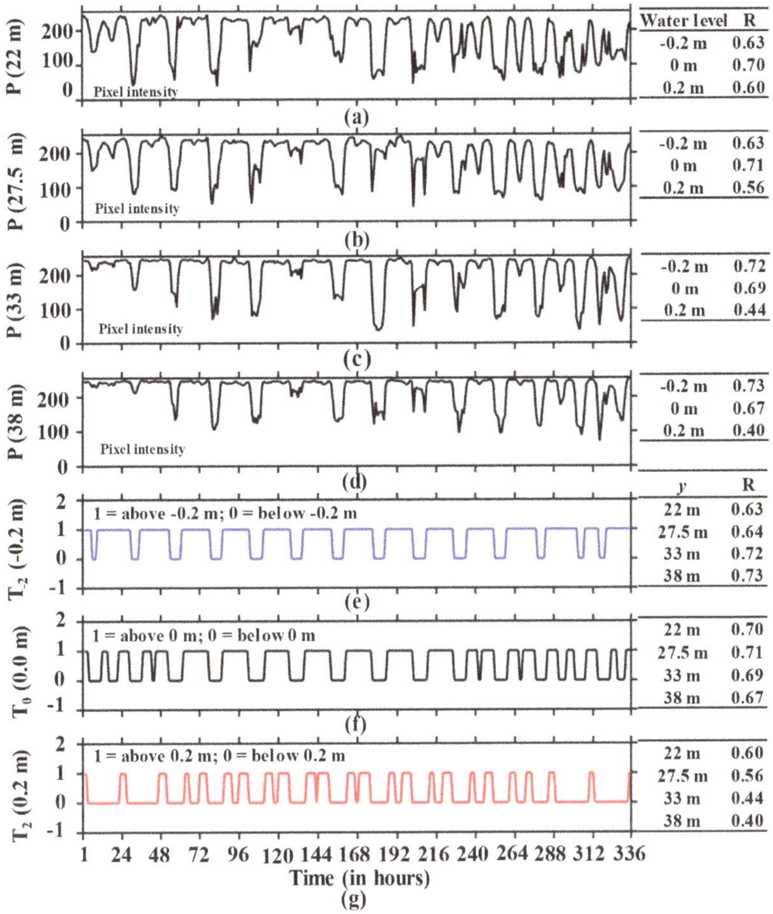

Figure 7. The pixel intensities over two weeks (17–30 June 2005) at $x = -49$ m and (**a**) $y = 22$ m, (**b**) $y = 27.5$ m, (**c**) $y = 33$ m, and (**d**) $y = 38$ m. (**e**) The tidal binary signal indicating above or below water level -0.2 m: (**f**) water level 0.0 m and (**g**) water level 0.2 m. The tables on the right-hand side of a–g illustrate the correlation coefficients between the pixel intensity variation and tidal binary signals. Higher correlation values show strong similarities between pixel intensity variation and binary signal representation of tide records.

The tidal elevation range between -0.8 and 0.8 m is considered with 0.1 m increments to generate binary pulse sequences, which are hereafter indicated with water level number $N(N = -8, -7, \ldots, 8)$. The study site exhibits a micro-tidal influence with a tidal variability range of approximately 1.5 m. Each tidal elevation within a given tidal range of -0.8–0.8 m with 0.1 m increments has a unique binary pulse signal that representing 1 and 0, where 1 and 0 indicates a binary pulse sequence above and below water level at a given water level of the tidal cycle. Each elevation produced a unique binary pulse sequence, which is defined by $T_N(n)$, as illustrated in Figure 7e–g for $T_{N=-2}(-0.2\text{ m})$, $T_{N=0}(0\text{ m})$, and $T_{N=2}(0.2\text{ m})$, respectively. As shown in Figure 7, the extracted pixel intensities from time-averaged radar images and binary pulse signals at water levels have signal similarities.

Now, the correlation coefficient $R(y,N)$ is calculated at each possible elevation to measure the signal similarities between the pixel intensity variation $P(y,N)$ at different cross-shore extents, in addition to the tidal binary pulse sequence $T_N(n)$ for different water level numbers:

$$R(y,N) = \frac{\sum_{n=1}^{24\times14}\{(P(y,n) - \overline{P}(y))(T_N(n) - \overline{T}_N)\}}{\sqrt{\sum_{n=1}^{24\times14}(P(y,n) - \overline{P}(y))^2}\sqrt{\sum_{n=1}^{24\times14}((T_N(n) - \overline{T}_N))^2}} \quad (1)$$

where $\overline{P}(y)$ is the mean pixel intensity and $\overline{T}_N$ is the mean tidal binary pulse sequence over two weeks. A high correlation value indicates the strong signal similarities between the pixel intensity variation and binary representation of the tidal record sequence, which describes the implication for bottom elevation.

The tables at the right-hand side of Figure 7a–d illustrate the correlation coefficients between each tidal binary signal pulse (−0.2 m, 0 m and 0.2 m) with individual pixel intensity variation, and Figure 7e–g illustrate the correlation coefficient between the pixel intensity variation at $y = 22$ m, 27.5 m, 33 m, and 38 m with individual tidal binary signal pulses. The maximum correlation coefficient is used to indicate the cross-shore position at a certain water level, and the process is repeated for all binary signal tidal elevation extracted from tide records. The outcomes are then utilized to plot a correlation map for a given longshore extent ($x = -49$ m) at different water levels (−0.8 to 0.8 m with 0.1 m increments) within the intertidal range ($y = 0$–80 m with 5.4 m increments), as shown in Figure 8. Here, no correlation coefficients are available at water levels of 0.6–0.8 m because the correlation values less than 0.2 at the 0.6 m level and the tide level did not reach 0.7 m and 0.8 m levels.

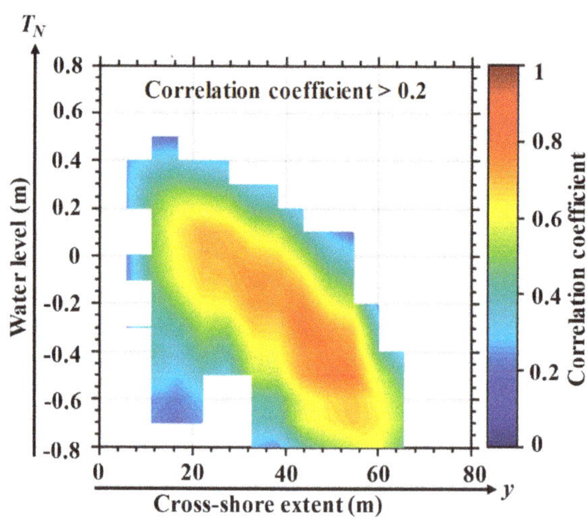

Figure 8. The distribution of correlation coefficients for different tidal signals (−0.8 to 0.8 m) and cross-shore positions $0 \leq y \leq 80$ m at $x = -49$ m during 17–30 June 2005.

By using the correlation coefficient map in Figure 8, the intertidal shore profile can be determined in the following two ways:

(i) To detect the water level from a cross-shore position correlation coefficient distribution, where the distribution is maximum. The detected point is regarded as the bottom elevation. Similarly, the maximum correlation at other water levels are estimated throughout the remaining cross-shore positions, and the intertidal cross-shore profile is determined. This approach is known as TWM, which was initially introduced by Bell et al. [28].

(ii) To detect the cross-shore position from the distribution of correlation coefficients across a water level, where the distribution is maximum. The detected point is also regarded as the bottom elevation. Proceeding the same way, the maximum correlation is established at other cross-shore positions across the remaining water levels, and the intertidal cross-shore profile is produced. This approach is called mTWM, which is the modified form of Bell et al.'s [28] approach.

Figure 9a shows an example of the correlation map for a given longshore extent ($x = -49$ m) at different water levels (−0.8–0.8 m with 0.1 m increments) within the cross-shore extent (y = 0–80 m with 5.4 m increments). The red arrow line indicates an example of the cross-shore position transect at $y = 27.5$ m. Figure 9b shows the correlation coefficient distribution at $y = 27.5$ m extracted from the map within the water levels range (−0.5–0.4 m). The maximum correlation is established at 0.0 m water level, which is regarded as the bottom elevation of this point. Similarly, the maximum correlation at other water levels are determined among the remaining cross-shore positions, and the intertidal cross-shore profile is estimated. Then, a linear fitting model is applied to determine the shoreline position at 0.0 m water level, and the intertidal foreshore slope is calculated from regression line coefficient within the intertidal range. As shown in Figure 9c, the shoreline position is determined at $y = 24.3$ m, and the intertidal foreshore slope is 0.017. Compared to the surveyed shore profile, we observed that the TWM failed to estimate the shore profile accurately from radar images.

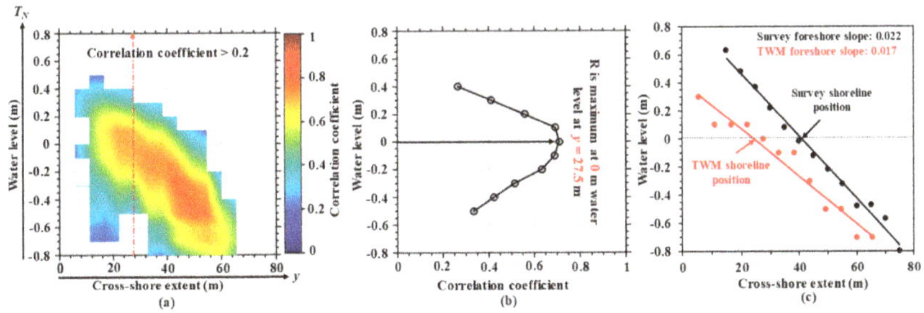

Figure 9. An example Temporal Waterline Method (TWM) detected beach profile at $x = -49$ m during 17–30 June 2005. (**a**) The distribution of correlation coefficients for different tidal signals (−0.8 to 0.8 m) and cross-shore positions ($0 \leq y \leq 80$ m). (**b**) The correlation coefficients for different tidal binary signals at $y = 27.5$ m. The maximum correlation is established at water level 0.0 m. (**c**) The intertidal beach profile determined by linear regression. The shoreline position ($y = 24.3$ m) is defined at water level 0.0 m, and the black regression line indicates the survey intertidal beach profile on 17 June 2005.

In TWM, to look for water levels from the distribution of correlation coefficients along the cross-shore positions, the maximum correlation coefficient is frequently established between −0.8 m and 0.4 m water levels. Due to the strong similarities between the cross-shore position pixel intensities extracted from images and tidal binary pulse signals (−0.8 to 0.4 m), a high correlation coefficient frequently appears at these water levels. In another sense, the original TWM cannot detect the water levels (0.5–0.8 m) due to the presence of lower correlation coefficients than in other water levels (−0.8 to 0.4 m). As a result, the TWM failed to estimate the actual intertidal shore profile and was unable to determine the shoreline position and intertidal foreshore slope with a significant bias.

Landwards cross-shore position pixel intensities have no periodic oscillation like tidal binary pulse signals, which is due to the irregularity of water line with time for each transition. In such reason, cross-shore position pixel intensities and tidal binary pulse signals provide poor and almost identical correlation coefficient values. Therefore, cross-shore direction-wise water level detection technique (TWM) is failed to estimate the expected objects. Under a changed search direction (mTWM) to find the bottom elevation, the extracted intertidal shore profile becomes closer to the survey profile. The above-described bottom elevation finding approach is considered as the modification in this study. Prior to estimating the intertidal shore profile, the input of the intertidal range was selected from the tidal elevation data. To ensure each of water levels in the intertidal shore profile, it is expected that the water level wise cross-shore position detection is more appropriate than TWM. The performance of this approach will be described later.

Figure 10a displays a correlation map for a given longshore extent ($x = -49$ m) at different water levels (-0.8–0.8 m with 0.1 m increments) within the intertidal range (y = 0–80 m with 5.4 m increments). The red arrow line indicates an example of a 0.0 m water level transect. Figure 10b shows the correlation coefficient distribution at a water level of 0.0 m extracted from the map within the cross-shore extent range ($0 \leq y \leq 80$ m). The maximum correlation is established at the $y = 27.5$ m cross-shore position, which is regarded as the bottom elevation of this point. Similarly, the maximum correlation at other cross-shore positions are determined among the remaining water levels and the intertidal cross-shore profile is estimated. Then, a linear fitting model is applied to determine the shoreline position at the 0.0 m water level, and the intertidal foreshore slope is calculated from regression line coefficients within the intertidal range. The shoreline position in Figure 10c is determined at $y = 30$ m, and the intertidal foreshore slope is 0.024. We compared the estimated slope with a survey intertidal profile slope on 17 June 2006. Figure 10c shows that the mTWM estimated intertidal foreshore slope is very close to the survey slope; however, the extracted shoreline positions have some landward shifts compared to the survey data. This shift is induced by the effect of wave set-up or wave run-up. The correlation coefficients greater than 0.2 is used for the both cases.

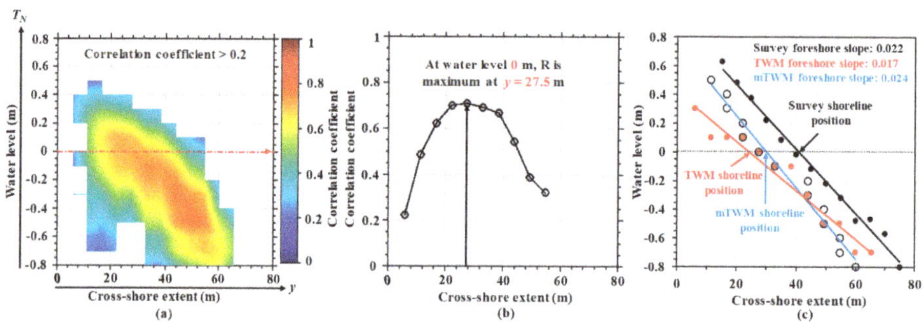

Figure 10. An example of a modified Temporal Waterline Method (mTWM) detected intertidal beach profile at $x = -49$ m during 17–30 June 2005. (**a**) The distribution of correlation coefficients for different tidal signals (-0.8 to 0.8 m) and cross-shore positions ($0 \leq y \leq 80$ m). (**b**) The distribution of correlation coefficients for cross-shore positions at water level 0.0 m. The maximum correlation is established at $y = 27.5$ m. (**c**) The intertidal beach profile determined by linear regression. The shoreline position ($y = 30$ m) is defined at water level 0.0 m, where red and black regression lines indicate the TWM and survey intertidal beach profile, respectively, on 17 June 2005.

The data collected over a two-week period were processed to obtain shoreline position and intertidal foreshore slope data. This process was repeated between 12 April 2005 and 31 December 2008 to estimate variations in the shoreline position and intertidal foreshore slope. Figure 11a,b depict variations in the shoreline position derived by TWM and mTWM compared to those obtained by survey data collected at the research pier; however, some random gaps are observed owing to the lack of strong waterline signals or the saturation of echo signals. The bias between the results obtained by survey data and mTWM derived results is presented in Figure 11c, showing 14 m mean absolute bias (MAB). A similar trend is observed between the shoreline positions estimated by mTWM and the survey data, which is confirmed by the correlation coefficient ($R = 0.86$), shown in Figure 11d; however, a horizontal shift is detected between the mTWM derived results and surveyed shoreline positions. This shift is induced by the effect of wave set-up or wave run-up or by a combination of correction factors and will be discussed in Sections 4.2 and 4.3. On the other hand, the calculated MAB between survey and TWM derived shoreline positions is 19.2 m, which is larger than the obtained MAB between survey and mTWM. The large number of random gaps also appear in the TWM estimation and are effects of linear regression conditions. This is clearly visible in Figure 11a.

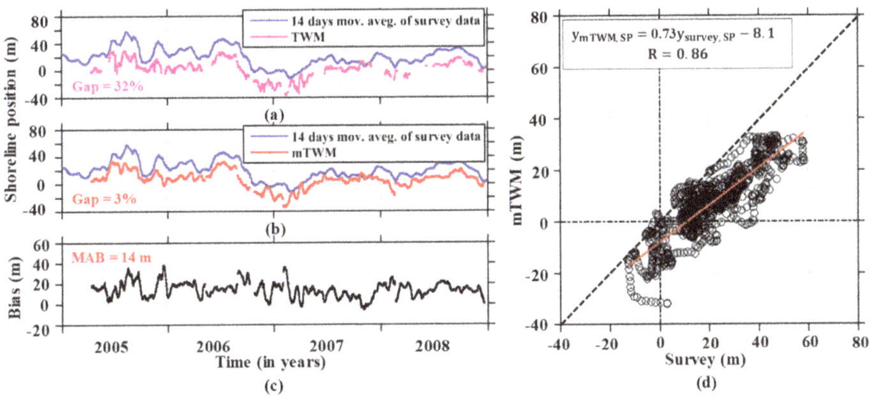

Figure 11. The temporal variations in the shoreline positions (SP) estimated from (**a**) the survey data and TWM and (**b**) the survey data and mTWM derived results (at $x = -49$ m), (**c**) the bias between the survey and mTWM results, and (**d**) the scatter plot of survey vs. mTWM. The red line represents the linear fitting line.

Figure 12a,b present the temporal variations in intertidal foreshore slopes derived by TWM and mTWM during the study period and present the survey data, and the bias between those obtained from the survey and mTWM outcomes are shown in Figure 12c. Compared to the surveyed slopes, TWM-estimated slopes are approximately three times milder. That means TWM cannot estimate the slopes accurately from radar images. The reason for this discrepancy is described above. In order to estimate the bias, the radar-detected values at each time point were subtracted from the survey detected values. The surveyed slope is sometimes steeper than that estimated by the mTWM owing to the comparison between different cross section transect estimated observations. This discrepancy can be attributed to the fact that survey data are collected along the research pier, while data used for the mTWM are extracted 49 m far from the research pier. In addition, a fixed gap is observed along the research pier ($-49 < x < 65$ m) due to saturated echo signals.

Water flows typically faster around piers and abutments, making them susceptible to local scour or the removal of sediment [46]. Due to local scour, sand is eroded around the pier and accumulated to the near side. As a consequence, the cross-shore beach profile changes gradually along the pier, and beach slopes can be steeper than the surroundings. Authors have visited the study site several times and sometimes found steeper foreshore close to the pier. This may be one possible reason for the discrepancy in intertidal foreshore slope estimations.

The mTWM is applied to all alongshore locations; the spatiotemporal variations in the shoreline positions and intertidal foreshore slopes for the entire area are obtained for the period between 12 April 2005 and 31 December 2008, as shown in Figure 13. The x-axis denotes alongshore extent, and the y-axis denotes the time history between 2005 and 2008. The color bar of Figure 13a indicates the landward (blue) and seaward (red) shoreline positions from its origin. On the other hand, the color bar pattern of Figure 13b shows the milder (blue) and steeper (red) intertidal beach slopes. However, 3% of the estimates are missing owing to the previously mentioned reason.

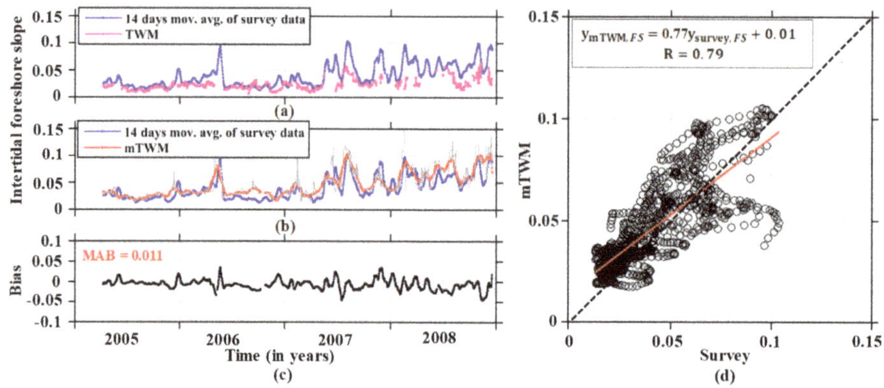

Figure 12. The temporal variations in intertidal foreshore slopes (FS) estimated from (**a**) the survey data and TWM and (**b**) the survey data and mTWM derived results (at $x = -49$ m), where the off-white and red lines indicate mTWM-derived intertidal foreshore slopes and its 14-day moving average, respectively. (**c**) The bias between the survey and mTWM results. (**d**) The scatter plot of the survey vs. mTWM.

4.2. Wave Set-up Correction

Wave set-up is defined as an increase in the mean water level above the still water level owing to momentum transfer to the water column with the presence of breaking waves, which is a common dynamic process in the nearshore zone [47]. Within the surf zone where wave breaking dissipates energy, the wave thrust decreases as the breaking surge travels shoreward; consequently, the mean water level rises, and thus wave set-up occurs. Figure 14a is a schematic illustration of the wave set-up that indicates the corresponding rise in waterline. It shows that the mTWM estimated results are shifted landwards compared to the survey data. Hence, the correction of wave set-up on the mTWM is essential to reduce the horizontal shift in shoreline positions. Based on this schematic scenario, the following strategy is adopted to correct the wave set-up and to obtain more accurate results by mTWM.

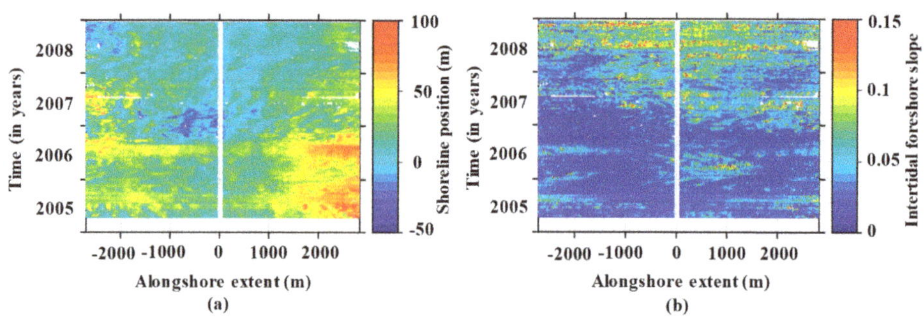

Figure 13. Spatial and temporal variations in (**a**) the shoreline positions and (**b**) the intertidal foreshore slopes estimated by mTWM during 12 April 2005 to 31 December 2008.

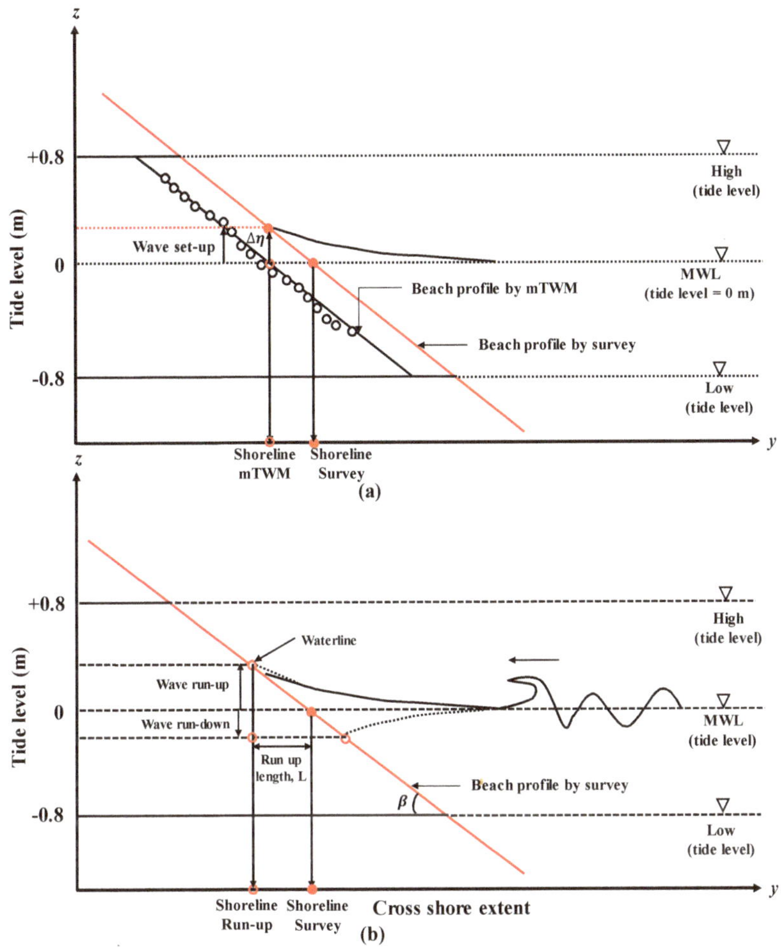

Figure 14. A schematic illustration of the shift in the estimated shoreline position due to (**a**) wave set-up or (**b**) wave run-up.

Several studies have been conducted on the wave set-up on sloping beaches. In this study, the following empirical formula is used to estimate the amount of wave set-up (η) at Hasaki beach caused by normal wave incidence [48]:

$$\frac{\eta}{H_0} = A_0 + A_1 log(H_0/L_0) + A_2(log(H_0/L_0))^2 \quad (2)$$

and

$$\left.\begin{aligned} A_0 &= 0.0063 + 0.768\, tan\beta \\ A_1 &= -0.0083 - 0.011\, tan\beta \\ A_2 &= 0.00372 + 0.0148\, tan\beta \end{aligned}\right\} \quad (3)$$

where $tan\beta$ is the bottom slope; H_0 is the offshore wave height, which is determined by the linear wave theory equation $H_1 = H_0 K_r K_s$ where K_r is the refraction coefficient, K_s is the shoaling coefficient, and H_1 is the NOWPHAS wave height; and L_0 is the offshore wave length, which is determined by $L_0 = 1.56T^2$ and where T is the wave period.

The amount of wave set-up $\Delta\eta$ at Hasaki beach due to oblique wave incidence is given by [48]

$$\Delta\eta = \eta \times (cos\theta_0)^{0.545+0.038\ log(H_0/L_0)} \quad (4)$$

where, θ_0 is the offshore wave direction estimated by Snell's law.

The amount of wave set-up is calculated for every two-hour time frame using Equations (2)–(4), as shown in Figure 15. In order to determine the wave set-up, the mean beach slope (0.04) is used, which is extracted by the mTWM from radar images during 2005–2008. A set-up corrected water level for the mTWM estimates is set by adding the wave set-up to the tide record, as shown in Figure 16. The corrected water level is shifted upward by approximately 0.17 m compared to the original tide record. A new correlation map is established with the corrected water level at $x = -49$ m shown in Figure 17a. The correlation coefficients obtained at the 0.0 m water level at cross-shore positions ($0 \leq y \leq 80$ m) are shown in Figure 17b with solid red circles. The maximum correlation is now established at $y = 33$ m. Similarly, the obtained intertidal beach profile with corrected tide signal is shown in Figure 17c. The results of the linear regression analysis indicate the estimated shoreline position at $y = 34$ m with intertidal foreshore slope $\beta = 0.024$. The shoreline position is now shifted 4 m landwards, while the intertidal foreshore remains almost the same. Hence, it may be expected that we can apply the tidal variation-based intertidal foreshore slopes extracted by mTWM for further analysis.

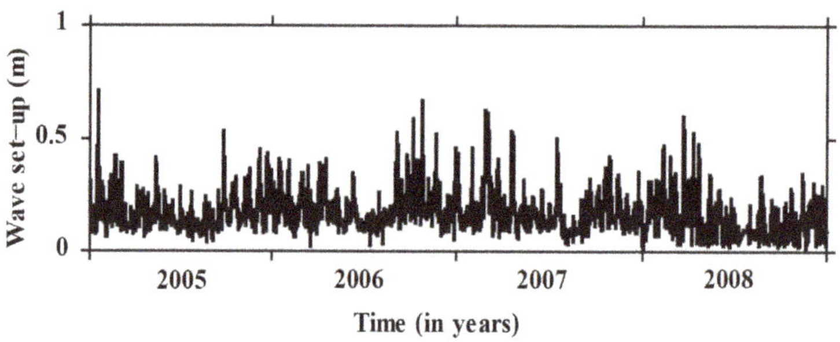

Figure 15. The variations in wave set-up during 2005–2008.

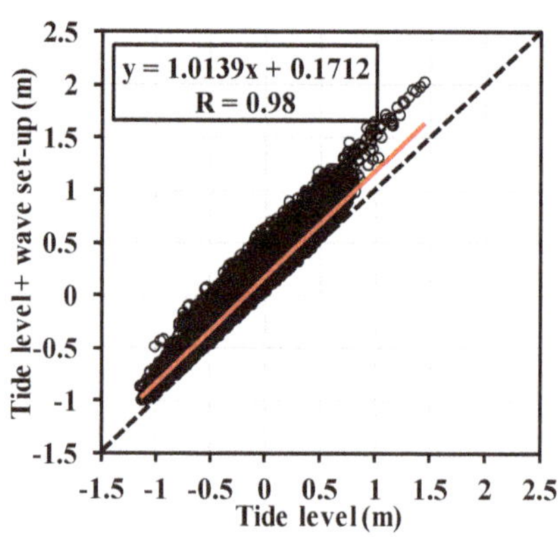

Figure 16. The scatter plot of the tide level and tide level + wave set-up (2005–2008). The red line represents the linear fitting line.

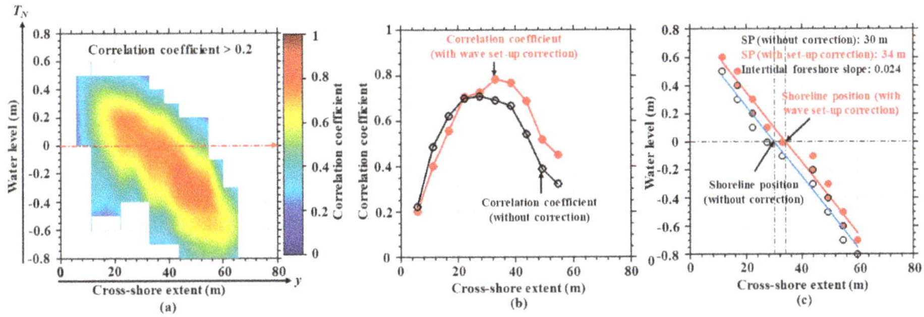

Figure 17. An example of the mTWM detected beach profile at $x = -49$ m with a corrected wave set-up during 17–30 June 2005. (**a**) The distribution of correlation coefficients between different tidal signals (–0.8 to 0.8 m) and cross-shore positions ($0 \leq y \leq 80$ m). (**b**) The distribution of correlation coefficients for cross-shore positions at a water level 0.0 m (without and with corrected wave set-up). The maximum correlation is established at $y = 33$ m for a corrected wave set-up. (**c**) The intertidal beach profile determined by linear regression. The shoreline position is defined at water level 0.0 m.

Figure 18a depicts a comparison between the survey and mTWM-extracted shoreline positions with corrected set-up, and the vertical elevation bias between the survey and mTWM results with a corrected wave set-up is shown in Figure 18b. As shown in Figure 18c, the correlation coefficient ($R = 0.84$) in both cases is almost equivalent to the previously obtained results. Similar trends are observed between shoreline positions obtained by the mTWM with a corrected wave set-up and the survey data; however, the MAB of the estimated results dropped from 14 m to 10.5 m.

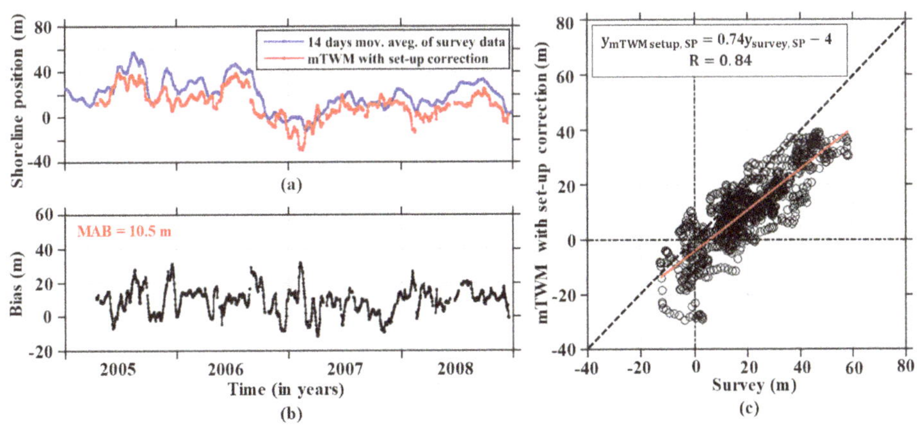

Figure 18. (**a**) The temporal variations in shoreline position (SP) estimated from survey data and mTWM derived results with corrected wave set-up (at $x = -49$ m). (**b**) The bias between the survey data and mTWM-derived results with corrected wave set-up. (**c**) The scatter plot of the survey vs. mTWM with corrected wave set-up. The red line represents the linear fitting line.

The variation in intertidal foreshore slope with a corrected wave set-up that shown in Figure 19 is similar to the previously estimated results (Figure 12).

Figures 18 and 19 confirm the effectiveness of the corrected wave set-up to determine the shoreline positions using the mTWM accurately. It allows shifting of the estimated shoreline position landward without changing the shape of the intertidal beach profile.

The mTWM with a corrected wave set-up is applied to all alongshore locations; the spatiotemporal variations in shoreline positions and intertidal foreshore slopes can be obtained for the entire study site from 12 April 2005 to 31 December 2008. For the sake of simplicity, wave set-up corrected spatiotemporal variations in shoreline positions and intertidal foreshore slopes are not shown in the manuscript.

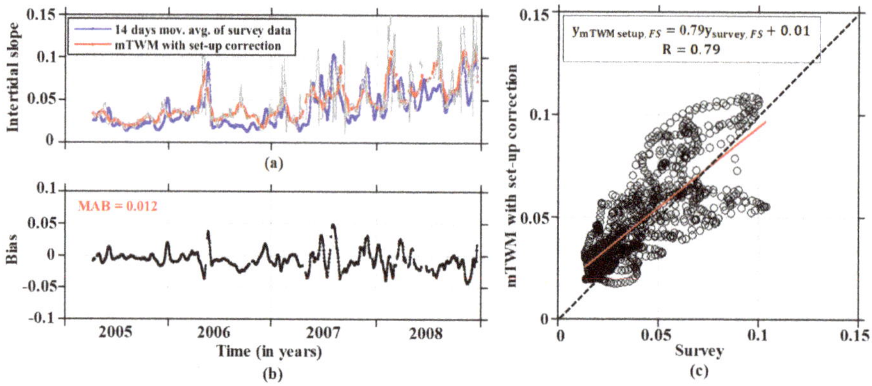

Figure 19. (**a**) The temporal variations in intertidal foreshore slope (FS) estimated from the survey and mTWM-derived results with corrected wave set-up (at $x = -49$ m), where off-white and red lines indicate the mTWM-derived intertidal foreshore slope and its 14-day moving average, respectively. (**b**) The bias between the survey data and mTWM-derived results with corrected wave set-up. (**c**) The scatter plot of the survey data vs. mTWM derived results with corrected wave set-up. The red line represents the linear fitting line.

4.3. Wave Run-up Correction

Wave run-up is defined as "the landward extent of wave uprush measured vertically from the still water level" [49]. Accordingly, during high wave conditions, the waterline positions shift more landward owing to the wave run-up effect. Basically, the run-up characteristics change with the beach slope and offshore wave properties. Figure 14b is a schematic illustration showing that the results obtained by the mTWM extraction with a corrected wave run-up are further shifted landward compared to the survey data. Hence, it may expect that the corrected wave run-up is essential to accurately determine the shoreline positions for the study site. The following strategy is adopted to correct the wave run-up to obtain more accurate estimates by the mTWM:

Based on both field and laboratory observations, Hunt [50] initially proposed the following wave run-up formula, which is typically defined by

$$\frac{R}{H_0} = c\xi_0 + d \tag{5}$$

where R is the vertical run-up normalized by the offshore wave height H_0 and c and d are dimensionless constants.

Several researchers [51–55] have investigated the amount of wave run-up on intermediate to reactive beaches by using video camera analysis. Hasan and Takewaka [44] first analyzed wave run-up using data extracted from X-band radar images at Hasaki beach and established a formula to calculate the infragravity wave run-up:

$$\frac{R}{H_0} = 1.025\xi_0 + 0.03 \tag{6}$$

where ξ_0 is the surf similarity parameter [56] and is expressed by

$$\xi_0 = \frac{tan\beta}{\sqrt{H_0/L_0}} \tag{7}$$

where $tan\beta$ is the beach slope, L_0 is the deep-water wave length defined as $L_0 = gT^2/2\pi$, g is the gravitational acceleration, and T is the wave period. The low surf similarity parameters ($\xi_0 \leq 0.3$) usually specify a dissipative beach condition while higher values suggest a more reactive condition. Figure 20 shows the variation in wave run-up height along a transect at $x = -49$ m over the period 12 April 2005 to 31 December 2008 using Equations (6) and (7). H_0 and L_0 are calculated from the wave records, and the intertidal foreshore slope ($tan\beta$) is derived from mTWM estimation considering tidal variation.

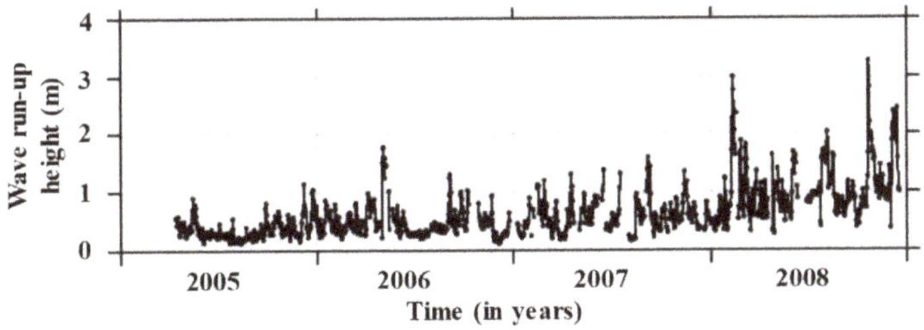

Figure 20. The wave run-up height (at $x = -49$ m) estimated using an empirical wave run-up formula [44].

The following formula is employed to correct the wave run-up effect on the shoreline position derived by the mTWM:

$$mTWM\ runup\ corrected\ SP = mTWM\ estimated\ SP + \left(\frac{Wave\ runup\ height}{mTWM\ estimated\ slope}\right) \tag{8}$$

where the second term on the right side denotes the landward horizontal shift due to wave run-up, which is known as the wave run-up length, and the 14-day moving average filter is used for the wave run-up length calculation.

The variations in shoreline position obtained by the mTWM with corrected wave run-up compared to the survey data are illustrated in Figure 21a. Correspondingly, the vertical elevation bias between the survey and mTWM-estimated results with corrected wave run-up is shown in Figure 21b, and the MAB is further reduced to 5.9 m (compared to the original value of 14 m), which is theoretically smaller than the spatial resolution of the radar measurements described in Section 2.3. The correlation coefficient between the shoreline position obtained by the mTWM with corrected wave run-up and survey data is 0.85, as shown in Figure 21c, which is almost equivalent to the previously estimated results. The results show that the utilized method can successfully estimate the shoreline position from radar images after corrected wave run-up.

The mTWM with corrected wave run-up is applied to all alongshore locations to determine the spatio-temporal variation in shoreline positions for the entire study area over the period between April 12, 2005 and December 31, 2008, as shown in Figure 22.

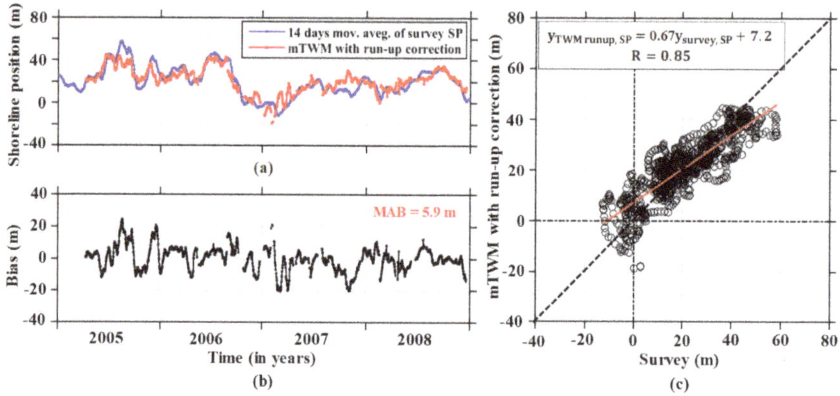

Figure 21. (**a**) The temporal variations in shoreline positions (SP) estimated from the survey data and mTWM-derived results with corrected wave run-up (at $x = -49$ m). (**b**) The estimated bias between the survey data and mTWM-derived results with corrected wave run-up. (**c**) The scatter plot of the survey data vs. mTWM-derived results with corrected wave run-up. The red line represents the linear fitting line.

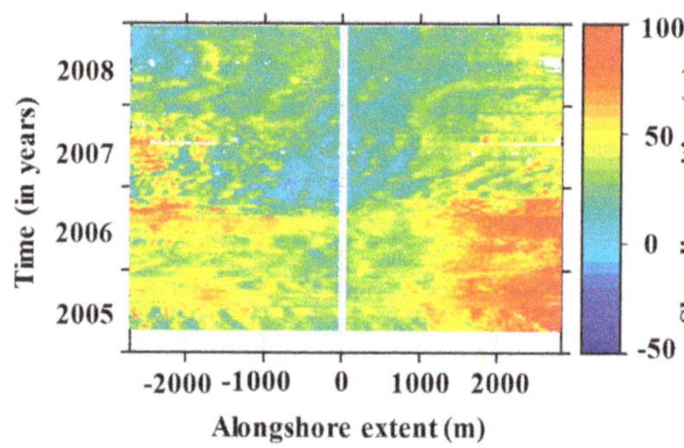

Figure 22. Spatial and temporal variations in the shoreline position derived by mTWM with corrected wave run-up during 12 April 2005 to 31 December 2008.

4.4. Shoreline Position Data Gaps Filled by Garcia's Method

The white blanks in Figure 22 indicate the 3% gaps in the shoreline data derived by mTWM owing to the lack of strong waterline signals caused by overflooding, the existence of strong radar reflectors, etc. Garcia's smoothing and gap filling method [57] was employed to fill these gaps, and the validity of this method will be checked.

Garcia's method is formulated based on a Penalized Least Squares regression by means of the Discrete Cosine Transform (PLS-DCT), which expresses the data in terms of a sum of cosine functions oscillating at different frequencies and is suitable for equally spaced data in one dimension and higher. As the DCT can be multidimensional, the DCT-based PLS regression can be instantly extended to multidimensional datasets. Furthermore, this method was adopted by Wang et al. [58], and its performances to fill the gaps in a global soil moisture dataset was analyzed. Recently, Bell et al. [28] and Bird et al. [29] have employed this algorithm to fill the gaps in beach profile data. This method will be adopted in this study to fill the random gaps in the temporal or spatiotemporal shoreline datasets.

To evaluate the performance of Garcia's method, the numerous random gaps in the mTWM derived shoreline positions with corrected wave run-up are considered. Figure 23a depicts the temporal variation in gap-filled shoreline positions with corrected wave run-up. The gaps in shoreline positions are filled by Garcia's method. The bias distribution between the survey data and mTWM extracted shoreline position results with corrected wave run-up is shown in Figure 23b with a MAB of 5.9 m, which is equivalent to that obtained before applying Garcia's method. As shown in Figure 23c, the gap-filled result implies that the variations in shoreline position are synchronized, which is confirmed by the good correlation coefficient (R = 0.85). This value is almost equivalent to that obtained before applying Garcia's method on the wave run-up correction shoreline positions. Therefore, the estimated variation of shoreline positions with corrected wave run-up are quite similar to the survey data. The efficiency of Garcia's filling method will be explained in the discussion section.

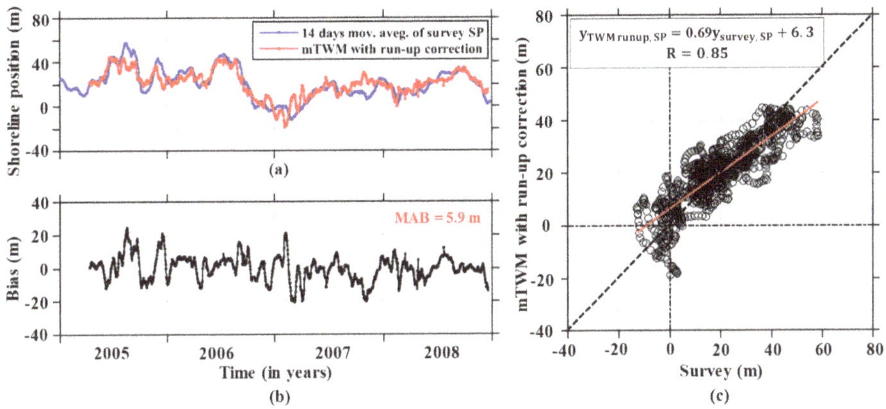

Figure 23. (**a**) The temporal variations in shoreline position (SP) estimated from the survey data and mTWM-derived results with corrected wave run-up (at $x = -49$ m) integrated with Garcia's filling method. (**b**) The estimated bias between the survey data and mTWM-derived results with corrected wave run-up. (**c**) The scatter plot of the survey data vs. mTWM-derived results with corrected wave run-up. The red line represents the linear fitting line.

Based on the above results, Garcia's method is implemented to fill the gaps in the entire spatiotemporal shoreline position data derived by mTWM with corrected wave run-up (Figure 22). Figure 24a shows the gap-filled spatiotemporal distribution of wave run-up corrected shoreline positions extracted from time-averaged X-band radar images over four years (2005–2008). The rainbow patterns of the color bar indicate the landwards (blue) and seawards (red) positions of shoreline from its origin. Correspondingly, Figure 24b,c show the spatial mean of the shoreline positions and temporal variations in significant wave height, respectively. The adopted gap-filling method demonstrates a reasonably good performance in terms of the smoothness of the shoreline data. Figure 24 suggests that Garcia's method is capable of filling the numerous random gaps in the shoreline position data. Moreover, we may conclude that Garcia's method combined with mTWM performs reasonably well to fill the random gaps in the shoreline position data, generating results nearly similar to the survey data. The results show seaward shifts of shoreline positions due to low wave conditions and landward shifts due to high wave conditions. The obtained result is in good agreement with Galal and Takewaka's [59] reported outcome. The obtained features were initially reported by Kuriyama and Lee [60] and is called beach cycle. Huge erosion was observed in October 2006, which was a result of high waves and storm surge attacking the Kashima Coast. Based on the results obtained by mTWM, we can be assured that the mTWM is a robust approach to detect intertidal shore profiles automatically from time-averaged X-band radar images, and it can be used any beach in the world to help the authorities to understand long- or short-term shoreline changes in coastal zones.

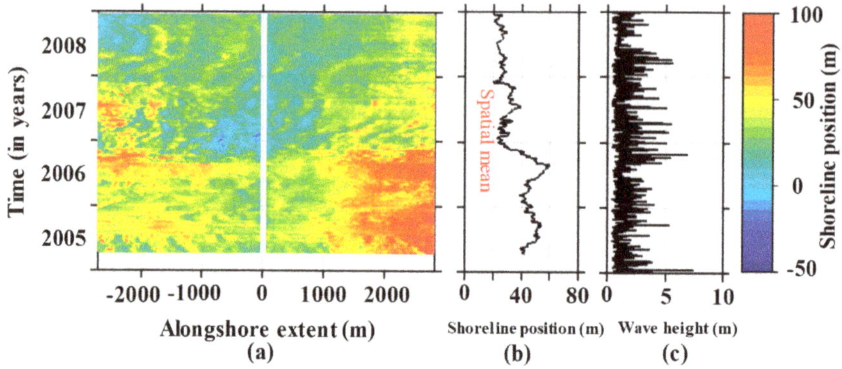

Figure 24. (**a**) Gaps in the shoreline positions derived by mTWM with corrected wave run-up filled by Garcia's smoothing and gap-filling method. (**b**) The spatial means of shoreline positions. (**c**) The temporal variations of significant wave height.

5. Discussion

This section discusses the mTWM extracted results, its adopted correction, and the possible reasons for the large bias. The purpose of this study was to automatically detect the shoreline positions and intertidal foreshore slopes using the mTWM. The automatic extraction of shoreline positions and intertidal foreshore slopes has been successfully performed, but systematic biases in the shoreline positions have been observed. The shoreline position estimated by the mTWM shifted landward, with a MAB of 14 m.

Similar shifts in estimated shoreline positions were reported in previous studies [6,28]. Takewaka [6] initially analyzed the time-averaged X-band radar images at Hasaki beach, Japan to estimate the shoreline positions and intertidal foreshore slopes by manual inspection. In this study, the horizontal shift between the estimated and surveyed shoreline positions was measured as 10 m. Recently, Bell et al. [28] automatically acquired intertidal shore profile datasets for Hilbre Island at the mouth of Dee Estuary, UK. To validate the accuracy of the calculated results, they compared the TWM-generated results with airborne LIDAR surveyed data for the same study site over the radar survey period. The vertical elevation bias between the two intertidal shore profiles is approximately $\pm$0.5 m. This can lead to horizontal shifts in the estimated shoreline positions depending on the beach slopes and wave conditions. If the beach slopes (1:20–1:100) vary with high wave conditions, the horizontal shift between the two observations can also vary (10–50 m). In both studies, the effects of wave set-up and run-up on reducing the horizontal shift were not considered.

The schematic illustration in Figure 14 suggests that wave set-up and run-up may be causing this type of horizontal shifts. Two correction methods were tested to reduce such systematic bias. The correction of wave set-up evaluates a mean waterline above a still water level and allows a horizontal distance between such waterline and tidal level [10]. Figure 18b shows the distribution of the bias between the survey data and shoreline positions derived by mTWM with corrected wave set-up. The MAB dropped by 3.5 m compared to the results obtained before applying a wave set-up correction to the shoreline position derived by mTWM (from 14 to 10.5 m); however, horizontal shifts were still present. This may suggest that Equations (2)–(4) are used appropriately to reduce some of the detected horizontal shifts in the study site; however, it was not bound in the spatial resolution of radar measurement.

To further reduce the horizontal shifts in the estimated shoreline positions and intertidal beach slopes, a wave run-up empirical formula is applied to achieve an upward shift of waterline position as well as a horizontal shift to the landward direction. The bias distribution between the survey and shoreline positions with corrected wave run-up is shown in Figure 21b, with a MAB of 5.9 m, which is smaller than the theoretical resolution of the radar measurements described earlier and the horizontal shift of bias that obtained by Takewaka [6]. Considering the use of X-radar images with a theoretical resolution of 7.5 m/pixel, the obtained MAB of 5.9 m is acceptable for an automatic application. Sometimes, the bias of estimated shoreline positions is higher than the spatial resolution of the radar measurements.

Figure 25a shows the frequency distribution of the estimated bias of shoreline positions. The estimated bias is the difference between the survey and mTWM-extracted run-up corrected gap-filled shoreline positions for each measurement. The distribution of the bias seems to follow the normal distribution, with a mean value of −0.05 m and a variance of 5.9 m. Hence, the overall results indicate that about 84% of the bias of the estimated shoreline positions are limited in the spatial resolution of the radar measurement and that the rest are slightly higher than the spatial resolution. The reason for the significant bias of the estimated shoreline positions will be described below. On the other hand, Figure 25b displays the frequency distribution of the estimated bias between the survey and only Garcia's method gap-filled shoreline positions. This estimated bias of shoreline positions is bound in [−10 m, 10 m]. That means Garcia's method interpolates nearly similar shoreline positions like the mTWM-derived shoreline positions. As shown in Figure 25b, the distribution of the bias seems to follow the normal distribution, with a mean value of −0.1 m and a variance of 3 m, where the 32 data are used in this case.

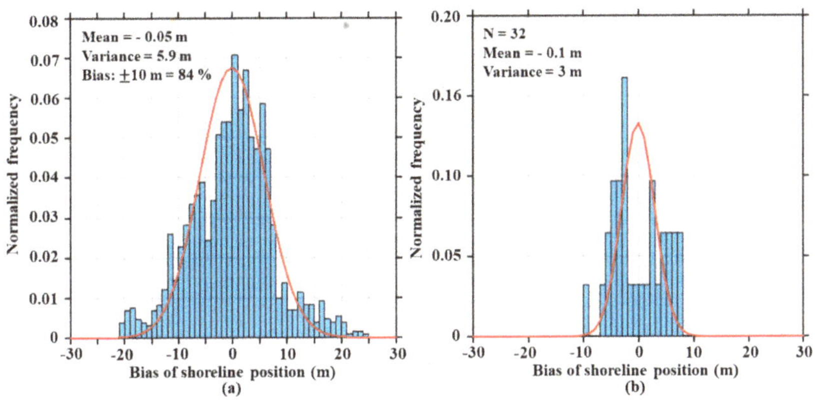

Figure 25. (**a**) A histogram of the bias between the survey and mTWM-detected shoreline positions during 2005–2008 (1 m bins) with the normal adjusted curve super-imposed in red. (**b**) A histogram of the bias between the estimated shoreline positions with the survey and only Garcia's filling method (1 m bins) with the normal adjusted curve superimposed in red.

Figure 26a shows the scatter plot between the mTWM-estimated intertidal slope and the estimated bias of the shoreline positions. It seems that the bias become large when the mTWM-estimated intertidal foreshore slopes are in the milder range (0.01–0.06). In such a beach slope condition, the waterline position easily shifted landward and the shoreline position also moved landward. As a result, it sometimes surpasses the survey shoreline position and produces negative shoreline bias. On the contrary, sometimes it falls behind the survey shoreline position for the same beach slope condition and produces a positive shoreline bias. The milder beach slope is not the only reason for this bias; it also depends on wave action. It is noteworthy that the estimated bias is still within the acceptable limits for steeper beach slope conditions (0.06–0.1). In such conditions, the shoreline position cannot be moved easily in a landward direction without an extreme weather event such as a typhoon. As shown in Figure 26b, these features are verified by the relationship between the frequency distribution of slopes with 0.1 equal intervals and the corresponding estimated MAB of shoreline positions. It is also seen in Figure 26b that the high frequency of a milder slope produces the large MAB of shoreline positions. As mentioned before, the action of waves (low and high) might also play an influential role in significant bias. The cumulative frequencies indicate the 75% milder slopes (0.01–0.06) can follow the abovementioned characteristics.

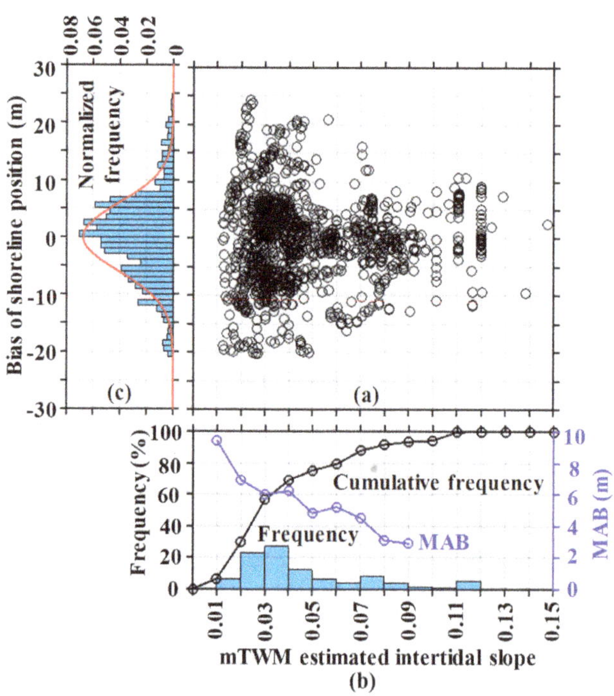

Figure 26. (**a**) The relationship between the mTWM-estimated slope and the estimated bias of shoreline positions. (**b**) The frequency distribution of mTWM-estimated slopes and the corresponding MAB estimation of shoreline positions. (**c**) The frequency distribution of estimated bias of shoreline positions.

Figure 27a shows the relationship between the 14-days moving averaged wave height and the estimated bias of shoreline positions. It is remarkable to observe that the bias of shoreline positions (positive or negative) becomes large when the waves are in low or high conditions with same beach slope. These features are confirmed in Figure 27b by the relationship between the frequency distribution of a wave height with 0.5 m equal intervals and the corresponding estimated MAB of shoreline positions. It can also be seen that the action of high waves produces significant MABs of the shoreline positions and vice versa. The cumulative frequencies indicate that 90% of waves can follow the abovementioned characteristics. Overall, the obtained results also show that the intertidal foreshore slopes (steeper or milder) and waves (low or high) are influential parameters for shoreline bias (positive or negative). It is further noted that some significant bias is also to be expected due to the different cross-sectional transect comparison of radar and survey-derived data.

Based on the results obtained by the gap-filled corrected wave run-up, we may conclude that more accurate shoreline positions are derived by mTWM with corrected wave run-up, which is very close to the survey data. This confirms that the correction of wave run-up and the application of Garcia's method are the most reasonable strategies to reduce horizontal shifts in shoreline positions at Hasaki beach, Japan.

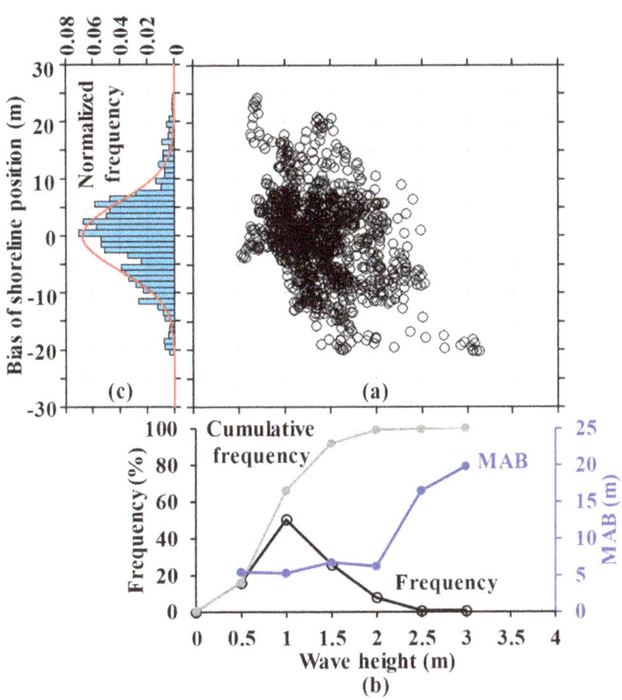

Figure 27. (**a**) The relationship between the wave height and estimated bias of the shoreline positions. (**b**) The frequency distribution of the wave height and corresponding MAB estimations of the shoreline positions. (**c**) The frequency distribution of the estimated bias of shoreline positions.

6. Conclusions

The mTWM is presented as a way to detect shoreline positions and intertidal foreshore slopes from X-band radar images. The method is slightly modified from the Bell et al. [28] approach (TWM). Cross-shore direction-wise bottom elevation estimation was considered as the TWM. Due to the presence of low signal similarities between pixel intensities and tidal binary signals at the landward cross-shore location, the TWM failed to estimate an accurate intertidal shore profile. Hence, determining the water level direction-wise bottom elevation and detecting intertidal shoreline profile are considered as the mTWM. The mTWM is successfully employed to detect accurate shoreline positions and intertidal foreshore slopes from X-band radar images collected over the course of two-week tidal cycle variation at microtidal sandy Hasaki beach, Japan, during 12 April 2005 to 31 December 2008. Compared to the survey data, the MAB of the detected shoreline positions was 14 m.

To reduce the horizontal shift between the shoreline positions derived by mTWM and the survey data, the corrected wave set-up was applied to the tidal record to compensate for the shift in the estimated results. The MAB between shoreline positions derived by mTWM with corrected wave set-up and the survey data was reduced to 10.5 m. Furthermore, the correction of the wave run-up was applied to the results obtained by the mTWM. This reduced the MAB to 5.9 m, which is smaller than the theoretical spatial resolution of X-band radar images; however, sometimes the estimated bias was larger than the spatial resolution. This larger bias is the effect of milder intertidal foreshore slopes and the action of waves. The frequency distribution results indicate that 84% of the estimated bias of the shoreline positions is limited in the spatial resolution of the radar measurement. On the other hand, numerous random gaps still existed in the dataset. These random gaps are due to a lack of strong waterline signals caused by overflooding, the existence of strong radar reflectors, etc.

To fill the random gaps, Garcia's method was applied in the mTWM-derived shoreline positions with the corrected wave run-up. The MAB between these estimated shoreline positions and the survey data was 5.9 m, and the frequency distribution of the estimated bias of shoreline positions was 84%, which are almost equivalent to that obtained before applying the gap-filling method. On the other hand, the bias between the survey and only Garcia's method filling shoreline positions are limited with [−10 m, 10 m]. This indicates the reliability of Garcia's method. We can conclude that the mTWM integrated with this method is an efficient and robust approach to automatically detect shoreline positions from time-averaged X-band radar images with the consideration of a wave run-up correction at sandy beaches during various periods and to demonstrate the practicability of the utilized method. Therefore, the temporal and spatial variations of a shoreline can be automatically and continuously monitored over the long term to help authorities understand coastal changes, facilitating coastal protection and sustainable development in coastal zones.

Author Contributions: D.K. contributed the design and analysis and wrote the paper; S.T. supervised the work and checked the manuscript.

Funding: This research received no external funding.

Acknowledgments: The radar observations are supported by the members of Littoral Drift Division and Port and Airport Research Institute. This study is financially supported by the Grants-in-Aid of the Japan Society for the Promotion of Science (JSPS), the Social Implementation Program on Climate Change Adaptation Technology (SI-CAT), and the Ministry of Education, Culture, Sports, Science and Technology (MEXT).

Conflicts of Interest: The authors declare no conflict of interest.

References

1. Dolan, R.; Hayden, B.P.; May, P.; May, S. The reliability of shoreline changes measurements from aerial photographs. *Shore Beach* **1980**, *48*, 22–29.
2. Boak, E.H.; Turner, I.L. Shoreline definition and detection: a review. *J. Coast. Res.* **2005**, *21*, 688–703. [CrossRef]
3. Crowell, M.; Leatherman, S.P.; Buckley, M.K. Historical shoreline change: Error analysis and mapping accuracy. *J. Coast. Res.* **1991**, *7*, 839–852.
4. Camfield, F.E.; Morang, A. Defining and interpreting shoreline change. *Ocean Coast. Manag.* **1996**, *32*, 129–151. [CrossRef]
5. Kraus, N.C.; Rosati, J.D. *Interpretation of Shoreline-Position Data for Coastal Engineering Analysis*; Coastal Engineering Technical Note II-39 (No. CERC-CETN-II-39); U.S. Army Engineer Research and Development Center: Vicksburg, MS, USA, 1997.
6. Takewaka, S. Measurements of shoreline positions and intertidal foreshore slopes with X-band marine radar system. *Coast. Eng. J.* **2005**, *47*, 91–107. [CrossRef]
7. Hanslow, D.J. Beach erosion trend measurement: a comparison of trend indicators. *J. Coast. Res.* **2007**, *50*, 588–593.
8. Ryu, J.H.; Kim, C.H.; Lee, Y.K.; Won, J.S.; Chun, S.S.; Lee, S. Detecting the intertidal morphologic change using satellite data. *Estuar. Coast. Shelf Sci.* **2008**, *78*, 623–632. [CrossRef]
9. Chen, W.W.; Chang, H.K. Estimation of shoreline position and change from satellite images considering tidal variation. *Estuar. Coast. Shelf Sci.* **2009**, *84*, 54–60. [CrossRef]
10. Chang, H.K.; Chen, W.W.; Liou, J.C. Shifting the waterlines of satellite images to the mean water shorelines considering wave runup, setup, and tidal variation. *J. Appl. Remote Sens.* **2015**, *9*, 096004. [CrossRef]
11. García-Rubio, G.; Huntley, D.; Russell, P. Evaluating shoreline identification using optical satellite images. *Mar. Geol.* **2015**, *359*, 96–105. [CrossRef]
12. Aarninkhof, S.G.; Turner, I.L.; Dronkers, T.D.; Caljouw, M.; Nipius, L. A video-based technique for mapping intertidal beach bathymetry. *Coast. Eng.* **2003**, *49*, 275–289. [CrossRef]
13. Aarninkhof, S.G.J.; Ruessink, B.G.; Roelvink, J.A. Nearshore subtidal bathymetry from time-exposure video images. *J. Geophys. Res. Oceans* **2005**, *110*, C06011. [CrossRef]

14. Holman, R.A.; Sallenger, A.H.; Lippmann, T.C.; Haines, J.W. The application of video image processing to the study of nearshore processes. *Oceanography* **1993**, *6*, 78–85. [CrossRef]
15. Holland, K.T.; Holman, R.A. Video estimation of foreshore topography using trinocular stereo. *J. Coast. Res.* **1997**, *13*, 81–87.
16. Plant, N.G.; Holman, R.A. Intertidal beach profile estimation using video images. *Mar. Geol.* **1997**, *140*, 1–24. [CrossRef]
17. Uunk, L.; Wijnberg, K.M.; Morelissen, R. Automated mapping of the intertidal beach bathymetry from video images. *Coast. Eng.* **2010**, *57*, 461–469. [CrossRef]
18. De Santiago, I.; Morichon, D.; Abadie, S.; Castelle, B.; Liria, P.; Epelde, I. Video monitoring nearshore sandbar morphodynamics on a partially engineered embayed beach. *J. Coast. Res.* **2013**, *65*, 458–463. [CrossRef]
19. Sobral, F.; Pereira, P.; Cavalcanti, P.; Guedes, R.; Calliari, L. Intertidal bathymetry estimation using video images on a dissipative beach. *J. Coast. Res.* **2013**, *65*, 1439–1444. [CrossRef]
20. Valentini, N.; Saponieri, A.; Molfetta, M.G.; Damiani, L. New algorithms for shoreline monitoring from coastal video systems. *Earth Sci. Inform.* **2017**, *10*, 495–506. [CrossRef]
21. Austin, M.J.; Masselink, G. Observations of morphological change and sediment transport on a steep gravel beach. *Mar. Geol.* **2006**, *229*, 59–77. [CrossRef]
22. Gaudin, D.; Delacourt, C.; Allemand, P.; Jaud, M.; Ammann, J.; Tisseau, C.; Véronique, C.U.Q. High resolution DEM derived from thermal infrared images: Example of Aber Benoit (France). In Proceedings of the 2009 IEEE International Geoscience and Remote Sensing Symposium, Cape Town, South Africa, 12–17 July 2009.
23. Dankert, H.; Horstmann, J. A marine radar wind sensor. *J. Atmos. Ocean. Technol.* **2007**, *24*, 1629–1642. [CrossRef]
24. Holman, R.; Haller, M.C. Remote sensing of the nearshore. *Annu. Rev. Mar. Sci.* **2013**, *5*, 95–113. [CrossRef] [PubMed]
25. Bell, P.S. Shallow water bathymetry derived from an analysis of X-band marine radar images of waves. *Coast. Eng.* **1999**, *37*, 513–527. [CrossRef]
26. Galal, E.M.; Takewaka, S. Longshore migration of shoreline mega-cusps observed with X-band radar. *Coast. Eng. J.* **2008**, *50*, 247–276. [CrossRef]
27. An, S.; Takewaka, S. A Study on the morphological characteristics around artificial headlands in Kashima Coast, Japan. *J. Coast. Res.* **2016**, *32*, 508–518. [CrossRef]
28. Bell, P.S.; Bird, C.O.; Plater, A.J. A temporal waterline approach to mapping intertidal areas using X-band marine radar. *Coast. Eng.* **2016**, *107*, 84–101. [CrossRef]
29. Bird, C.O.; Bell, P.S.; Plater, A.J. Application of marine radar to monitoring seasonal and event-based changes in intertidal morphology. *Geomorphology* **2017**, *285*, 1–15. [CrossRef]
30. Koopmans, B.N.; Wang, Y. Satellite radar data for topographic mapping of the tidal flats in the Wadden Sea, The Netherlands. In Proceedings of the Second Thematic Conference on Remote Sensing for Marine and Coastal Environments, New Orleans, LA, USA, 31 January–2 February 1994.
31. Mason, D.C.; Davenport, I.J.; Robinson, G.J.; Flather, R.A.; McCartney, B.S. Construction of an inter-tidal digital elevation model by the "Water-Line" Method. *Geophys. Res. Lett.* **1995**, *22*, 3187–3190. [CrossRef]
32. Heygster, G.; Dannenberg, J.; Notholt, J. Topographic mapping of the German tidal flats analyzing SAR images with the waterline method. *IEEE Trans. Geosci. Remote Sens.* **2010**, *48*, 1019–1030. [CrossRef]
33. Ryu, J.H.; Won, J.S.; Min, K.D. Waterline extraction from Landsat TM data in a tidal flat: A case study in Gomso Bay, Korea. *Remote Sens. Environ.* **2002**, *83*, 442–456. [CrossRef]
34. Zhao, B.; Guo, H.; Yan, Y.; Wag, Q.; Li, B. A simple waterline approach for tidelands using multi-temporal satellite images: A case study in the Yangtze Delta. *Estuar. Coast. Shelf Sci.* **2008**, *77*, 134–142. [CrossRef]
35. Xu, Z.; Kim, D.J.; Kim, S.H.; Cho, Y.K.; Lee, S.G. Estimation of seasonal topographic variation in tidal flats using waterline method: A case study in Gomso and Hampyeong Bay, South Korea. *Estuar. Coast. Shelf Sci.* **2016**, *183*, 213–220. [CrossRef]
36. Dellepiane, S.; De Laurentiis, R.; Giordano, F. Coastline extraction from SAR images and a method for the evaluation of the coastline precision. *Pattern Recognit. Lett.* **2004**, *25*, 1461–1470. [CrossRef]
37. Fuse, T.; Ohkura, T. Development of shoreline extraction method based on spatial pattern analysis of Satellite SAR images. *Remote Sens.* **2018**, *10*, 1361. [CrossRef]
38. Paravolidakis, V.; Ragia, L.; Moirogiorgou, K.; Zervakis, M. Automatic coastline extraction using edge detection and optimization procedures. *Geosciences* **2018**, *8*, 407. [CrossRef]

39. Pardo-Pascual, J.E.; Almonacid-Caballer, J.; Ruiz, L.A.; Palomar-Vázquez, J. Automatic extraction of shorelines from Landsat TM and ETM+ multi-temporal images with subpixel precision. *Remote Sens. Environ.* **2012**, *123*, 1–11. [CrossRef]
40. Katoh, K. Changes of sand grain distribution in the surf zone. In Proceedings of the Coastal Dynamics, New York, NY, USA, 1 January 1995; pp. 335–364.
41. Kuriyama, Y. Medium-term bar behavior and associated sediment transport at Hasaki, Japan. *J. Geophys. Res. Oceans* **2002**, *107*, 15–19. [CrossRef]
42. Kuriyama, Y.; Ito, Y.; Yanagishima, S. Medium-term variations of bar properties and their linkages with environmental factors at Hasaki, Japan. *Mar. Geol.* **2008**, *248*, 1–10. [CrossRef]
43. Hasan, G.J.; Takewaka, S. Observation of a stormy wave field with X-band radar and its linear aspects. *Coast. Eng. J.* **2007**, *49*, 149–171. [CrossRef]
44. Hasan, G.J.; Takewaka, S. Wave run-up analyses under dissipative condition using X-band radar. *Coast. Eng. J.* **2009**, *51*, 177–204. [CrossRef]
45. Takewaka, S.; Taishi, Y. Rip current observation with X-band radar. *Coast. Eng. Proc.* **2012**, *1*, 43. [CrossRef]
46. Zhai, Y. Time-Dependent Scour Depth under Bridge-Submerged Flow. Master's Thesis, University of Nebraska-Lincoln, Lincoln, NE, USA, May 2010.
47. Lentz, S.; Raubenheimer, B. Field observations of wave setup. *J. Geophys. Res. Oceans* **1999**, *104*, 25867–25875. [CrossRef]
48. Goda, Y. *Random Seas and Design of Maritime Structures*, 3rd ed.; World Scientific Publishing Company: Singapore, 2010.
49. Melby, J.A. *Wave Runup Prediction for Flood Hazard Assessment*; Technical Report ERDC/CHL TR-12-24; Coastal and Hydraulics Laboratory, U.S. Army Engineer Research and Development Center: Vicksburg, MS, USA, 2012; p. 126.
50. Hunt, I.A. Design of sea-walls and breakwaters. *Trans. Am. Soc. Civ. Eng.* **1959**, *126*, 542–570.
51. Mase, H. Random wave runup height on gentle slope. *J. Waterw. Port Coast. Ocean Eng.* **1989**, *115*, 649–661. [CrossRef]
52. Ruessink, B.G.; Kleinhans, M.G.; Van den Beukel, P.G.L. Observations of swash under highly dissipative conditions. *J. Geophys. Res. Oceans* **1998**, *103*, 3111–3118. [CrossRef]
53. Ruggiero, P.; Holman, R.A.; Beach, R.A. Wave run-up on a high-energy dissipative beach. *J. Geophys. Res. Oceans* **2004**, *109*, C06025. [CrossRef]
54. Stockdon, H.F.; Holman, R.A.; Howd, P.A.; Sallenger, A.H., Jr. Empirical parameterization of setup, swash, and runup. *Coast. Eng.* **2006**, *53*, 573–588. [CrossRef]
55. Holman, R.A.; Bowen, A.J. Longshore structure of infragravity wave motions. *J. Geophys. Res. Oceans* **1984**, *89*, 6446–6452. [CrossRef]
56. Iribarren Cavanilles, R.; Casto Nogales, M. *Protection Des Ports*; Technical Report; PIANC: Lisbon, Portugal, 1 January 1949.
57. Garcia, D. Robust smoothing of gridded data in one and higher dimensions with missing values. *Comp. Stat. Data Anal.* **2010**, *54*, 1167–1178. [CrossRef]
58. Wang, G.; Garcia, D.; Liu, Y.; De Jeu, R.; Dolman, A.J. A three-dimensional gap filling method for large geophysical datasets: Application to global satellite soil moisture observations. *Environ. Model. Softw.* **2012**, *30*, 139–142. [CrossRef]
59. Galal, E.M.; Takewaka, S. Temporal and spatial shoreline variability observed with a X-band radar at Hasaki coast, Japan. In Proceedings of the Coastal Sediments, San Diego, CA, USA, 11–15 May 2015; pp. 1–9.
60. Kuriyama, Y.; Lee, J.H. Medium-term beach profile change on a bar-trough region at Hasaki, Japan, investigated with complex principal component analysis. In Proceedings of the Fourth Conference on Coastal Dynamics, Lund, Sweden, 11–15 June 2001; pp. 959–968.

Journal of
Marine Science and Engineering

MDPI

Article

A Quality Control Method for Broad-Beam HF Radar Current Velocity Measurements

Belinda Lipa *, Donald Barrick and Chad Whelan

Codar Ocean Sensors, 1914 Plymouth St., Mountain View, CA 94043, USA; Don@codar.com (D.B.); Chad@codar.com (C.W.)

* Correspondence: Belinda@codar.com; Tel.: +1-650-851-5517; Fax: +1-408-773-0514

Received: 6 March 2019; Accepted: 15 April 2019; Published: 19 April 2019

Abstract: This paper describes a method to provide quality control for radial velocity maps derived from radar echo voltage cross spectra measured by broad-beam high frequency radars. The method involves the comparison of voltage cross spectra measured at Doppler frequencies in the Bragg region with values predicted from basic equations defining the complex voltage cross spectra in terms of the measured antenna patterns and the radar cross section. Poor agreement at a given Doppler frequency indicates contamination of the spectra, usually due to interference; velocity results from that Doppler frequency are then eliminated. Examples are given of its application to broad-beam radars operating at four sites.

Keywords: current velocity measurement; high-frequency (HF) radar oceanography; remote sensing; quality control

1. Introduction

Crombie [1] first observed and identified the distinctive features of sea-echo in radar Doppler spectra. Barrick [2,3] derived expressions for the high frequency (HF) sea-echo Doppler spectrum in terms of the surface current velocities and the ocean wave height directional spectrum. Methods were then developed to interpret radar sea-echo spectra in terms of Barrick's equations to give surface current velocity maps [4,5], directional ocean wave spectra and wind direction [6–8]. These methods are presently used by commercial, operational systems to give current maps and directional wave information.

The first HF radar systems to be developed were phased arrays, which ideally have narrow radar beams [9,10]. Although the interpretation of narrow-beam signals backscattered from the sea is simpler, the disadvantages of phased-array systems for applications are their large physical size with consequent difficult installation requirements and high operating cost. It is mainly for these reasons that compact, transportable systems were developed in the seventies [11], followed by the SeaSonde (Codar Ocean Sensors, Mountain View, CA, USA) [12,13], which has been available commercially since 1990. The SeaSonde has three small antennas, two crossed loops and a vertical monopole/dipole, all aligned along a single vertical axis. Interpretation of the signal voltages using Barrick's equations yields both the surface current field and directional ocean wave parameters. Lipa and Barrick derived analysis methods for broad-beam HF radars based on equations defining the complex voltage cross spectra in terms of the antenna patterns and the radar cross section; these can be used to obtain the radial current velocity as a function of range and direction [4]. In this paper, we describe how these equations can provide real-time quality control (QC) for radial current velocities derived from the cross spectra. This is based on internal consistency checks that are used to assess the quality of the radial velocities. Analysis procedures presently implemented in SeaSonde system software are based on the multiple signal classification (MUSIC) algorithm for direction-finding [5], which provides reduced analysis time. The solutions for current velocities from MUSIC are consistent with those from the standard least-squares procedure, which is based on equations for the measured voltage cross spectra that can also be used for QC, as described in this paper.

Operational coastal HF ocean surface wave radar networks have proliferated such that regional and national scale operational networks have been installed in the United States [14], Asia [15], Europe [16] and India [17]. The primary products of these networks are surface current maps used for a variety of scientific and operational applications, including field wave measurement [18,19], search and rescue [20], oil spill response [21–24], vessel navigation [25], fisheries [26], tsunami detection [27]. With the increase in both extent and operational uses of the HF Radar networks, there is now a greater need for more automated QC of the real-time data products.

A recommended best practice for HF radar operators is to measure and use the antenna response pattern for each station's receive antenna [28]. Kohut et al. [29] showed that applying measured antenna patterns to the HF radar Doppler spectra analysis improved the surface current data accuracy. Methods for measuring the antenna pattern at SeaSonde stations more easily and more often have progressed to include using aerial drones and passing ships of opportunity [30–32]. Most SeaSonde HF Radar systems now operate using a measured antenna pattern.

Several QC methods have been developed for use in real-time software processing, as outlined and referenced in manuals issued by the United States Integrated Ocean Observing System (U.S. IOOS) [33]. Recent studies for broad-band radars have demonstrated site-specific methods real-time QC based on area-averaging, SNR limits and MUSIC signal quality metrics [34,35]. QC methods based on SNR limits for phased array radars are described in [36]. The QC method presented here differs from previous methods as it is based on the fundamental equations for the radar sea echo and measured antenna patterns; previous methods are based on analysis of derived parameters and statistical fluctuations of the echo data.

The paper is organized as follows: Section 2 summarizes the radar spectral theory on which the methods are based; Section 3 outlines the method used for the derivation of radial current velocities; Section 4 outlines the method used to provide QC for radial velocities, based on internal consistency checks supplied by software and applied to measured data from a radar system at Shelter Cove; Section 5 shows typical results from application of these methods to radar data from three radar sites located at Trinidad, Big Creek (California) and Sea Bright (New Jersey).

2. Radar Spectral Theory

In this paper, we focus the analysis on SeaSonde broad-beam systems [13], which operate at several hundred sites. The basic SeaSonde data set consists of complex voltage cross spectra (see for example [4]) measured by the monopole and two crossed-loop antennas. The radar echo spectra from circular range cells over the coverage area consist of dominant peaks produced by first-order Bragg scatter from waves with one-half the radar wavelength, from which current velocities are derived. The first-order peaks are surrounded by sidebands produced by higher-order scatter, as shown in Figure 1. First-order peaks at positive Doppler frequencies are produced by echo from advancing Bragg waves; those at negative Doppler frequencies come from receding Bragg waves.

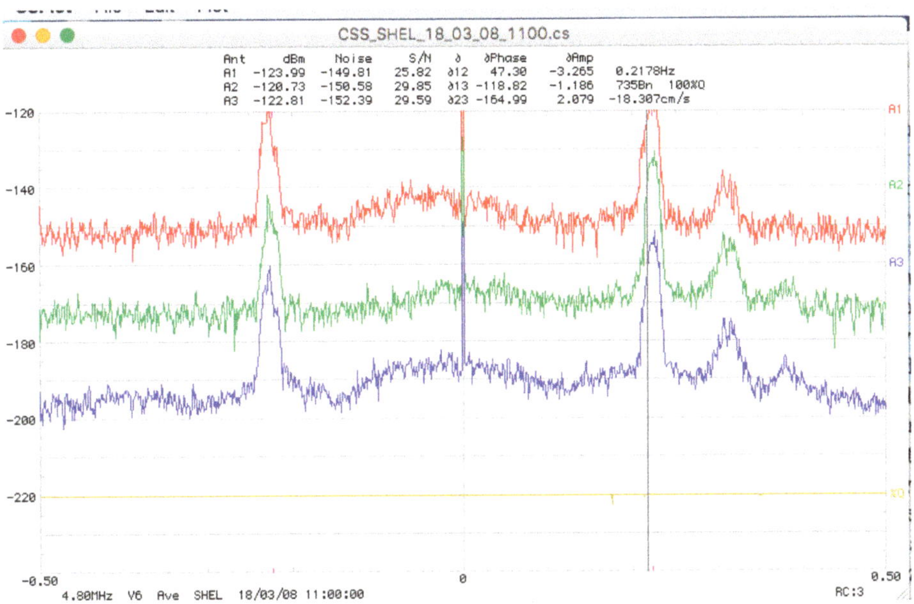

Figure 1. An example of an uncontaminated radar spectrum obtained from a SeaSonde system, measured at Shelter Cove, California, 8 March, 2018, 11a.m. range 18 km. Radar-echo spectral power (dB) from the three SeaSonde antennas is plotted vs. Doppler frequency for the Loop 1 antenna (A1, red), the Loop 2 antenna (A2, green) and the monopole (A3, blue). The curves are offset by 20 dB for better visibility. The first-order Bragg peaks at Doppler ± 0.22 Hz are surrounded by second-order echo, with the second-order peak especially visible to the right of the positive Bragg peak. The yellow line is an indicator of spectral quality.

Barrick [2] derived the following expression for the narrow-beam first-order radar cross section $\sigma^1(\omega,\phi)$ at radian frequency ω and azimuthal direction ϕ:

$$\sigma^1(\omega,\phi) = 2^6\pi k_0^4 \sum_{m'=\pm 1} S\left(2k_0,\ \phi + (m'+1)\frac{\pi}{2}\right)\partial(\omega - m'\omega_B - 2k_0 v_r(\phi)) \tag{1}$$

where k_0 is the radar wavenumber, $S(k, \phi)$ is the directional ocean wave spectrum for angle ϕ and wavenumber k, $v_r(\phi)$ is the radial current velocity and ω_B is the radian Bragg frequency defined by $\sqrt{2gk_0}$, where g is the gravitational constant.

For a broad-beam system, complex voltage cross spectra from the three antennas $\langle \widetilde{V}_i \widetilde{V}_j^* \rangle$, where i, j = 1,2,3, at a given frequency can be expressed as follows in terms of the antenna patterns and the radar cross section:

$$\langle |\widetilde{V}_1|^2 \rangle = \int_{\phi_1}^{\phi_2} \left|\widetilde{Z}_1\right|^2 \sigma(\phi)d\phi$$

$$\langle |\widetilde{V}_2|^2 \rangle = \int_{\phi_1}^{\phi_2} \left|\widetilde{Z}_2\right|^2 \sigma(\phi)d\phi$$

$$\langle |\widetilde{V}_3|^2 \rangle = \int_{\phi_1}^{\phi_2} \sigma(\phi)d\phi$$

$$\langle \widetilde{V}_1 \widetilde{V}_3^* \rangle = \int_{\phi_1}^{\phi_2} \widetilde{Z}_1(\phi)\sigma(\phi)d\phi$$

$$\langle \widetilde{V}_2 \widetilde{V}_3^* \rangle = \int_{\phi_1}^{\phi_2} \widetilde{Z}_2(\phi)\sigma(\phi)d\phi$$

$$\langle \widetilde{V}_1 \widetilde{V}_2^* \rangle = \int_{\phi_1}^{\phi_2} \widetilde{Z}_1(\phi)\widetilde{Z}_2^*(\phi)\sigma(\phi)d\phi \tag{2}$$

where the angular brackets denote infinite ensemble averages, * indicates a complex conjugate and where $\widetilde{Z}_1$ (ϕ), $\widetilde{Z}_2$ (ϕ) are the complex antenna pattern functions for Loops 1 and 2, normalized by the monopole value; $\sigma(\phi)$ is the narrow-beam radar cross section defined by Equation (1); ϕ_1 and ϕ_2 are the angles at which the radar beam is cut off by the coastline.

3. Analysis for Radial Current Velocities

Lipa and Barrick [4] described analysis of broad-beam HF radar Doppler echo spectra to give radial current velocities, assuming ideal antenna patterns. It follows from Equation (1) that the Doppler shift from the positive and negative Bragg frequencies uniquely defines a value of the radial current velocity; interpretation of the data using least-squares methods yields the azimuth angles at which this velocity occurs. It is assumed that each radial velocity can occur at either one or two azimuth angles; standard least-squares methods are used to determine the velocity values and choose between a one-angle or a two-angle solution.

This least-squares analysis is here extended to use measured antenna patterns in the interpretation of the radar cross spectra, and a QC method is identified.

Assuming that the velocity defined by a given Doppler frequency occurs at only one or two azimuth angles leads to the following equations, where the first/second entries define a one/two-angle solutions.

$$\int_{\phi_1}^{\phi_2} \left|\widetilde{Z}_1\right|^2 \sigma(\phi)d\phi = p(\phi)\left|\widetilde{Z}_1(\phi)\right|^2 \ (One-angle) = p(\phi_1)\left|\widetilde{Z}_1(\phi_1)\right|^2 + p(\phi_2)\left|\widetilde{Z}_1(\phi_2)\right|^2 \ (Two-angle)$$

$$\int_{\phi_1}^{\phi_2} \left|\widetilde{Z}_2\right|^2 \sigma(\phi)d\phi = p(\phi)\left|\widetilde{Z}_2(\phi)\right|^2 \ (One-angle) = p(\phi_1)\left|\widetilde{Z}_2(\phi_1)\right|^2 + p(\phi_2)\left|\widetilde{Z}_2(\phi_2)\right|^2 \ (Two-angle)$$

$$\int_{\phi_1}^{\phi_2} \sigma(\phi)d\phi = p(\phi) \ (One-angle) = p(\phi_1) + p(\phi_2) \ (Two-angle)$$

$$\int_{\phi_1}^{\phi_2} \widetilde{Z}_1(\phi)\sigma(\phi)d\phi = p(\phi) \ \widetilde{Z}_1(\phi) \ (One-angle) = p(\phi_1)\widetilde{Z}_1(\phi_1) + p(\phi_2)\widetilde{Z}_1(\phi_2) \ (Two-angle)$$

$$\begin{aligned} \int_{\phi_1}^{\phi_2} \widetilde{Z}_2(\phi)\sigma(\phi)d\phi &= p(\phi) \ \widetilde{Z}_2(\phi) \ (One-angle) \\ &= p(\phi_1)\widetilde{Z}_2(\phi_1) + p(\phi_2)\widetilde{Z}_2(\phi_2) \ (Two-angle) \end{aligned}$$

$$\begin{aligned}\int_{\phi_1}^{\phi_2} \widetilde{Z}_1(\phi)\widetilde{Z}_2^*(\phi)\sigma(\phi)d\phi &= p(\phi)\ \widetilde{Z}_1(\phi)\check{Z}_2^*(\phi)\ (One-angle)\\ &= p(\phi_1)\widetilde{Z}_1(\phi_1)\check{Z}_2^*(\phi_1)+p(\phi_2)\widetilde{Z}_1(\phi_2)\check{Z}_2^*(\phi_2)\ (Two-angle)\end{aligned} \tag{3}$$

In Equation (3), the multiplicative factors $p(\phi)$ have the same values for all the equations.

The normalized least-squares sum $S(\omega)$ at each frequency is then formed. For a one angle solution, $S(\omega)$ is given by

$$S(\omega) = \left[\langle|\widetilde{V}_1|_2\rangle - p(\phi)|\widetilde{Z}_1(\phi)|^2\right]^2 + \left[\langle|\widetilde{V}_2|_2\rangle - p(\phi)|\widetilde{Z}_2(\phi)|^2\right]^2 + \cdots + \left[\langle\widetilde{V}_1\widetilde{V}_2^*\rangle - p(\phi)\widetilde{Z}_1(\phi)\check{Z}_2^*(\phi)\right]^2 / \langle|\widetilde{V}_3|^2\rangle \tag{4}$$

and the normalized RMS (Root-mean-squared) deviations defined by:

$$\mathrm{RMS}(\omega) = \sqrt{S(\omega) / \langle|\widetilde{V}_3|^2\rangle} \tag{5}$$

Minimizing $S(\omega)$ with respect to $p(\phi)$ and ϕ provides the optimum fit values, leading to radial current velocities.

We then proceed in a similar fashion for the two-angle solution, minimizing the sum of least-squares deviations to give optimum values of $\phi_1, \phi_2, p(\varnothing_1), p(\varnothing_2)$. Statistical methods to select between a one- or a two-angle solution are described in [4] Section IIIB.

4. Application of Quality Control for Derived Radial Velocities

The minimum value of the RMS deviations can be used to provide QC for the derived radial velocities: If the deviations are large, this indicates that the model is not a good fit to the data, and the derived velocities for that Doppler bin should be discarded. Large deviations can be due to interference or incorrect boundaries set to define the first-order region. First-order boundaries are set by locating the limits of the strong signal region surrounding the Bragg frequencies. We now illustrate this process by analysis of the contaminated radar spectrum shown in Figure 2, for which interference is clearly masking the negative Bragg region.

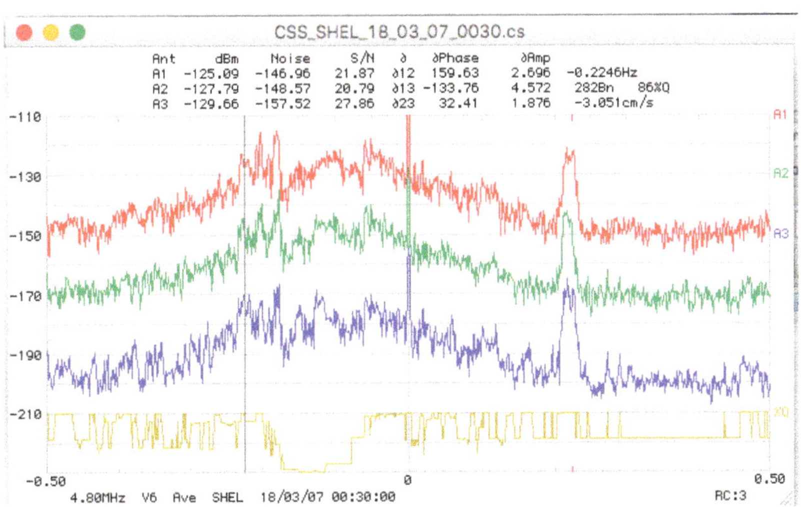

Figure 2. An example of a contaminated radar spectrum, measured at Shelter Cove, California: 7 March 2018, 00:30 a.m., range 18 km. The radar transmit frequency was 4.8 MHz. For plot details, see the caption of Figure 1.

To provide a comparison between current velocity results from contaminated and uncontaminated Bragg regions, this spectrum was analyzed to give current velocities, using first only negative- and then only positive-Doppler frequencies. Results are shown in Figure 3.

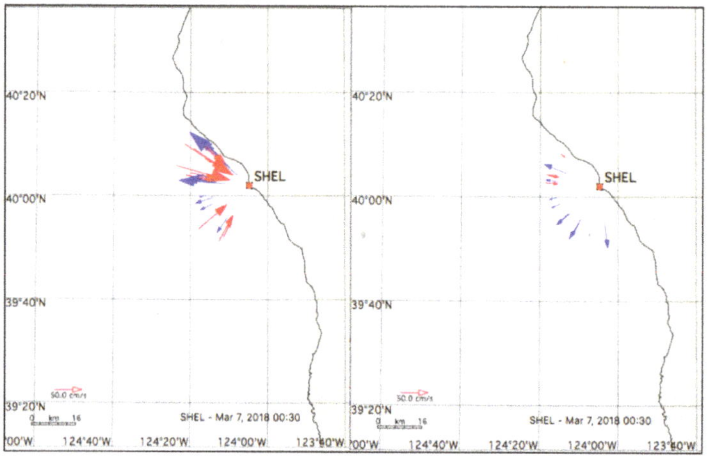

(**a**) (**b**)

Figure 3. Radial velocities obtained from the radar data 7 March, 2018, 00:30 a.m., range 18 km from (**a**) negative and (**b**) positive Doppler first-order regions. Red/blue radial vectors approach/recede from radar.

The radial current velocities shown in Figure 3a are quite erratic due to the interference, whereas Figure 3b indicates a regular pattern. Corresponding normalized RMS deviations are plotted vs. Doppler frequency in Figure 4 for both positive and negative Doppler frequencies.

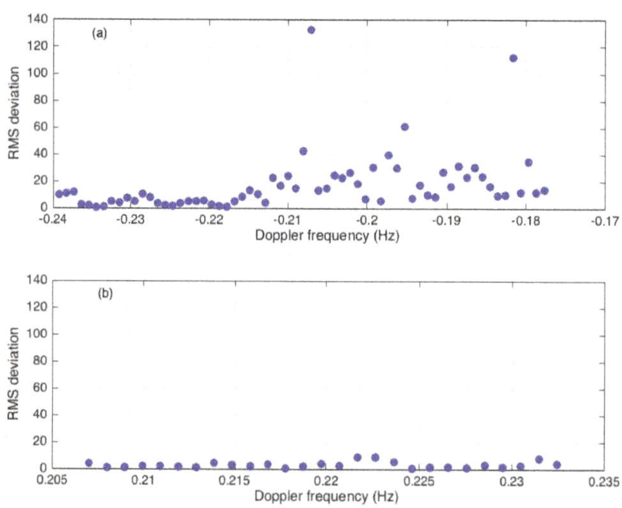

Figure 4. RMS deviations plotted vs. Doppler frequency for Shelter Cove on 7 March, 2018, 00:30 a.m., range 18 km (**a**) negative (**b**) positive Doppler frequencies.

From Figure 4, the values of the RMS deviations for the uncontaminated spectral points with positive Doppler frequency are usually less than 5. Setting this value as a limit (S_{lim} = 5) radial current velocities obtained for which the RMS deviations do not exceed S_{lim} are plotted in Figure 5.

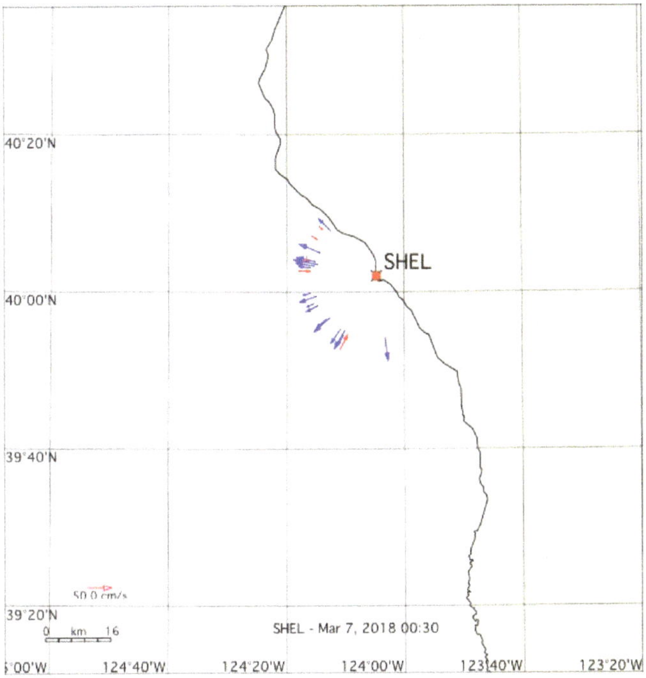

Figure 5. Radial velocities for which the RMS deviations do not exceed 5, from the radar data 7 March, 2018, 00:30 a.m., range 18 km.

In practice, a limit S_{lim} for a given site is determined by applying this method to an extensive array of non-contaminated spectra and identifying a reasonable lower limit. Methods are under development to automate this procedure. Radial current velocities are output only if the RMS deviations are less than S_{lim}. This eliminates results from the corresponding contaminated Doppler spectral points.

5. Examples of Results

In this section, we give examples of the application of the method described in Section 4 to provide QC for radial velocities from three other radar sites. Current velocities shown are obtained from measured radar spectra, unaveraged over time or distance. Figures 6–8 show current velocity maps obtained with and without QC. A quantitative measure of the improvement provided by QC is given by the resulting reduction in the normalized RMS deviations, see Equation (5). The average of this quantity over the map RMS_{av} is quoted for each case.

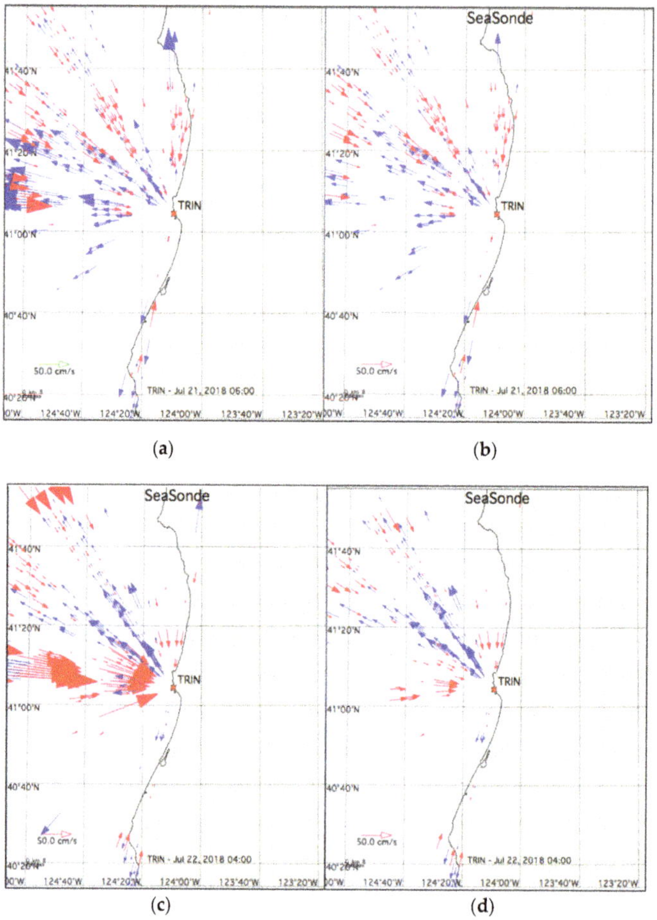

Figure 6. Comparison of current velocity maps obtained with and without QC (Quality Control) at Trinidad, California. Radar transmit frequency: 4.8 MHz. The threshold S_{lim} was set to 25. 21 July, 2018, 6:00 a.m.: (**a**) without QC, RMS_{av} = 32.2 (**b**) with QC, RMS_{av} = 11.2. 22 July, 2018, 4:00 a.m.: (**c**) without QC, RMS_{av} = 83.9 (**d**) with QC, RMS_{av} = 12.8.

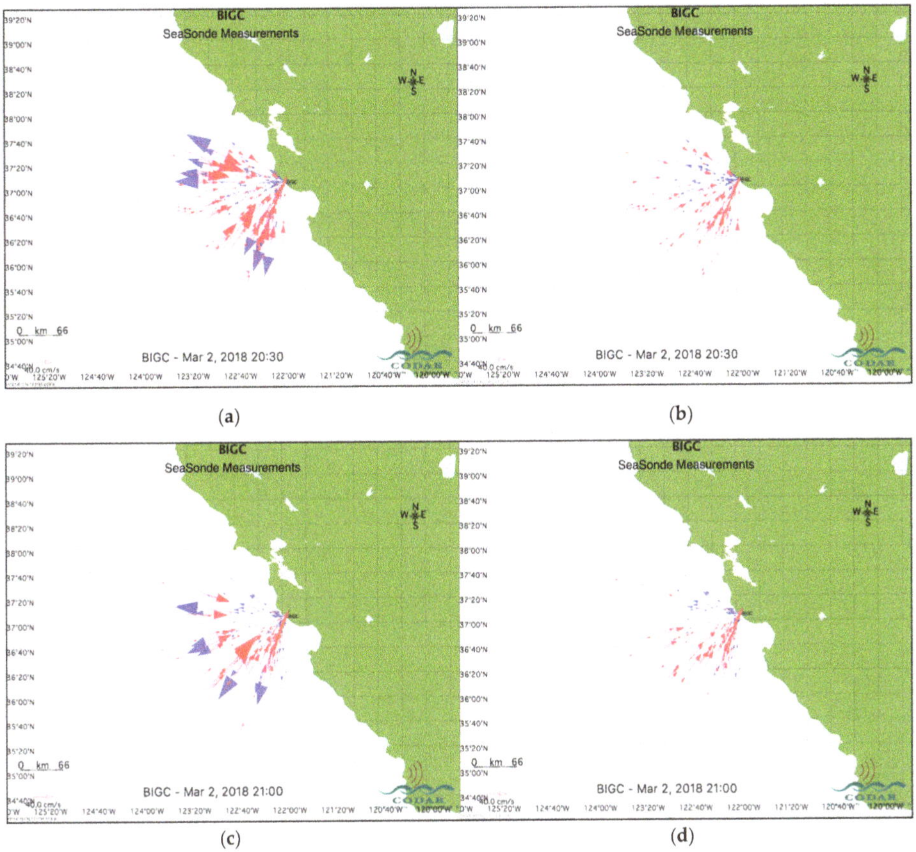

Figure 7. Comparison of current velocity maps obtained with and without QC at Big Creek, California. Radar transmit frequency 4.5 MHz. The threshold S_{lim} was set to 8.0. 2 March, 2018, 8:30 p.m.: (**a**) without QC, RMS_{av} = 858.6 (**b**) with QC, RMS_{av} = 1.9. 2 March, 2018, 21:00: (**c**) without QC RMS_{av} = 414.4, (**d**) with QC RMS_{av} = 1.6.

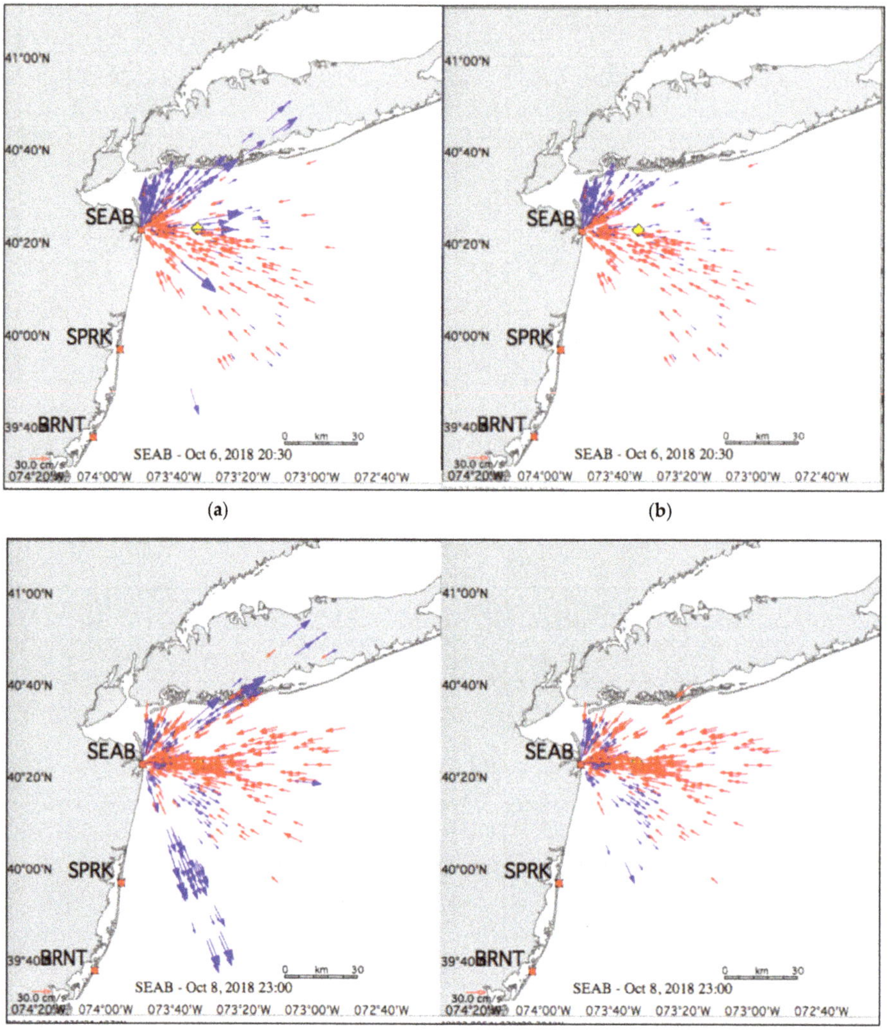

Figure 8. Comparison of current velocity maps obtained with and without QC at Sea Bright, New Jersey. Radar transmit frequency 13.5 MHz. The threshold S_{lim} was set to 1.4. 6 October, 2018, 20:30, 6:00 a.m.: (**a**) without QC, RMS_{av} = 2.9 (**b**) with QC, RMS_{av} = 0.48. 8 October, 2018, 11:00 p.m.: (**c**) without QC RMS_{av} = 3.2 (**d**) with QC RMS_{av} = 0.49.

6. Conclusions

This paper summarizes a method for improving the quality of radial velocities from broad-beam sea-echo voltage cross spectra, based on the application of a least-squares direction-finding algorithm. This method provides QC for eliminating poor-quality radar spectral points from the analysis, when the least-squares deviations exceed a preset limit. The method is illustrated using data from a radar at Shelter Cove, California. Results are given from application of this method to data from three additional radar sites, including both long-range and standard-range broad-beam systems.

The next step is to automate the method, which can then run on systems in the field to improve output current velocity data quality. Current velocities shown in this paper are obtained from short-term radar spectra, involving minimal averaging at the spectral level. Further processing involves averaging over time and distance. Resulting current velocities from radar sites with significant interference can be validated by comparison with drifters and in situ devices.

A similar method is currently under development to provide QC for output from directional ocean wave spectra.

The majority of HF radars in operational use for ocean-monitoring employ compact broad-beam antennas; other systems rely on beam-forming.In the future it may be possible to determine a common QC procedure for all ocean-monitoringHF radars, since all output results are based on Barrick's Equation (1) for the narrow-beam first-order radar cross section.

Author Contributions: Conceptualization, B.L., C.W., D.B.; Methodology, B.L.; Software, B.L.; Validation, B.L. Formal Analysis, B.L.; Investigation, B.L.; Resources, B.L., C.W.; Data Curation, B.L., C.W.; Writing-Original Draft Preparation, B.L.; Writing-Review & Editing, D.B., C.W.

Funding: This research received no external funding.

Acknowledgments: We are grateful to Hardik Parikh and Maeve Daugharty for providing data and maps used in the analysis. Radar spectral data came from the Central and Northern California Ocean Observing System (CeNCOOS) and the Mid-Atlantic Regional Association Coastal Ocean Observing System (MARACOOS), both of which receive funding from the U.S. Integrated Ocean Observing System (IOOS®). We are grateful to the individuals who maintain those systems and provided the data: Deedee Shideler and Marcel Losekoot of Bodega Marine Laboratory; Hugh Roarty and Ethan Handel of Rutgers University.

Conflicts of Interest: The authors are involved with the development and sale of SeaSonde systems.

References

1. Crombie, D.D. Doppler spectrum of sea echo at 13.56 Mc/s. *Nature* **1955**, *175*, 681–682. [CrossRef]
2. Barrick, D.E. First-order theory and analysis of MF/HF/VHF scatter from the Sea. *IEEE Trans. Antennas Propag.* **1972**, *AP-20*, 2–10. [CrossRef]
3. Barrick, D.E. Remote sensing of sea-state by radar. In *Remote Sensing of the Troposphere*; Derr, V.E., Ed.; U.S. National Oceanic and Atmospheric Administration: Silver Spring, MD, USA, 1972; pp. 12-1–12-46.
4. Lipa, B.J.; Barrick, D.E. Least-squares methods for the extraction of surface currents from CODAR crossed-loop data: Application at ARSLOE. *IEEE J. Ocean. Eng.* **1983**, *8*, 226–253. [CrossRef]
5. Lipa, B.; Nyden, B.; Ullman, D.; Terrill, E. SeaSonde Radial Velocities: Derivation and Internal Consistency. *IEEE J. Ocean. Eng.* **2006**, *31*, 850–861. [CrossRef]
6. Lipa, B.; Barrick, D. *Analysis Methods for Narrow-Beam High-Frequency Radar Sea Echo*; Technical Report 420-WP; U.S. Department of Commerce, National Oceanic and Atmospheric Administration, Environmental Research Laboratories: Washington, DC, USA, 1982; Volume 56.
7. Lipa, B.; Barrick, D. Extraction of sea state from HF radar sea echo: Mathematical theory and modeling. *Radio Sci.* **1986**, *21*, 81–100. [CrossRef]
8. Lipa, B.; Nyden, B. Directional wave information from the seasoned. *IEEE J. Ocean. Eng.* **2005**, *30*, 221–231. [CrossRef]
9. Stewart, R.; Joy, J. HF radio measurements of surface currents. *Deep Sea Res.* **1974**, *21*, 1039–1049. [CrossRef]
10. Barrick, D.; Headrick, J.; Bogle, R.; Crombie, D. Sea backscatter at HF: Interpretation and utilization of the echo. *Proc. IEEE* **1974**, *62*, 673–680. [CrossRef]
11. Barrick, D.; Evans, M.; Weber, B. Ocean surface currents mapped by radar. *Science* **1977**, *198*, 138–144. [CrossRef]
12. Barrick, D.; Lipa, B. A compact transportable HF radar system fordirectional coastal wave field measurements. In *Ocean Wave Climate*; Earle, M.D., Malahoff, A., Eds.; Plenum: New York, NY, USA, 1979; pp. 153–201.
13. The SeaSonde. Available online: http://www.codar.com/SeaSonde.shtml (accessed on 10 April 2019).
14. Harlan, J.; Terrill, E.; Hazard, L.; Keen, C.; Barrick, D.; Whelan, C.; Howden, S.; Kohut, J. The Integrated Ocean Observing System High-Frequency Radar Network: Status and Local, Regional, and National Applications. *Mar. Technol. Soc. J.* **2010**, *44*, 122–132. [CrossRef]
15. Fujii, S.; Heron, M.; Kim, K.; Lai, J.-W.; Lee, S.-H.; Wu, X.; Wu, X.; Wyatt, L.; Yang, W.-C. An Overview of Developments and Applications of Oceanographic Radar Networks in Asia and Oceania Countries. *Ocean Sci. J.* **2013**, *48*, 69–97. [CrossRef]

16. Rubio, A.; Mader, J.; Corgnati, L.; Mantovani, C.; Griffa, A.; Novellino, A.; Quentin, C.; Wyatt, L.; Schulz-Stellenfleth, J.; Horstmann, J.; et al. HF Radar Activity in European Coastal Seas: Next Steps toward a Pan-European HF Radar Network. *Front. Mar. Sci.* **2017**, 4–8. [CrossRef]
17. Jena, B.; Arunraj, K.; Suseentharan, V.; Tushar, K.; Karthikeyan, T. Indian coastal ocean radar network. *Curr. Sci.* **2019**, *116*, 372–378.
18. Saviano, S.; Kalampokis, A.; Zambianchi, E.; Uttieri, M. A Year-Long Assessment of Wave Measurements Retrieved from an HF Radar Network in the Gulf of Naples (Tyrrhenian Sea, Western Mediterranean Sea). *J. Oper. Oceanogr.* **2019**, 1–15. [CrossRef]
19. Daugharty, M.; Lipa, B.; Barrick, D.; Roarty, H.; Dicopoulos, J. Mapping Wave Variation in MARACOOS with SeaSonde Compact HF Radar. In Proceedings of the ORCA Conference, Okinawa, Japan, 2–4 June 2018.
20. Abascal, A.; Castanedo, S.; Fernandez, V.; Medina, R. Backtracking Drifting Objects using Surface Currents from High-Frequency (HF) Radar. *Ocean Dyn.* **2012**, *58*, 1–18. [CrossRef]
21. Bellomo, A.; Griffa, S.; Cosoli, P.; Falco, R.; Gerin, I.; Iermano, A.; Kalampokis, Z.; Kokkini, A.; Lana, M.; Magaldi, I.; et al. Toward an integrated HF radar network in the Mediterranean Sea to improve search and rescue and oil spill response: The TOSCA project experience. *J. Oper. Oceanogr.* **2015**, *8*, 95–107. [CrossRef]
22. Abascal, A.; Sanchez, J.; Chiri, H.; Ferrer, M.; Cárdenas, M.; Gallego, A.; Castanedo, S.; Medina, R.; Alonso-Martirena, A.; Berx, B.; et al. Operational Oil Spill Trajectory Modelling Using HF Radar Currents: A Northwest European Continental Shelf Case Study. *Mar. Pollut. Bull.* **2017**, *119*, 336–350. [CrossRef]
23. Cardenas, M.; Abascal, A.; Castanedo, S.; Chiri, S.; Ferrer, M.; Sanchez, J.; Medina, R.; Turrell, W.; Hughes, S.; Gallego, A.; Berx, B. Spill Trajectory Modeling Based on HF Radar Currents in the North Sea: Validation with Drifter Buoys. In Proceedings of the Thirty Eighth AMOP Technical Seminar, Vancouver, BC, Canada, 2–5 June 2015; pp. 123–142.
24. Whelan, C.; Barrick, D.; Lilleboe, P.; Breivik, Ø.; Kjelaas, A.; Fernandez, V.; Alonso-Martirena, A. Rapid Deployable HF RADAR for Norwegian emergency spill operations. In Proceedings of the Oceans 2010 IEE, Sydney, Australia, 24–27 May 2010; pp. 1–3.
25. Roarty, H.; Smith, M.; Glenn, S.; Barrick, D.; Page, E.; Statscewich, H.; Weingartner, T. Expanding Maritime Domain Awareness Capabilities in the Arctic: High Frequency Radar Vessel-tracking. In Proceedings of the IEEE Radar Conference, Ottawa, ON, Canada, 29 April–3 May 2013; pp. 1–5.
26. Sciascia, R.; Berta, M.; Carlson, D.; Griffa, A.; Panfili, M.; La Mesa, M.; Corgnati, L.; Mantovani, C.; Domenella, E.; Fredj, E.; et al. Linking Sardine Recruitment in Coastal Areas to Ocean Currents Using Surface Drifters and HF Radar: A Case Study in the Gulf of Manfredonia, Adriatic Sea. *Ocean Sci.* **2018**, *14*, 1461–1482. [CrossRef]
27. Lipa, B.; Barrick, D.; Isaacson, J. *Coastal Tsunami Warning with Deployed HF Radar Systems*; Mokhtari, M., Ed.; InTechOpen: Rijeka, Croatia, 2016; Available online: http://www.intechopen.com/books/tsunami/coastal-tsunami-warning-with-deployed-hf-radar-systems (accessed on 10 April 2019). [CrossRef]
28. Cook, T.; Hazard, L.; Otero, M.; Zelenke, B. (Eds.) *Deployment and Maintenance of a High-Frequency Radar (HFR) for Ocean Surface Current Mapping: Best Practices*; University of California San Diego, Scripps Institution of Oceanography for SCCOOS: La Jolla, CA, USA, 2008; 19p. Available online: https://www.oceanbestpractices.net/handle/11329/368 (accessed on 10 April 2019).
29. Kohut, J.; Glenn, S. Improving HF Radar Surface Current Measurements with Measured Antenna Beam Patterns. *J. Atmos. Ocean. Technol.* **2003**, *20*, 1303–1316. [CrossRef]
30. Emery, B.; Washburn, L.; Whelan, C.; Barrick, D.; Harlan, J. Measuring Antenna Patterns for Ocean Surface Current HF Radars with Ships of Opportunity. *J. Atmos. Ocean. Technol.* **2014**, *31*, 1564–1582. [CrossRef]
31. Washburn, L.; Romero, E.; Johnson, C.; Emery, B.; Gotschalk, C. Measurement of Antenna Patterns for Oceanographic Radars Using Aerial Drones. *J. Atmos. Ocean. Technol.* **2017**, *34*, 971–981. [CrossRef]
32. Whelan, C.; Hubbard, M.; Trockel, D.; Parikh, H. Benefits of Multiple Antenna Pattern Measurement Methods for Maintaining a Regional HF Radar Network. In Proceedings of the ORCA Conference, Okinawa, Japan, 2–4 June 2018.
33. U.S. Department of Commerce, National Oceanic and Atmospheric Administration, National Ocean Service, Integrated Ocean Observing System. Manual for Real-Time Quality Control of High Frequency Radar Surface Currents Data: A Guide to Quality Control and Quality Assurance of High Frequency Radar Surface Currents Data Observations. Volume 54, 2016. Available online: https://cdn.ioos.noaa.gov/media/2017/12/HFR_QARTOD_Manual_05_26_16.pdf (accessed on 10 April 2019).

34. Kirincich, A.; de Paolo, T.; Terrill, E. Improving HF Radar Estimates of Surface Currents Using Signal Quality Metrics, with Application to the MVCO High-Resolution Radar System. *J. Atmos. Ocean. Technol.* **2012**, *29*, 1377–1390. [CrossRef]
35. De Paolo, T.; Terrill, E.; Kirincich, A. Improving SeaSonde radial velocity accuracy and variance using radial metrics. In Proceedings of the OCEANS MTS/IEEE Conference, Genoa, Italy, 18–21 May 2015; pp. 1–9.
36. Cosoli, S.; Grcic, B.; de Vos, S.; Hetzel, Y. Improving Data Quality for the Australian High Frequency Ocean Radar Network through Real-Time and Delayed-Mode Quality-Control Procedures. *Remote Sens.* **2018**, *10*, 1476. [CrossRef]

Journal of
Marine Science and Engineering

MDPI

Article

Estimation of Coastal Currents Using a Soft Computing Method: A Case Study in Galway Bay, Ireland

Lei Ren [1,2], Jianming Miao [1,*], Yulong Li [1,*], Xiangxin Luo [1,*], Junxue Li [3,*] and Michael Hartnett [4]

1 School of Marine Engineering and Technology, Sun Yat-sen University, Guangzhou 510275, China; renleicomeon@163.com
2 Southern Marine Science and Engineering Guangdong Laboratory (Zhuhai), Zhuhai 519082, China
3 College of Science and Technology, Hebei Agricultural University, Cangzhou 061100, China
4 College of Engineering and Informatics, National University of Ireland Galway, H91 TK33 Galway, Ireland; michael.hartnett@nuigalway.ie
* Correspondence: miaojm@mail.sysu.edu.cn (J.M.); liylong7@mail.sysu.edu.cn (Y.L.); luoxx6@mail.sysu.edu.cn (X.L.); lijunxue1125@163.com (J.L.)

Received: 1 April 2019; Accepted: 14 May 2019; Published: 20 May 2019

Abstract: In order to obtain forward states of coastal currents, numerical models are a commonly used approach. However, the accurate definition of initial conditions, boundary conditions and other model parameters are challenging. In this paper, a novel application of a soft computing approach, random forests (RF), was adopted to estimate surface currents for three analysis points in Galway Bay, Ireland. Outputs from a numerical model and observations from a high frequency radar system were used as inputs to develop soft computing models. The input variable structure of soft computing models was examined in detail through sensitivity experiments. High correlation of surface currents between predictions from RF models and radar data indicated that the RF algorithm is a most promising means of generating satisfactory surface currents over a long prediction period. Furthermore, training dataset lengths were examined to investigate influences on prediction accuracy. The largest improvement for zonal and meridional surface velocity components over a 59-h forecasting period was 14% and 37% of root mean square error (RMSE) values separately. Results indicate that the combination of RF models with a numerical model can significantly improve forecasting accuracy for surface currents, especially for the meridional surface velocity component.

Keywords: coastal surface currents; soft computing; radar; sensitivity experiments; numerical model

1. Introduction

Interaction between atmospheric forces such as wind, river inflows and tide drive the movement of coastal water bodies. The horizontal phenomenon related to this movement is surface current. Good understanding of coastal surface currents is of great importance for many coastal economic operations, in particular marine renewable energy exploration/exploitation. In general, numerical models and observation platforms are powerful and conventional tools to study patterns of coastal surface currents, providing useful information [1]. However, model approximations and simplifications in defining initial and boundary conditions, model grid structure and other factors lead to a reduction in model performance and accuracy. Model prediction accuracy over the long term is hard to guarantee. Oceanic observation tools such as satellites and radars based on electro communication technologies are useful means to record surface flow information over a large coastal domain [2]. However, these observation tools cannot provide forecasts.

In order to accurately forecast states of parameters of interest by making the best use of available observations, soft computing approaches such as artificial neural networks (ANNs), support vector machines (SVMs) and random forests (RF) have been widely used as an alternative in a number of fields [3–5]. Soft computing models are computationally more efficient since they construct the relationship between input variables and output variables directly. In soft computing models, predictions are generated from a trained soft computing model, which is based on relationships between inputs variables and output variables. According to some case studies such as the prediction of renewable energy sources [6], forecasting urban water demand [7] and ecological prediction [8], RF is a powerful and efficient soft computing approach used to establish internal relationships among datasets and ultimately provide forecasting states with high accuracy. Since the RF algorithm is capable of dealing with large numbers of predictor variables even in the presence of complex interaction, it has been successfully applied in genetics, clinical medicine and bioinformatics within the past decades [9]. Currently, it has become increasingly a more popular approach in oceanographic engineering. For example, Ibarra-Berastegi et al. [6] applied the RF algorithm to produce short-term forecasting of the wave energy flux in comparison to using a physical model—wave model (WAM) and analogues. Their comparative results indicate that the RF model outperforms other statistical techniques when compared with the WAM, and a window of forecasting horizons emerges in which the use of RF outperforms any other solutions. Lahouar and Slama [10] applied the RF algorithm to predict electrical load demand of the day ahead, finding that RF coupled with expert selection was able to capture complex load behavior and solve some special cases that are specific to culture, high temperature, religious events and moving holidays owing to appropriate input variable structure. Catani et al. [11] used a RF algorithm to estimate landslide susceptibility and found that the dimension of parameter space, the mapping scale and the training process strongly influenced the classification accuracy and the prediction process. They also showed that the choice of the training set was of key importance for obtaining accurate results. Mahjoobi and Mosabbeb [12] used present time and previous time data as input variables to predict significant wave heights using a support vector machine. Moreover, Balas et al. [13] used historical data to predict missing wave parameters using a soft computing approach. Soft computing models are commonly established based only on observational data. In this research, a three-dimensional coastal hydrodynamic model i.e., environment fluid dynamics code (EFDC) of Galway Bay has been set up by Ren et al. [14]. This is the first time that the RF algorithm has been used to predict surface currents and the first time RF has been used in combination with high frequency coastal radar data. The predictive model described herein was developed in order to further improve forecasting capability of surface currents in this area and to provide continuous and operational predictions in the future. Outputs of surface currents $u(t)$ and $u(t-1)$ from the EFDC model were chosen as main input variables to establish RF models in this research. Reasons for including previous observations lies in that the variation of surface current patterns is continuous in time at each physical location, development of surface currents at the present time step is related to previous states. Galway Bay is located on the west coast of Ireland, tidal water elevation and winds are known to be the main driving forces that generate surface currents in the bay. Tidal water elevation, wind speeds and wind directions were examined as input variables to examine the input variable structure of the RF model developed.

In this research, the soft computing approach RF was adopted to predict surface currents at the 1/4-scale marine energy test site Galway Bay area using surface current data measured by a high frequency coastal ocean dynamic application radar (CODAR) system and outputs from a numerical model. Surface current speeds at the present time step are denoted by $u(t)$, and surface currents i hour(s) before the present time are denoted as $u(t-i)$.

The main objective of this research was to use model data and observations from radars to run a soft computing model for coastal surface current prediction. The layout of this paper is as follows: Section 2 describes the methodologies used and outlines the study domain; this section summarizes the numerical model, the high frequency radar (HFR) system, the random forests algorithm and criteria assessment skills. Results are presented in Section 3, followed by a discussion in Section 4. Research conclusions are presented in Section 5.

2. Methodologies

2.1. Study Domain

Galway Bay is located on the west coast of Ireland with an entrance opening onto the northeast Atlantic Ocean. The bay is semi-enclosed as it is partially shielded from the harsh Atlantic conditions by three small islands. Hydrodynamics within the bay are mainly affected by oceanic flows to the bay from the adjacent continental shelf and wind-driven currents. Figure 1 shows the extents of the model domain and the measurement deployment locations. The water depth of the inner bay covered by the radar system ranged from 10–40 m. Meteorological conditions in Galway Bay are mainly influenced by Atlantic weather systems [15]. Effects of wind forcing on surface flow patterns has been studies in details by Ren et al. [16].

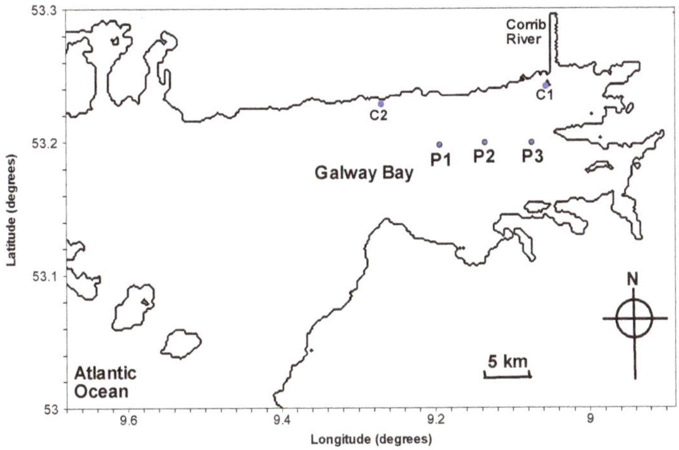

Figure 1. Galway Bay domain; (C1 and C2 indicate radar stations; P1–P3 indicate three analysis points).

2.2. Numerical Model

EFDC is a finite difference based model which can simulate three-dimensional flows and transport processes in surface water systems, rivers, lakes, estuaries, wetlands and coastal areas [17]. The model structure includes four major modules: (1) A hydrodynamic module; (2) a water quality module; (3) a sediment transport module; and (4) a toxics module. The EFDC model solves the three dimensional, vertically hydrostatic, free surface, turbulent-averaged equations of motion for a variable density fluid and can simulate both barotropic and baroclinic circulation. The model uses a stretched (sigma) vertical coordinate system and a Cartesian or curvilinear orthogonal horizontal coordinate system [18–20]. The EFDC model has been successfully applied in a number of areas [21–23].

A three-dimensional computational model of Galway Bay had been developed previously. A bathymetric model of the Galway Bay was developed using data from the Integrated Mapping for the Sustainable Development of Ireland's Marine Resource (INFOMAR) program. The deepest area is approximately 70 m near the south west corner of the open boundaries. The mean water depth in the domain is 30.7 m. Tidal elevation for boundary conditions obtained from the Oregon Tide Prediction Software (OTPS), wind data obtained from the Europe Centre for Mid-Range Weather Forecasts (ECMWF) and River Corrib flow data provided by the Irish Open Public Work (OPW) were used to drive the models. Detailed EFDC modeling setup for Galway Bay such as wind forcing and vertical layer structure is described in Ren et al. [14], O'Donncha et al. [24] and Ren et al. [25].

2.3. Radar Data

There are two commonly used types of HFR observation systems: Beam-forming (e.g., CODAR and ocean state measuring and analyzing radar (OSMAR) system) and direction-finding (e.g., WERA HFR system) [26–28]. The former HFR system uses a distributed array of elements to electronically scan ocean surface with a relatively narrow beam; while the latter HFR system uses a small-aperture antenna to from a quite broad beam for obtaining sea echoes [29]. A land-based CODAR HFR observation system consisting of two radars has been deployed in the Galway Bay area since 2011, as shown in Figure 1. Each radar station is capable of monitoring radial surface currents toward or away from the station. The raw ocean surface information is obtained from radar signals, which are scattered in 360°, measurement information returns to the radar receiver when the radar signal scatters off a wave that is exactly half of the transmitted signal wavelength [30–32]. The information in the radio-wave backscatter exploited from the ocean surface is used to infer movement of the near surface layer, including parameters of surface currents, waves and winds [33–36]. Surface flow fields in the area covered by both radars were combined based on the radial current maps for at least two stations [37,38]. Measurements obtained from the HFR systems are available in near real time. Temporal and spatial resolution of surface flow fields are sixty minutes and 300 m, respectively. The operating frequency is 25 MHz for both radars. The measured surface currents at the three analysis points (P1, P2 and P3) by the HFR system were taken as target fields to train and assess the RF models. Validation of radar surface currents have been undertaken by Ren et al. [14], O'Donncha et al. [24] and Ren et al. [25], and reasonably good agreement between radar data and ADCP observations provided confidence in using radar data for other applications such as establishing RF models.

2.4. Random Forests

The RF algorithm proposed by Breiman [39] is an improved version of the decision tree learning approach, which integrates the prediction of multiple uncorrelated decision trees [40]. The RF algorithm is based on bagging that builds a large collection of de-correlated trees, and then averages them [41,42]. The RF algorithm has been implemented in a number of fields due to its robustness and satisfactory predictions with high accuracy.

The RF methodology does not only produce one decision tree, but produces a variety of trees using subsets of a training dataset. The RF algorithm can be expressed in the following steps [43].

(1) Select n_{RF} multiple bootstrap samples from the dataset;
(2) Develop an unpruned regression tree for each of the bootstrap samples by randomly sampling m_{try} of the predictors and select the best split among those variables;
(3) Forecast new data by aggregating the predictions of the n_{RF} trees; for a regression analysis, the averaged value is taken as output.

For the classification case, a random forest obtains a class vote from every tree, and then classifies them using a majority vote; for the regression case, predictions from every tree are averaged as final outputs [41]. The RF approach is to build n_{RF} multiple decision trees and merge them together to generate a prediction with higher accuracy. In each decision tree, n_{RF} features were selected and used.

In this research, the RF algorithm was applied to establish regression models, input variables including historical outputs of surface currents from both the EFDC model and radar observations were taken as the target field was viewed as a full dataset.

The n_{RF} trees were separately developed. Each tree includes m samples, which were randomly selected. In the selection process for all trees, the number of predictors m_{try} was the same. The recommend values of m_{try} for the classification and regression were given as [11]:

Classification:

$$\mathrm{m}_{try} = \sqrt{\mathrm{p}} \tag{1}$$

Regression:

$$m_{try} = \frac{p}{3} \tag{2}$$

where, p is the number of predictors.

Each tree was developed by training the selected samples. R packages named *"randomForest"* from the Comprehensive Archive R Network (CRAN) was used in this research [8,43]. An estimate of the error rate can be obtained in RF approach based on the training data: Firstly, out of bag (OOB) data not used in the training were used for prediction using the tree grown with the bootstrap sample; secondly, the OOB predictions were aggregated and the error rate computed which was called the OOB estimate of error rate [43,44].

Advantages of the RF algorithm are: (a) RF is capable of dealing with high-dimensional datasets, twice randomly sampling reduces the dimension of the datasets; (b) bootstrapping sampling process results in around one third data are out of bag (OOB) samples, OOB samples are used to compute the unbiased error rate and variable importance; (c) the prediction only depends on one user-selected parameter, the number of predictors are chosen randomly at each node [8].

In this research, outputs of surface currents from EFDC model, tidal water elevation, wind speed and wind direction were considered as input variables to establish RF models. River discharge was not considered here because it had weak impacts on the three analysis points located far from the estuary. Selection of input variable structure was presented in detail in Section 3.1.

2.5. Criteria Skills

In order to quantify performances of the RF models developed in a quantitative way, multiple statistics including correlation (R), bias, root mean square error (RMSE) and scatter index (SI) were computed using the following Equations (3)–(6) [45]. The correlation coefficient is an indicator of the linear relationship between two datasets; bias indicates the trend of a measurement process to systematically over- or underestimate the magnitude of a predicted parameter; RMSE is an error index presenting an overall error distribution; SI gives the percentage of expected error for the parameter.

$$bias = \overline{y} - \overline{x} \tag{3}$$

$$\mathrm{RMSE} = \sqrt{\frac{\sum_{i=1}^{n}(y_i - x_i)}{n}} \tag{4}$$

$$\mathrm{SI} = \frac{RMSE}{\overline{x}} \tag{5}$$

$$\mathrm{R} = \frac{\sum_{i=1}^{n}(x_i - \overline{x})(y_i - \overline{y})}{\sqrt{\sum_{i=1}^{n}(x_i - \overline{x})^2}\sqrt{\sum_{i=1}^{n}(y_i - \overline{y})^2}} \tag{6}$$

where,

x_i is an observed value;
y_i is a predicted value;
n is the number of observations;
$\overline{x}$ and $\overline{y}$ is the mean of x and y, respectively.

Statistics were computed for each RF model at every analysis point respectively.

3. Results

3.1. RF Models

A full dataset consisting of 305 hourly values starting from Julian Day 220 01:00, 2013 was used in this research. Based on division principles proposed by Aydogan et al. [46], the full dataset was subdivided as follows: (i) A training dataset accounting for 60% of the total; (ii) a test dataset accounting for 20% of the total and (iii) a prediction dataset accounting for 20% of the total. Surface currents from the EFDC model were outputted at the same time steps as the HFR observations. Tidal elevations and wind stresses were used as boundary conditions for the numerical model simulations; in order to further explore the relationship between surface flows and the dominant forcing factors tidal elevations, wind speeds and wind directions were adopted as input variables in establishing the RF models as well as surface current outputs from the numerical model. RF models for predicting surface current components at each of the three analysis points were established separately. The functional equation for RF models is expressed as:

$$\mathrm{URF}(t) = f(UEF(t), UEF(t-1), TWL(t), WS(t), WD(t)) \tag{7}$$

where,

$\mathrm{URF}(t)$ is the output of surface current components from RF models at time step t (cm/s);
$f(\bullet)$ indicates the random forests function;
$UEF(t)$ is the EFDC model surface current output at time step t (cm/s);
$UEF(t-1)$ is the EFDC model surface current output at time step $(t-1)$ (cm/s);
$TWL(t)$ is the tidal water level from OTPS at time step t (m);
$WS(t)$ is the ECMWF wind speed at time step t (m/s);
$WD(t)$ is the ECMWF wind direction at time step t (°).

RF models were established for predicting surface currents based on training the relationship between two datasets. A schematic of the development process of proposed models is shown in Figure 2. HFR data were only used as target variables during the procedure of training RF models; HFR data were used to evaluate the RF modeling performance when the test dataset was adopted; HFR data were not applied in the RF models during the forecasting period.

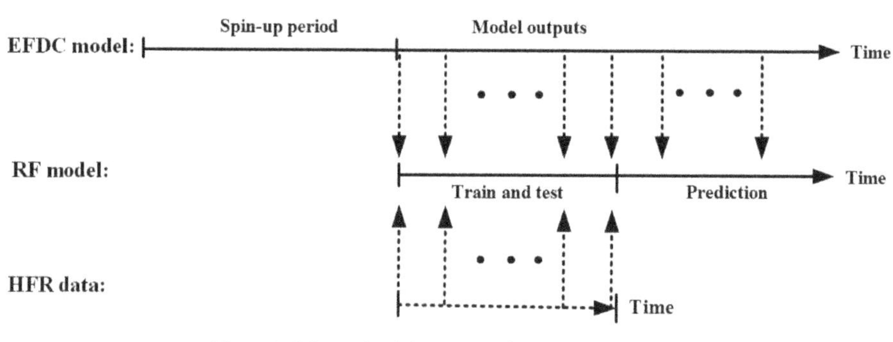

Figure 2. Schematic of the proposed prediction method.

To define an appropriate input variable structure, sensitivity experiments were performed to determine the appropriate number of input variables. Tidal water elevation, wind speed and wind direction were linearly interpolated to the radar data observation time step. Mean and range of these variables of the training dataset, test dataset and forecasting dataset are presented in Tables 1–3.

Table 1. Mean and range of training dataset (60% of total dataset).

Point	Variable	Mean	Minimum	Maximum
P1	U(HFR) (cm/s)	16.29	0.81	35.63
	UEF (cm/s)	14.43	1.26	36.89
	TWL (m)	0	−2.12	2.08
	WS (m/s)	5.35	1.81	7.98
	WD (degrees)	186.39	102.40	255.01
P2	U(HFR) (cm/s)	17.19	1.20	35.80
	UEF (cm/s)	15.55	0.78	32.64
	TWL (m)	0.00	−2.12	2.08
	WS (m/s)	5.36	1.81	7.98
	WD (degrees)	185.77	97.35	255.01
P3	U(HFR) (cm/s)	16.71	2.41	34.54
	UEF (cm/s)	17.87	2.54	35.53
	TWL (m)	0.00	−2.12	2.08
	WS (m/s)	5.36	1.81	7.98
	WD (degrees)	185.77	97.35	255.01

Note that U(HFR) indicates total surface velocity from high frequency radar; TWL indicates tidal water elevation from the Oregon Tide Prediction Software (OTPS) prediction model; WS and WD indicates wind speed and wind direction from the European Centre for Medium-Range Weather Forecasts (ECMWF) forecasting model respectively.

Table 2. Mean and range of testing dataset (20% of total dataset).

Point	Variable	Mean	Minimum	Maximum
P1	U(HFR) (cm/s)	14.23	2.03	32.34
	UEF (cm/s)	10.87	1.15	25.61
	TWL (m)	−0.03	−1.51	1.44
	WS (m/s)	6.21	3.38	8.41
	WD (degrees)	198.72	171.64	247.65
P2	U(HFR) (cm/s)	14.69	2.76	27.18
	UEF (cm/s)	11.88	0.98	25.66
	TWL (m)	−0.02	−1.51	1.44
	WS (m/s)	6.13	3.38	8.41
	WD (degrees)	198.15	166.09	247.65
P3	U(HFR) (cm/s)	13.91	4.96	24.59
	UEF (cm/s)	12.14	2.01	21.89
	TWL (m)	−0.02	−1.51	1.44
	WS (m/s)	6.13	3.38	8.41
	WD (degrees)	198.15	166.09	247.65

Tables 1–3 show mean, minimum and maximum values of the three datasets at each analysis point (P1, P2, and P3). The mean values, minimum and maximum values varied in these datasets at each analysis point. Variations among these datasets is of great significance when examining and assessing the effectiveness of the proposed RF models. The datasets are important in determining whether or not the proposed RF models are robust enough to generate accurate predictions.

Sensitivity experiments of the input variable structure were performed for surface velocity components separately at each analysis point, as presented in Tables 4 and 5. Table 4 presents statistics of each RF model on testing input variable structure for the zonal surface velocity component; statistics for the meridional surface velocity component are presented in Table 5. The test dataset, accounting for 20% of the total dataset, was used to evaluate the performance of each RF model based on statistics between RF model results and radar data.

Table 3. Mean and range of prediction dataset (20% of total dataset).

Point	Variable	Mean	Minimum	Maximum
P1	U(HFR) (cm/s)	17.44	2.94	34.91
	UEF (cm/s)	16.51	1.92	32.79
	TWL (m)	−0.04	−2.53	2.36
	WS (m/s)	6.09	4.21	8.84
	WD (degrees)	211.88	167.31	250.24
P2	U(HFR) (cm/s)	17.65	3.20	31.83
	UEF (cm/s)	17.74	1.55	30.75
	TWL (m)	0.07	−2.43	2.36
	WS (m/s)	6.02	4.21	8.84
	WD (degrees)	210.12	167.31	250.24
P3	U(HFR) (cm/s)	16.49	1.68	28.13
	UEF (cm/s)	17.74	3.14	32.42
	TWL (m)	0.07	−2.43	2.36
	WS (m/s)	6.02	4.21	8.84
	WD (degrees)	210.12	167.31	250.24

Table 4. Statistics of zonal surface velocity component (testing dataset).

Model	Input Variables	R1	R2	Bias	RMSE1 (cm/s)	RMSE2 (cm/s)	SI
UP1RF1	UEF(t), UEF(t − 1)	0.9259	0.9429	−3.3166	5.8804	7.1152	0.7323
UP1RF2	UEF(t), UEF(t − 1),TWL(t)	0.9204	0.9429	−3.5441	6.1199	7.1152	0.7621
UP1RF3	UEF(t), UEF(t − 1),WS(t)	0.9457	0.9429	−3.0766	5.2315	7.1152	0.6515
UP1RF4	UEF(t), UEF(t − 1),WD(t)	0.9337	0.9429	−2.3496	5.1282	7.1152	0.6396
UP1RF5	UEF(t), UEF(t − 1),TWL(t),WS(t)	0.9420	0.9429	−3.4323	5.5389	7.1152	0.6897
UP1RF6	UEF(t), UEF(t − 1),TWL(t),WD(t)	0.9271	0.9429	−2.8268	5.5479	7.1152	0.6909
UP1RF7	**UEF(t), UEF(t − 1),WS(t),WD(t)**	**0.9450**	**0.9429**	**−2.2880**	**4.7701**	**7.1152**	**0.5940**
UP1RF8	UEF(t), UEF(t − 1),TWL(t),WS(t),WD(t)	0.9428	0.9429	−2.7379	5.0978	7.1152	0.6348
UP2RF1	UEF(t), UEF(t − 1)	0.8733	0.8446	0.9801	6.2389	7.1729	0.8376
UP2RF2	UEF(t), UEF(t − 1),TWL(t)	0.8729	0.8446	1.6002	6.1501	7.1729	0.8256
UP2RF3	**UEF(t), UEF(t − 1), WS(t)**	**0.8862**	**0.8446**	**1.0590**	**5.9567**	**7.1729**	**0.7997**
UP2RF4	UEF(t), UEF(t − 1),WD(t)	0.8831	0.8446	1.4787	6.0854	7.1729	0.8170
UP2RF5	UEF(t), UEF(t − 1),TWL(t),WS(t)	0.8823	0.8446	1.1141	6.0047	7.1729	0.8061
UP2RF6	UEF(t), UEF(t − 1),TWL(t),WD(t)	0.8744	0.8446	1.4671	6.2523	7.1729	0.8394
UP2RF7	UEF(t), UEF(t − 1),WS(t),WD(t)	0.8851	0.8446	1.4156	6.0472	7.1729	0.8118
UP2RF8	UEF(t), UEF(t − 1),TWL(t),WS(t),WD(t)	0.8821	0.8446	1.1639	6.1089	7.1729	0.8201
UP3RF1	UEF(t), UEF(t − 1)	0.8661	0.8764	−4.3970	6.9525	6.2198	1.0892
UP3RF2	UEF(t), UEF(t − 1),TWL(t)	0.8682	0.8764	−4.8322	7.2423	6.2198	1.1346
UP3RF3	UEF(t), UEF(t − 1),WS(t)	0.9029	0.8764	−4.2053	6.2616	6.2198	0.9810
UP3RF4	UEF(t), UEF(t − 1),WD(t)	0.8696	0.8764	−3.7031	6.4570	6.2198	1.0116
UP3RF5	UEF(t), UEF(t − 1),TWL(t),WS(t)	0.9004	0.8764	−4.6492	6.6429	6.2198	1.0407
UP3RF6	UEF(t), UEF(t − 1),TWL(t),WD(t)	0.8700	0.8764	−4.1360	6.7427	6.2198	1.0564
UP3RF7	**UEF(t), UEF(t − 1),WS(t), WD(t)**	**0.8988**	**0.8764**	**−3.4843**	**5.8521**	**6.2198**	**0.9168**
UP3RF8	UEF(t), UEF(t − 1),TWL(t),WS(t),WD(t)	0.8995	0.8764	−3.9344	6.1614	6.2198	0.9653

Note that R1 indicates the correlation coefficient between radar data and RF model results; R2 indicates the correlation coefficient between radar data and numerical model outputs only; RMSE1 indicates root mean square error between radar data and RF model results; RMSE2 indicates root mean square error between radar data and numerical model outputs; SI indicates the scatter index between radar data and RF model results; bias is the difference between radar data and RF model results. The same parameters are used as follows.

Table 4 shows that model UP1RF7, which used EFDC model outputs, wind speed and wind direction as input variables, produced the minimum root mean square error (RMSE) value (RMSE1) for the zonal surface velocity component at point P1; the same input variable structure for point P3 generated the minimum RMSE (RMSE1); while the best input variables for point P2 were EFDC model outputs and wind speed. This indicates that inclusion of wind speed was a highly significant parameter and improves RF model performance for the zonal surface velocity component. However, the effect of wind directions varied from location to location. Improvement of RMSE1 values for P1, P2 and P3 were 33%, 17% and 6% in comparison with RMSE2 respectively. Additionally, correlation (R1) between RF model results and radar data was enhanced as well at these locations comparing with correlation (R2) between EFDC model results and radar data. The maximum improvement in correlation was at point P2, by 5%. Table 4 shows that using (i) EFDC model outputs at two time-steps; (ii) wind speeds and (iii) wind directions to establish RF models can further improve estimation of zonal surface velocity components.

Table 5. Statistics of meridional surface velocity component (testing dataset).

Model	Input Variables	R1	R2	Bias	RMSE1 (cm/s)	RMSE2 (cm/s)	SI
VP1RF1	VEF(t), VEF(t − 1)	0.1722	0.3670	−0.2246	5.7549	5.5092	−23.8347
VP1RF2	VEF(t), VEF(t − 1),TWL(t)	0.3217	0.3670	0.1495	5.2536	5.5092	−21.7586
VP1RF3	VEF(t), VEF(t − 1),WS(t)	0.2996	0.3670	0.8936	5.5466	5.5092	−22.9719
VP1RF4	VEF(t), VEF(t − 1),WD(t)	0.6852	0.3670	0.7174	4.1340	5.5092	−17.1215
VP1RF5	VEF(t), VEF(t − 1),TWL(t),WS(t)	0.4922	0.3670	1.0752	4.9408	5.5092	−20.4630
VP1RF6	**VEF(t), VEF(t − 1),TWL(t),WD(t)**	**0.7941**	**0.3670**	**0.9206**	**3.6745**	**5.5092**	**−12.2185**
VP1RF7	VEF(t), VEF(t − 1),WS(t),WD(t)	0.5934	0.3670	1.9665	4.9099	5.5092	−20.3348
VP1RF8	VEF(t), VEF(t − 1),TWL(t),WS(t),WD(t)	0.7270	0.3670	1.8684	4.2173	5.5092	−17.4663
VP2RF1	VEF(t), VEF(t − 1)	0.1779	0.5097	−0.0881	7.4797	6.9956	17.7816
VP2RF2	VEF(t), VEF(t − 1),TWL(t)	0.1350	0.5097	1.5291	7.7613	6.9956	18.4510
VP2RF3	VEF(t), VEF(t − 1),WS(t)	0.4335	0.5097	0.9096	7.0880	6.9956	16.8503
VP2RF4	VEF(t), VEF(t − 1),WD(t)	0.7307	0.5097	0.2646	5.1449	6.9956	12.2310
VP2RF5	VEF(t), VEF(t − 1),TWL(t),WS(t)	0.4789	0.5097	1.8376	6.8769	6.9956	16.3485
VP2RF6	**VEF(t), VEF(t − 1),TWL(t),WD(t)**	**0.8180**	**0.5097**	**1.3854**	**4.7285**	**6.9956**	**11.2410**
VP2RF7	VEF(t), VEF(t − 1),WS(t),WD(t)	0.6424	0.5097	0.8763	5.8199	6.9956	13.8357
VP2RF8	VEF(t), VEF(t − 1),TWL(t),WS(t),WD(t)	0.7440	0.5097	1.5349	5.2029	6.9956	12.3688
VP3RF1	VEF(t), VEF(t − 1)	0.0127	0.4749	2.4416	8.9829	9.9814	6.4104
VP3RF2	VEF(t), VEF(t − 1),TWL(t)	0.0479	0.4749	1.6222	8.9451	9.9814	6.3835
VP3RF3	VEF(t), VEF(t − 1),WS(t)	0.4449	0.4749	−0.4086	7.4830	9.9814	5.3401
VP3RF4	VEF(t), VEF(t − 1),WD(t)	0.8133	0.4749	−1.3446	5.0184	9.9814	3.5812
VP3RF5	VEF(t), VEF(t − 1),TWL(t),WS(t)	0.4157	0.4749	−0.1224	7.4912	9.9814	5.3459
VP3RF6	**VEF(t), VEF(t − 1),TWL(t),WD(t)**	**0.8670**	**0.4749**	**−0.9184**	**4.4955**	**9.9814**	**3.2081**
VP3RF7	VEF(t), VEF(t − 1),WS(t),WD(t)	0.7648	0.4749	−0.5450	5.2767	9.9814	3.7656
VP3RF8	VEF(t), VEF(t − 1),TWL(t),WS(t),WD(t)	0.8191	0.4749	−0.4309	4.7590	9.9814	3.3962

Sensitivity experiments for the meridional surface velocity component, presented in Table 5, showed that RF models using EFDC model outputs, tidal water elevation and wind direction yielded the best estimations at each of the three locations. Improvements in RMSE values for the meridional surface velocity component between RF model estimations and radar data were 33%, 32% and 55% respectively at points P1, P2 and P3 in comparison with RMSE2. Correlations between RF model results and radar data (R1) were also significantly improved by 116%, 60% and 83% respectively for points P1, P2 and P3 in comparison with correlation between EFDC model results and radar data (R2).

Improvements were generally more significant in the meridional surface velocity component than the zonal surface velocity component at all points based on statistics as presented in Tables 4 and 5. The meridional component is difficult to predict as its signal generally is quite low and is more induced by transient wind stresses than deterministic tidal forcing. Results show that the proposed RF models can improve prediction accuracy for surface velocity components, and it is especially significant for difficult to predict meridional surface velocity components.

3.2. Comparison of Predictions

RF models at the three analysis points, using the best input variable structure, were applied to produce predictions using the forecasting dataset, and compared against "future" radar data. Time series of surface velocity components at the three analysis points are presented in Figures 3–8 respectively.

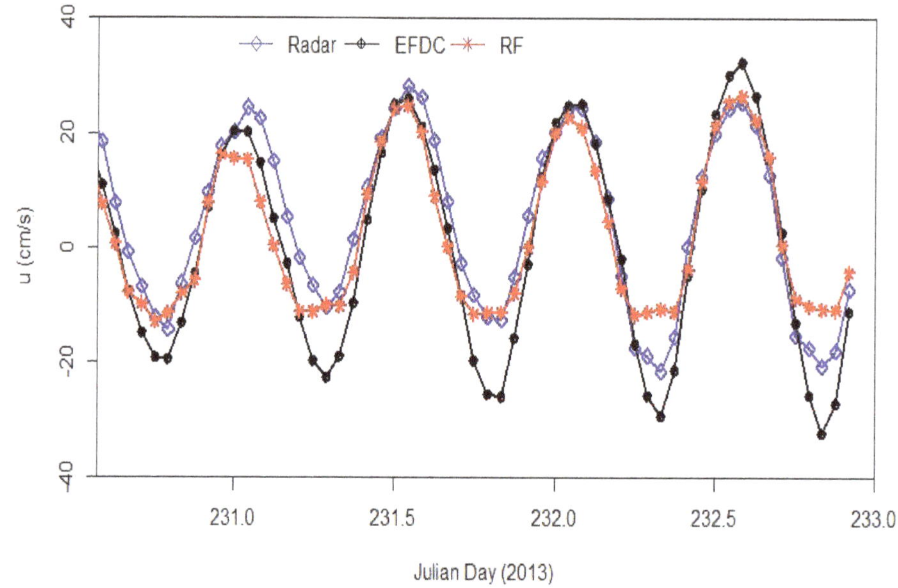

Figure 3. Predictions of zonal surface velocity component at point P1.

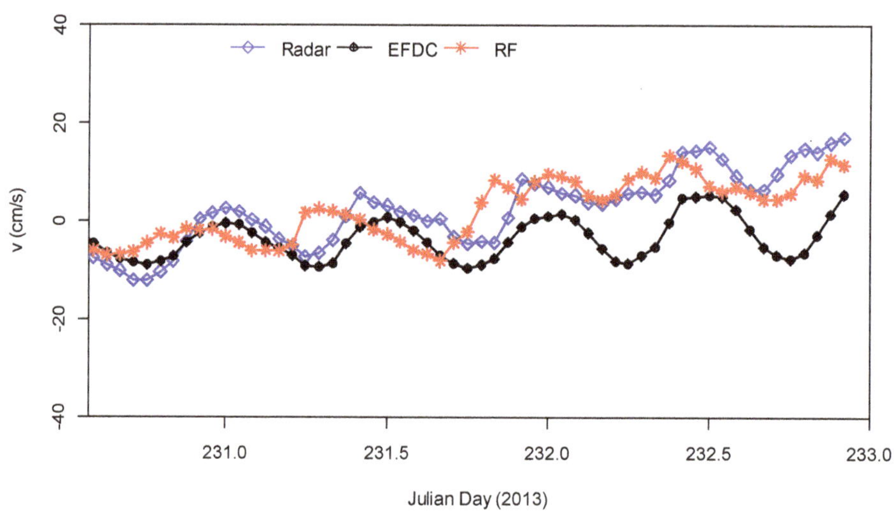

Figure 4. Predictions of meridional surface velocity component at point P1.

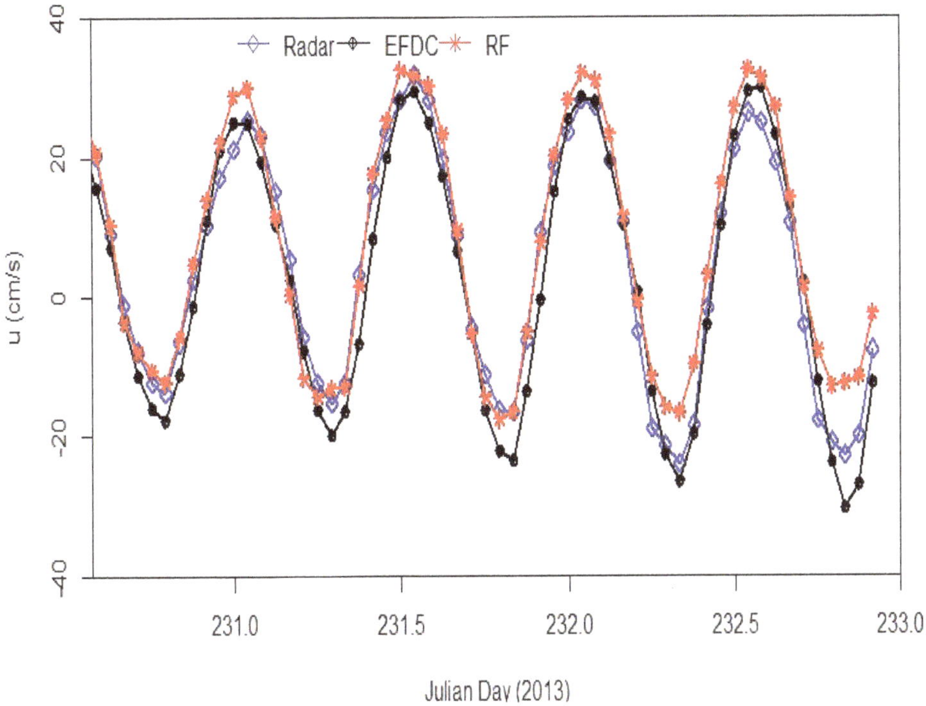

Figure 5. Predictions of zonal surface velocity component at point P2.

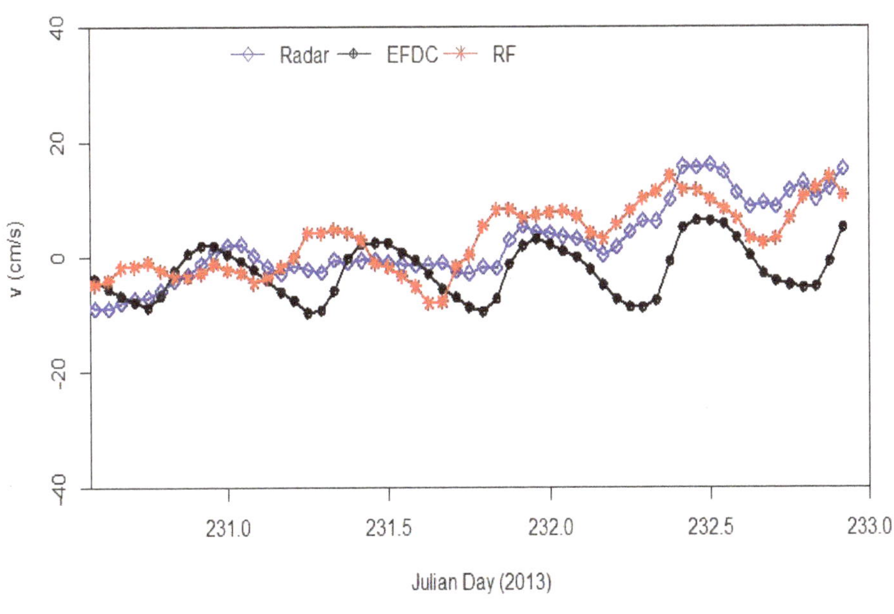

Figure 6. Predictions of meridional surface velocity component at point P2.

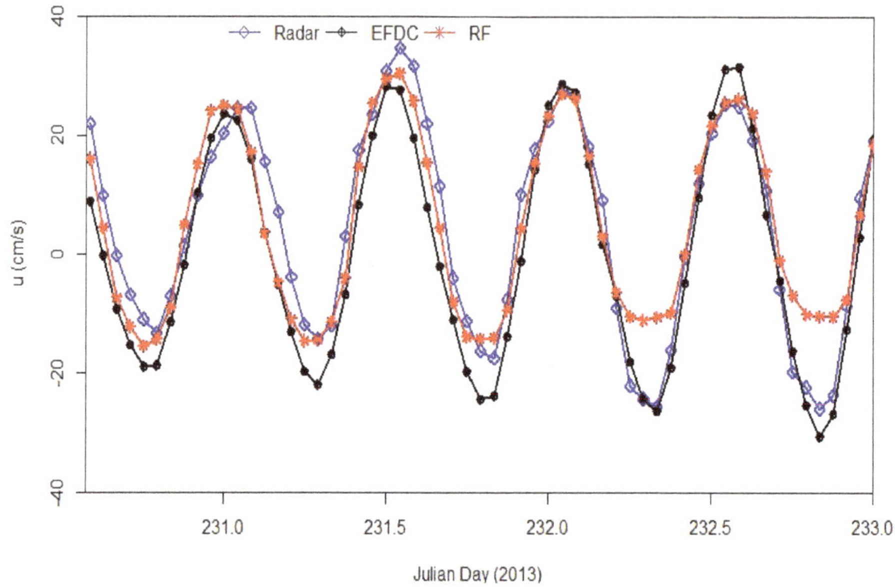

Figure 7. Predictions of zonal surface velocity component at point P3.

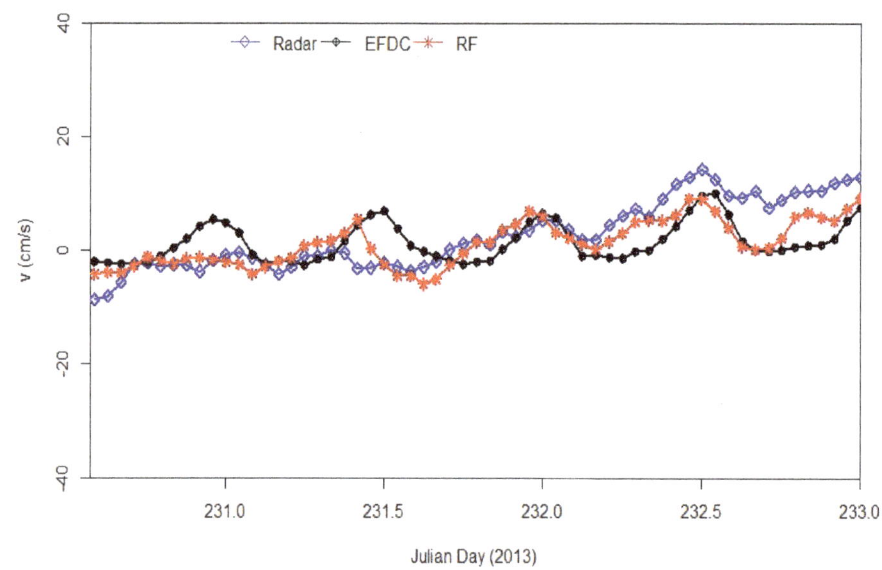

Figure 8. Predictions of meridional surface velocity component at point P3.

Figures 3–8 show that established RF models can produce predictions which have a better agreement with radar data in comparison with EFDC model results at the three analysis points. Statistics between HFR data and RF model predictions, and EFDC results were computed and are presented in Table 6.

Table 6. Statistics of forecasting dataset (59 h).

Location	Component	R1	R2	Bias	RMSE1 (cm/s)	RMSE2 (cm/s)	SI
P1	u	0.9326	0.9589	0.2477	6.3775	6.7774	1.4291
P1	v	0.8250	0.4551	−1.3209	3.7671	5.5251	1.3993
P2	u	0.9772	0.9795	2.7250	4.5843	4.2732	0.9378
P2	v	0.7650	0.4937	0.7659	4.4600	7.5313	2.0941
P3	u	0.9335	0.9710	−2.0553	5.9289	6.9026	1.1886
P3	v	0.7395	0.6713	−0.4820	5.1840	8.1931	2.1957

Table 6 shows that the RF models generally generated better prediction of surface velocity components than EFDC models based on RMSE values over 59 h forecasting period; the improvements are statistically highly significant for meridional surface velocity components. Improvement of the RMSE value for the meridional surface velocity components between the RF model predictions and radar data was greater than 32% at the three analysis points; the maximum RMSE improvement of the meridional surface velocity component was at point P2 at 41%, while improvement of RMSE value for the zonal surface velocity component was greater than 6% at locations P1 and P3. Although correlation between RF model predictions and radar data was quite similar to the correlation between EFDC results and radar data, the RMSE value between RF model predictions and radar data (RMSE1) increased by 7% in comparison with RMSE2. It is also clearly seen from Figures 5, 6 and 8 that RF models were much closer to HFR data than EFDC output at surface current maxima and minima; this is highly significant in terms of accurate prediction of tidal energy and transport effects for operational purposes at marine renewable energy sites. In general, established RF models can improve prediction of surface current components; improvement was more significant in the meridional surface velocity component than the zonal surface velocity component.

4. Discussion

The above analysis shows that the newly developed RF models were capable of producing improved surface velocity components, especially for the meridional surface velocity component. In the above RF models, the length of training dataset at the three analysis locations accounted for 60% of the total dataset i.e., 184 h. The RF models generally produced highly satisfactory predictions of surface velocity components over the 59-h forecasting period. The authors were also interested in assessing the effect of the length of RF training datasets on model predictions. Sensitivity analyses on the length of training dataset were performed at the three analysis locations respectively. Seven versions of training dataset lengths, from one day (24 h) to one week (168 h), were performed. Statistics over a 59-h prediction period between predicted results from RF models and HFR data were computed and are presented in Table 7.

Table 7 shows that RF models generated forecasts closer to HFR data than the EFDC model, in terms of RMSE values, except for models with a 24-h training length. In addition, correlation coefficients between RF model results and HFR data were comparable to those between EFDC model results and HFR data. High correlation existed in all RF models for the zonal surface velocity component, since the EFDC model produced satisfactory results for the zonal surface velocity component. The authors focused on improving forecasting accuracy for the meridional surface velocity component in this research. For the meridional surface velocity component, statistical values of the correlation coefficient between RF forecasts and HFR data generally had an increasing trend as the length of the training dataset increased. This indicated that larger training datasets produced better forecasts than using a small training dataset. This was because the larger variation range of each input variable used for training can be more comprehensively captured by RF models, and then the developed RF model is more robust to deal with the forecasting dataset which may have a large variation range. Additionally, all correlation coefficients R1 between RF models and radar data were greater than the correlation coefficient R2 between EFDC model results and radar data. This illustrated that the proposed approach that combines a numerical model with the RF model is effective in enhancing forecasts for the meridional surface velocity components. Moreover, RMSE values between RF model results and HFR data generally decreased as training datasets over longer periods were used. This proved that RF models generated forecasts closer to HFR data when a larger dataset was adopted during training phase. Improvement of the meridional surface velocity component by the RF model in comparison with EFDC model results was 36% based on RMSE values when the training dataset length was 168 h. Moreover, computational cost of using the RF algorithm based on R software was less than one hour for these models, which is much faster than a numerical model with data assimilation.

Table 7. Averaged statistics for training length tests (59-h prediction, averaged at three points).

Length (hours)	Variable	R1	R2	Bias	RMSE1 (cm/s)	RMSE2 (cm/s)	SI
24	u	0.95	0.97	1.60	6.40	5.98	1.35
	v	0.73	0.54	−1.36	5.20	7.08	2.20
48	u	0.95	0.97	0.86	5.63	5.98	1.18
	v	0.74	0.54	−0.74	4.84	7.08	2.05
72	u	0.95	0.97	−0.80	5.50	5.98	1.16
	v	0.76	0.54	−0.77	4.63	7.08	1.96
96	u	0.94	0.97	0.14	5.68	5.98	1.19
	v	0.76	0.54	−0.80	4.66	7.08	1.98
120	u	0.95	0.97	0.41	5.23	5.98	1.10
	v	0.76	0.54	−0.91	4.62	7.08	1.96
144	u	0.95	0.97	0.03	5.57	5.98	1.17
	v	0.69	0.54	−1.74	5.23	7.08	2.20
168	u	0.94	0.97	0.58	5.83	5.98	1.23
	v	0.77	0.54	−0.41	4.54	7.08	1.92

In short, application of RF method based on a numerical model and HFR data improve both forecasts of surface velocity components. But forecasting performance of meridional surface velocity component was more sensitive to training dataset length than the zonal surface velocity component.

5. Conclusions

Marine renewable energy sites require the most accurate data and forecasts possible for resources assessment, project planning and operational purposes. In this research, outputs of surface current components from a coastal hydrodynamic EFDC model, along with tidal elevations, wind speeds and directions were taken as input variables to establish RF models for components of surface water velocity. Input variable structures of each RF model was examined at three analysis points in the model domain within Galway Bay. Influences of training dataset lengths on model performances were also examined. Improvements in RF models were assessed based on several skills criteria: Correlation, RMSE, bias and SI in comparison with HFR data.

The main conclusions from this research are:

1. The best input variable structure in establishing RF models for the zonal surface velocity component was obtained using outputs u(t) and u(t − 1) from the EFDC model and wind speeds as input variables at the three analysis points, wind direction was needed at some locations. Correlation coefficients of zonal surface velocity components between RF model results and HFR data were improved and greater than 0.89 during testing.
2. The best RF models for the meridional surface velocity component used outputs v(t) and v(t − 1) from the EFDC model, tidal elevations and wind directions as input variables at the three analysis points. Correlation coefficients of the meridional surface velocity component between RF model

results and HFR data were significantly improved and greater than 0.79 for the three analysis points during testing. This is a significant result as the meridional component of velocity at Galway Bay is a low signal and strongly influenced by transient wind conditions.

3. The RF models produced comparable, or improved, forecasts over 59 h for both surface velocity components. Improvement was more significant for the meridional surface velocity component than for the zonal surface velocity component in terms of RMSE values at the three analysis points. The maximum improvement in RMSE values between RF model results and radar data for the zonal surface velocity component was 14% at point P3 and 37% for the meridional surface velocity component at point P2.
4. The RF models were robust and efficient enough to generate high-accuracy predictions using a very short-term training dataset, even less than 48 h. Correlation of both surface velocity components between HFR data and RF model prediction over a 59-h period was greater than 0.73 at the three analysis points when only a 24-h dataset was used to train RF models. This is of great importance for various realistic applications such as operations of a marine renewable energy site as improvement in the meridional surface velocity component, by RF model in comparison with EFDC model results, was 36% based on RMSE values when the training dataset length was 168 h.
5. Along with the improved forecast statistics, maximum and minimum velocities are also significantly improved using RF models; this is clearly seen by inspecting the zonal velocities of Figures 5, 6 and 8. At P1, Figure 5, the minimum values of HRF and RF were almost the same, whereas the minimum value from the EFDC model output was about 50% lower that the measured HFR data. This is a recurring feature of improved predictions using RF. Minimum and maximum values are very significant in terms of resource assessment and site operations, so this improved predictability is very important to developers of marine renewable energy extraction sites.

In short, the established RF models were capable of improving prediction accuracy for coastal surface currents. The RF model might become a robust, efficient and promising approach to provide timely and useful surface current information especially for the meridional surface velocity component of the Galway Bay area. This research is a test and case study to develop a soft computing model using an RF method to improve forecasting accuracy of surface currents. Future research will focus on combining the RF algorithm into the EFDC model to generate predictions over a large domain.

Author Contributions: L.R. and J.M. wrote the original draft; Y.L. visualized data; X.L. undertook analysis; J.L. prepared methodology and M.H. reviewed and edited the manuscript.

Funding: This research was funded by Sun Yat-sen University Basic Research Funding [76170-18841203].

Acknowledgments: We would like to thank ECMWF for providing wind data and University of Oregon for providing the OTPS tide prediction software. The authors gratefully acknowledge financial support of Sun Yat-sen University Basic Research Funding [76170-18841203].

Conflicts of Interest: The authors declare no conflict of interest.

References

1. Peng, M.; Schmalz, R.A., Jr.; Zhang, A.; Aikman, F., III. Towards the Development of the National Ocean Service San Francisco Bay Operational Forecast System. *J. Mar. Sci. Eng.* **2014**, *2*, 247–286. [CrossRef]
2. Ren, L.; Hartnett, M. Comparative Study on Assimilating Remote Sensing High Frequency Radar Surface Currents at an Atlantic Marine Renewable Energy Test Site. *Remote Sens.* **2017**, *9*, 1331. [CrossRef]
3. He, Z.; Wen, X.; Liu, H.; Du, J. A comparative study of artificial neural network, adaptive neuro fuzzy inference system and support vector machine for forecasting river flow in the semiarid mountain region. *J. Hydrol.* **2014**, *509*, 379–386. [CrossRef]
4. Mountrakis, G.; Im, J.; Ogole, C. Support vector machines in remote sensing: A review. *ISPRS J. Photogramm. Remote Sens.* **2011**, *66*, 247–259. [CrossRef]

5. Malek, S.; Gunalan, R.; Kedija, S.Y.; Lau, C.F.; Mosleh, M.A.A.; Milow, P.; Lee, S.A.; Saw, A. Random forest and Self Organizing Maps application for analysis of pediatric fracture healing time of the lower limb. *Neurocomputing* **2018**, *272*, 55–62. [CrossRef]
6. Ibarra-Berastegi, G.; Saénz, J.; Esnaola, G.; Ezcurra, A.; Ulazia, A. Short-term forecasting of the wave energy flux: Analogues, random forests, and physics-based models. *Ocean Eng.* **2015**, *104*, 530–539. [CrossRef]
7. Herrera, M.; Torgo, L.; Izquierdo, J.; Pérez-García, R. Predictive models for forecasting hourly urban water demand. *J. Hydrol.* **2010**, *387*, 141–150. [CrossRef]
8. Prasad, A.M.; Iverson, L.R.; Liaw, A. Newer Classification and Regression Tree Techniques: Bagging and Random Forests for Ecological Prediction. *Ecosystems* **2006**, *9*, 181–199. [CrossRef]
9. Strobl, C.; Malley, J.; Tutz, G. An introduction to Recursive Partitioning: Rationale, Application and Characteristics of Classification and Regression Trees, Bagging and Random Forests. *Psychol. Methods* **2009**, *14*, 323–348. [CrossRef]
10. Lahouar, A.; Slama, J.B.H. Day-ahead load forecast using random forest and expert input selection. *Energy Convers. Manag.* **2015**, *103*, 1040–1051. [CrossRef]
11. Catani, F.; Lagomarsino, D.; Segoni, S.; Tofani, V. Landslide susceptibility estimation by random forests technique: Sensitivity and scaling issues. *Nat. Hazards Earth Syst. Sci.* **2013**, *13*, 2815–2831. [CrossRef]
12. Mahjoobi, J.; Mosabbeb, E.A. Prediction of significant wave height using regressive support vector machines. *Ocean Eng.* **2009**, *36*, 339–347. [CrossRef]
13. Balas, C.E.; Koc, L.; Balas, L. Predictions of Missing Wave Data by Recurrent Neuronets. *J. Waterw. Portcoast. Ocean Eng.* **2004**, *130*, 256–265. [CrossRef]
14. Ren, L.; Nash, S.; Hartnett, M. Observation and modeling of tide- and wind-induced surface currents in Galway Bay. *Water Sci. Eng.* **2015**, *8*, 345–352. [CrossRef]
15. Wen, L. Three-Dimensional Hydrodynamic Modelling in Galway Bay. Ph.D. Thesis, University College Galway, Galway, Ireland, 1995.
16. Ren, L.; Nagle, D.; Hartnett, M.; Nash, S. The Effect of Wind Forcing on Modeling Coastal Circulation at a Marine Renewable Test Site. *Energies* **2017**, *10*, 2114. [CrossRef]
17. Hamrick, J.M. *A Three-Dimensional Environmental Fluid Dynamics Computer Code: Therotical and Computatonal Aspects*; Virginia Institute of Marine Science: Gloucester Point, VA, USA, 1992.
18. Hamrick, J.M. *User's Manual for the Environmental Fluid Dynamics Computer Code*; Department of Physical Sciences, School of Marine Science, Virginia Institute of Marine Science, College of William and Mary: Williamsburg, VA, USA, 1996.
19. Hamrick, J.M. *EFDC Technical Memorandum*; Tetra Tech: Fairfax, VA, USA, 2006.
20. Hamrick, J.M. *The Environmental Fluid Dynamics Code Theory and Computation Volume 1: Hydrodynamics and Mass Transport*; Tetra Tech, Inc.: Fairfax, VA, USA, 2007; p. 60.
21. Wang, C.; Sun, Y.; Zhang, X. Numerical Simulation of 3D Tidal Currents Based on the EFDC Model in Jiaozhou Bay. *Period. Ocean Univ. China* **2008**, *38*, 833–840.
22. Zhang, Q.; Tan, F.; Han, T.; Wang, X.; Hou, Z.; Yang, H. Simulation of sorting sedimentation in the channel of huanghua harbor by using 3D multi-sized sediment transport model of EFDC. In Proceedings of the 32nd Conference on Coastal Engineering, Shanghai, China, 30 June–5 July 2010; pp. 1–11.
23. Hayter, E.J. *Literature Review of EFDC Applications Demonstrating Capability for Use in the Jacksonville Harbor Feasibility Study*; US Army Engineer Research and Development Center: Vicksburg, MS, USA, 2014; pp. 1–36.
24. O'Donncha, F.; Hartnett, M.; Nash, S.; Ren, L.; Ragnoli, E. Characterizing observed circulation patterns within a bay using HF radar and numerical model simulations. *J. Mar. Syst.* **2015**, *142*, 96–110. [CrossRef]
25. Ren, L.; Nash, S.; Hartnett, M. Forecasting of Surface Currents via Correcting Wind Stress with Assimilation of High-Frequency Radar Data in a Three-Dimensional Model. *Adv. Meteorol.* **2016**, *2016*, 8950378. [CrossRef]
26. Gurgel, K.W.; Antonischski, G.; Essen, H.H.; Schlick, T. Wellen Radar (WERA): A new ground-wave HF radar for ocean remote sensing. *Coast. Eng.* **1999**, *37*, 219–234. [CrossRef]
27. Shay, L.K.; Martinez-Pedraja, J.; Cook, T.M.; Haus, B.K.; Weisberg, R.H. High-Frequency Radar Mapping of Surface Currents Using WERA. *J. Atmos. Ocean. Technol.* **2007**, *24*, 484–503. [CrossRef]
28. Zeng, Y.; Zhou, H.; Lai, Y.; Wen, B. Wind Direction Mapping with a Modified Wind Spreading Function by Broad-Beam High-Frequency Radar. *IEEE Geosci. Remote Sens. Lett.* **2018**, *15*, 679–683. [CrossRef]
29. Lai, Y.; Zhou, H.; Zeng, Y.; Wen, B. Quantifying and Reducing the DOA Estimation Error Resulting from Antenna Pattern Deviation for Direction-Finding HF Radar. *Remote Sens.* **2017**, *9*, 1285. [CrossRef]

30. Wang, J.; Dizaji, R.; Ponsford, A.M. Analysis of clutter distribution in bistatic high frequency surface wave radar. In Proceedings of the Conference on Electrical & Computer Engineering, Niagara Falls, ON, Canada, 2–5 May 2004; pp. 1301–1304.
31. Haus, B.K.; Wang, J.D.; Rivera, J.; Martinez-Pedraja, J.; Smith, N. Remote Radar Measurement of Shelf Currents off Key Largo, Florida, U.S.A. *Estuar. Coast. Shelf Sci.* **2000**, *51*, 553–569. [CrossRef]
32. Paduan, J.D.; Washburn, L. High-Frequency Radar Observations of Ocean Surface Currents. *Annu. Rev. Mar. Sci.* **2013**, *5*, 115–136. [CrossRef]
33. Shulman, I.; Paduan, J.D. Assimilation of HF radar-derived radials and total currents in the Monterey Bay area. *Deep Sea Res. Part II Top. Stud. Oceanogr.* **2009**, *56*, 149–160. [CrossRef]
34. Haines, S.; Seim, H.; Muglia, M. Implementing Quality Control of High-Frequency Radar Estimates and Application to Gulf Stream Surface Currents. *J. Atmos. Ocean. Technol.* **2017**, *34*, 1207–1224. [CrossRef]
35. Rubio, A.; Mader, J.; Corgnati, L.; Mantovani, C.; Griffa, A.; Novellino, A.; Quentin, C.; Wyatt, L.; Schulz-Stellenfleth, J.; Horstmann, J.; et al. HF Radar Activity in European Coastal Seas: Next Steps toward a Pan-European HF Radar Network. *Front. Mar. Sci.* **2017**, *4*, 8. [CrossRef]
36. Hisaki, Y.; Kashima, M.; Kojima, S. Surface current patterns observed by HF radar: Methodology and analysis of currents to the north of the Yaeyama Islands, East China Sea. *Ocean Dyn.* **2016**, *66*, 329–352. [CrossRef]
37. Sun, Y.; Chen, C.; Beardsley, R.C.; Ullman, D.; Butman, B.; Lin, H. Surface circulation in Block Island Sound and adjacent coatsal and shelf regions: A FVCOM-CODAR comparison. *Prog. Oceanogr.* **2016**, *143*, 26–45. [CrossRef]
38. Liu, Y.; Weisberg, R.H.; Merz, C.R.; Lichtenwalner, S.; Kirkpatrick, G.J. HF Radar Performance in a Low-Energy Environment: CODAR SeaSonde Experience on the West Florida Shelf. *J. Atmos. Ocean. Technol.* **2010**, *27*, 1689–1710. [CrossRef]
39. Breiman, L. *Random Forests-Random Features;* University of California: Berkeley, CA, USA, 1999; pp. 1–28.
40. Niu, D.; Pu, D.; Dai, S. Ultra-Short-TermWind-Power Forecasting Based on the Weighted Random Forest Optimized by the Niche Immune Lion Algorithm. *Energies* **2018**, *11*, 1098. [CrossRef]
41. Hastie, T.; Tibshirani, R.; Friedman, J. *The Elements of Statistical Learning:Data Mining, Inference, and Prediction;* Springer: Los Altos, CA, USA, 2009.
42. Liu, C.; Tang, F.; Bak, C.L. An Accurate Online Dynamic Security Assessment Scheme Based on Random Forest. *Energies* **2018**, *11*, 1914. [CrossRef]
43. Liaw, A.; Wiener, M. Classification and regression by randomForest. *R News* **2002**, *2*, 18–22.
44. Huang, N.; Lu, G.; Xu, D. A Permutation Importance-Based Feature Selection Method for Short-Term Electricity Load Forecasting Using Random Forest. *Energies* **2016**, *9*, 767. [CrossRef]
45. Mahjoobi, J.; Etemad-Shahidi, A. An alternative approach for the prediction of significant wave heights based on classification and regression trees. *Appl. Ocean Res.* **2008**, *30*, 172–177. [CrossRef]
46. Aydogan, B.; Ayat, B.; Öztürk, M.N.; Özkan Çevik, E.; Yüksel, Y. Current velocity forecasting in straits with artificial neural networks, a case study: Strait of Istanbul. *Ocean Eng.* **2010**, *37*, 443–453. [CrossRef]

Journal of
Marine Science and Engineering

MDPI

Article

Inversion of HF Radar Doppler Spectra Using a Neural Network

Rachael L. Hardman * and Lucy R. Wyatt

Schools of Mathematics and Statistics, University of Sheffield, Sheffield S3 7RH, UK
* Correspondence: rlhardman1@sheffield.ac.uk

Received: 20 May 2019; Accepted: 29 July 2019; Published: 6 August 2019

Abstract: For a number of decades, coastal HF radar has been used to remotely measure ocean surface parameters, including waves, at distances exceeding 100 km. The information, which has value in many ocean engineering applications, is obtained using the HF radar cross-section, which relates the directional ocean spectrum to the received radar signal, through a nonlinear integral equation. The equation is impossible to solve analytically, for the ocean spectrum, and a number of numerical methods are currently used. In this study, a neural network is trained to infer the directional ocean spectrum from HF radar Doppler spectra. The neural network is trained and tested on simulated radar data and then validated with data collected off the coast of Cornwall, where there are two HF radars and a wave buoy to provide the sea-truth. Key ocean parameters are derived from the estimated directional spectra and then compared with the values measured by both the wave buoy and an existing inversion method. The results are encouraging; for example, the RMSE of the obtained mean wave direction decreases from 20.6° to 15.7°. The positive results show that neural networks may be a viable solution in certain situations, where existing methods struggle.

Keywords: HF radar; artificial neural network; remote sensing; inversion; radar cross-section; monostatic radar; ocean wave directional spectrum; TensorFlow

---

1. Introduction

By employing the perturbation method of Rice [1], Barrick [2,3] derived the expected monostatic radar cross-section of the ocean surface $\sigma(\omega)$, given an ocean spectrum $S(\vec{k})$ and radar wavenumber k_0. The expression is written as:

$$\sigma(\omega) = \sigma^{(1)}(\omega) + \sigma^{(2)}(\omega), \tag{1}$$

where $\sigma^{(1)}(\omega)$ signifies the first order scatter and $\sigma^{(2)}(\omega)$ the second order. Explicitly, the first order contribution is:

$$\sigma^{(1)}(\omega) = 2^6 \pi k_0^4 \sum_{m=\pm 1} S(m\vec{k_B})\delta(\omega - m\omega_B), \tag{2}$$

and is due to Bragg scattering from ocean waves with half the radio wavelength, or equivalently wavenumber $k_B = 2k_0$, travelling directly towards or away from the radar. When there is no ocean current, the first order contribution appears as two dominant peaks, as described by the delta function $\delta()$, in the Doppler spectrum at the frequencies $\pm\omega_B$, where:

$$\omega_B = \sqrt{2gk_0 \tanh(2k_0 d)} \tag{3}$$

is known as the *Bragg frequency*, for ocean depth d and gravity g. If a current *is* present, the whole spectrum is subject to a Doppler shift, proportional to the current speed and direction. HF radar has been used as a remote measuring device of ocean currents for decades; see the works of Paduan and Washburn [4] and Abascal et al. [5] for further details.

The second order term,

$$\sigma^{(2)}(\omega) = 2^6 \pi k_0^4 \sum_{m,m'=\pm} \iint_{-\infty}^{\infty} |\Gamma_T|^2 S(m\vec{k_1}) S(m'\vec{k_2}) \delta(\omega - m\omega_1 - m'\omega_2)\, dp\, dq, \tag{4}$$

describes the second order combinations of the electromagnetic and hydrodynamic waves. In $\sigma^{(2)}(\omega)$, $\vec{k_1}$ and $\vec{k_2}$ are the two contributing wave vectors (with respective angular frequencies ω_1 and ω_2), which sum to give $\vec{k_B}$. $|\Gamma_T|^2$ is the *coupling coefficient*, which is a function of $\vec{k_1}$ and $\vec{k_2}$ and contains the mathematics of the nonlinear combinations. More detail on the coupling coefficient can be found in the work of Lipa and Barrick [6], amongst many others. The second order contribution surrounds the first order peaks and will henceforth be referred to as the *sidebands*.

Numerical methods, such as those described by Holden and Wyatt [7], can be used to calculate the value of $\sigma(\omega)$ for a measured, or modelled, $S(\vec{k})$. For this work, a Python code was written to simulate the radar data, using a vectorised version of the code presented by Lipa and Barrick [8]. An example of a simulated radar Doppler spectrum, along with a measured radar Doppler spectrum, is shown in Figure 1. To validate the radar cross-section expression, a value for $S(\vec{k})$, measured by the wave buoy described in Section 2.1.2, has been used as input to the simulation. The resulting Doppler spectrum has then been compared to a radar Doppler spectrum (measured by the radar described in Section 2.1.1), measured at the same time and location as the wave buoy. The two spectra are satisfactorily similar, with the most noticeable differences at frequencies near zero-Doppler. This is because values near zero-Doppler are not calculated in the simulation code, to save time, as they are not used in any inversion methods.

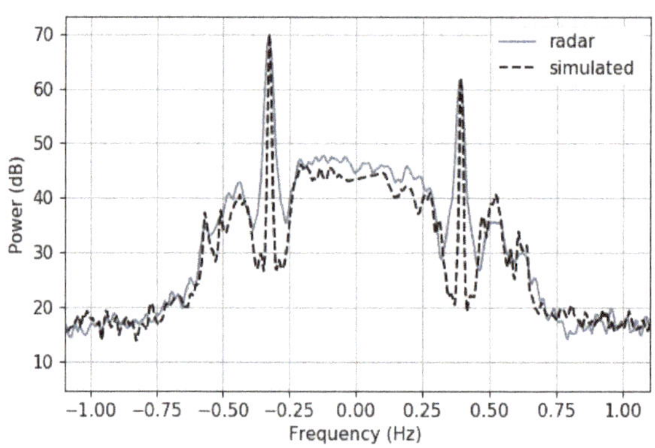

Figure 1. Comparison of measured and simulated monostatic Doppler spectra on 1 November 2012 20:05. The figure shows a comparison of a radar-measured Doppler spectrum from the Wave Hub HF dataset at the location of the wave buoy and a Doppler spectrum simulated using Equations (2) and (4) with $S(\vec{k})$ from the wave buoy input.

The inversion of Equation (4), for $S(\vec{k})$, allows us to measure the wave spectrum remotely. This has important applications in numerous coastal engineering topics, including the testing of and assimilation into operational wave models, supporting vessel navigation and monitoring both renewable energy sources and climate change. The works of Lipa [9,10], Wyatt [11], Green and Wyatt [12], Howell and Walsh [13] and Hisaki [14] are notable in the history of inverting Equation (4) numerically. Empirical methods such as that of Gurgel et al. [15], who empirically fit wave height to the second order sidebands using a wave buoy, are also possible, if one wants to avoid inverting the equation.

In this paper, a new inversion method is proposed, which uses a neural network to invert the Doppler spectrum (see Hardman [16] for full details). The neural network was trained on simulated Doppler spectra, generated using the radar cross-section expression given in Equations (2) and (4), using directional spectra measured from a local wave buoy and modelled using parameters from a WAVEWATCH III [17] model as the input. The performance of the method was measured against the existing method of Wyatt [11] and Green and Wyatt [12], henceforth referred to as the *Seaview method*, which has been validated in a number of works (e.g., [18–20]). To compare the two methods, both have been used to invert the same set of measured HF radar data, where a wave buoy is used to provide the sea-truth.

The majority of the existing inversion methods rely on the radar cross-section expression given in Equations (2) and (4). Thus, the methods produce good results when the data are accurately modelled by this expression. However, as was shown by Wyatt et al. [21], this was not always the case. For phased array receivers, using a beamforming method, large sidelobes can occur. Large sidelobes, coupled with a strong ocean current, or (as shown by Grosdidier et al. [22] for bistatic radar setups) low radar resolution can mean that the data are not modelled by the derived radar cross-section and, consequently, is not appropriate for inversion using the existing methods. A neural network learns by example, so if the effects of the radar data are included in the data that it learns from, then it should theoretically be able to model the relationship between the radar Doppler spectra and the corresponding ocean spectrum $S(\vec{k})$. Motivated by the goal of inverting data that cannot be modelled by the radar cross-section expression, this work is an experiment to see if a neural network can successfully invert high-quality HF Doppler spectra, i.e., when the data *are* representative of the radar cross-section equations. Therefore, no additional averaging or sidelobes were included in the simulations at this stage.

Although this is the first time that a neural network has been used to invert Doppler spectra to obtain directional ocean spectra, they have previously been used to measure wind speeds and water levels from HF radar data. Mathew and Deo [23] trained a neural network with HF-measured values of significant wave height, wave period, wave direction and wind direction input, to predict wind speed as measured by a wave buoy. Zeng et al. [24] did the same, but removed wave direction from the input data. Similarly, Shen et al. [25] trained a neural network to predict wind speed using the measurements of a local wave buoy and HF-measured first order Bragg peaks. To measure sea level, Wahle and Stanev [26] trained a neural network on tidal gauge data and HF radar current measurements. Seemingly all of the previous neural network applications in the field have used other measuring devices, such as wave buoys or tidal gauges, alongside measured radar data, to train their algorithms. In this work, by using simulations, the need for another device has been eradicated, and the HF radar system has no dependence on other devices.

We begin by describing the experimental setup and the necessary datasets in Section 2.1. Outlines of the new and existing methods for wave inversion, tested in this work, are described in Section 2.2, before the results of both methods are presented in Section 3. A discussion including the limitations of the method and scope for future work is given in Section 4, and concluding comments follow in Section 5.

2. Materials and Methods

2.1. Data

2.1.1. HF Radar

In the experiment, HF radar data, obtained by two monostatic WERAHF radar systems [27] on the north coast of Cornwall, SW England, were used to validate the inversion methods. The radars, which are operated by Plymouth University, were set up to aid the experiments at a test site for offshore renewable energy, known as Wave Hub. A full description of the site and results of the experiment were given by Lopez and Conley [28]. At the experiment site, shown in Figure 2, one radar is situated at

Pendeen (50.16° N, 5.67° W), and a second radar is situated, approximately 40 km away, at Perranporth (50.33° N, 5.18° W). Both radars operate at approximately 12.3 MHz. Radar data from November 2012 were used where hourly measurements were available. Each transmitter consisted of 4 elements arranged in a square, and each phased array receiver consisted of 16 antennas. The received signals were digitally beamformed onto a rectangular grid with a resolution of 1 km per cell. This dataset is referred to in this work as the *Wave Hub HF radar data*.

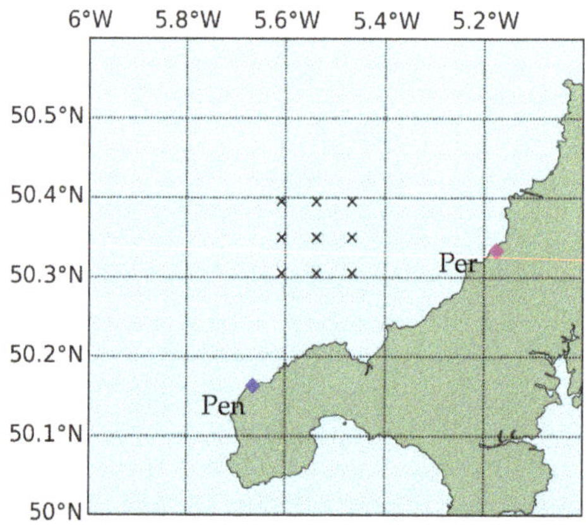

Figure 2. The Pendeen radar is shown by the blue diamond and the Perranporth radar by the magenta diamond. The wave buoy is shown by the white circle, and the location of the radar data used to validate the inversion method is shown by the black crosses.

2.1.2. Wave Buoy

A directional wave buoy situated in the radar coverage area (50.31° N, 5.61° W), at a location where the ocean depth is 53 m, was used to provide the sea-truth. Approximately 20 km from the Pendeen radar and 30 km from the Perranporth radar, measurements from the "Seawatch Mini II directional buoy" were in 30-min intervals, and the directional spectra were calculated using a maximum entropy method (see Lopez et al. [29] for details). Two different datasets from the wave buoy were used in this study; one for simulating radar Doppler spectra and one for validating the method. The two datasets overlapped, every hour, in November 2012, so to avoid bias in the neural network, the overlapping data were removed from the simulation dataset. The specifications of both wave buoy datasets are given in Table 1.

Table 1. Specifications of the wave buoy datasets.

	Simulation	Validation
Dates	1 September 2012–20 December 2012	1–30 November 2012
Time resolution	30 min (with data overlapping the validation set removed)	60 min
Frequency resolution	0.0078 Hz in [0–1 Hz] (129 measurements)	0.0078 Hz in [0–1 Hz]
Directional resolution	12° in [0–360] (30 measurements)	4° in [0–2π] (90 measurements)
Data points	4830	648

2.1.3. WAVEWATCH III

WAVEWATCH III (WW3) data (see Tolman et al. [17]), from the IOWAGAdatabase [30], were used in this experiment, alongside the wave buoy data, to simulate Doppler spectra. The WW3 dataset provides partition parameters for significant wave height, peak period and wave direction, on a grid with a resolution of $\sim$3 km. Data were available every 3 h, for January 2010–December 2012, although there were some gaps. A total of 7058 directional spectra were modelled using the WW3 data, using the model parameters to construct a Pierson Moskowitz spectrum with $\cos^{2s}$ directional spreading (where $s = 2$).

2.2. Methodology

2.2.1. The Seaview Inversion Method

First presented by Wyatt [11], the Seaview inversion (so named because it is being further developed by and is available from Seaview Sensing Ltd.) is an iterative method that begins with an estimated directional spectrum and progressively modifies it to minimise the difference between the calculated and measured Doppler spectra. As explained by Wyatt [20], the initially estimated spectrum was modelled using a Pierson Moskowitz spectrum with a sech directional model, using an empirically-estimated wave height and wind direction derived from the analysis of the first order peaks. The second order radar cross-section expression in Equation (4) was used to calculate the Doppler spectrum using the estimated $S(\vec{k})$, and the output was compared to the measured Doppler spectrum. The spectrum $S(\vec{k})$ was then updated by the rules of the algorithm, and the process was continued until a level of tolerance was achieved between the predicted and measured Doppler spectra. The method was restricted to the frequencies close to the Bragg peaks, where the shorter wave, $\vec{k_2}$, was assumed to be approximately equal to the Bragg wavevector and thus driven by local winds. The initially estimated spectrum could then be used to model $\vec{k_2}$, and then, Equation (4) was simplified and the resulting expression quasi-linear.

At each location, two Doppler spectra, measured by two monostatic radars, were used in order to resolve the directional ambiguity that existed for a single radar measurement. Additionally, for each Doppler spectrum, only values from the sidebands ($\pm$0.6 Hz) around the dominant Bragg peak were used as input to the inversion method. Furthermore, a signal-to-noise ratio of 15 dB of the largest sideband was a minimum requirement, and each spectrum was normalised by the power in the first order peak, to remove radar effects such as path loss.

2.2.2. The Neural Network Inversion Method

A neural network consists of an *input layer*, a set of *hidden layers* and an *output layer*, where each layer comprises a varying number of *nodes*. The number of nodes in the input and output layers is dependent on the specific problem, and the number in the hidden layers is set by the user. Furthermore, each node is connected to all of the nodes in the previous and following layers by some weight, $w_{kj}^{[l]}$, which signifies the weight between the jth node in the $(l-1)^{\text{th}}$ layer and the kth node in the lth layer. At each layer, a bias vector, $\vec{b}^{[l]}$, is also added. Figure 3 depicts a basic neural network.

The goal of a neural network is to learn appropriate values for the weights, connecting the neurons, in order to map accurately some input vector $\vec{x}$ to an output vector $\vec{y}$.

To train the neural network, a *training set* was used, which consisted of input and output data pairs, $(\vec{x}_i, \vec{y}_i)$. The training process included two steps: forward propagation and backward propagation.

In the forward propagation step, an input value $\vec{x}$ (which, for the following expression, is equivalent to $\vec{a}^{[0]}$), was propagated from left to right through the neural network. To do this, at hidden layer l, we calculated:

$$\vec{a}^{[l]} = g(\vec{z}^{[l]}), \tag{5}$$

where:

$$\vec{z}^{[l]} = \vec{w}^{[l]}\vec{a}^{[l-1]} + \vec{b}^{[l]}, \tag{6}$$

for $\vec{z}^{[l]} \in \mathbb{R}^{m\times 1}$, $\vec{w}^{[l]} \in \mathbb{R}^{m\times n}$, $\vec{a}^{[l-1]} \in \mathbb{R}^{n\times 1}$, and $\vec{b}^{[l]} \in \mathbb{R}^{m\times 1}$, where there are m nodes in layer l and n in layer $l-1$. In Equation (5), g is an *activation function*, which is usually some nonlinear function, which allows the algorithm to learn more complex nonlinear relationships; some common choices include the sigmoid, relu, elu [31] and tanh functions. This process was repeated until the output layer was reached, and then, the generated value, $\hat{\vec{y}}$, was compared to the training set output value, $\vec{y}$, by some defined metric, here called the *cost function*, which measures the error in the prediction.

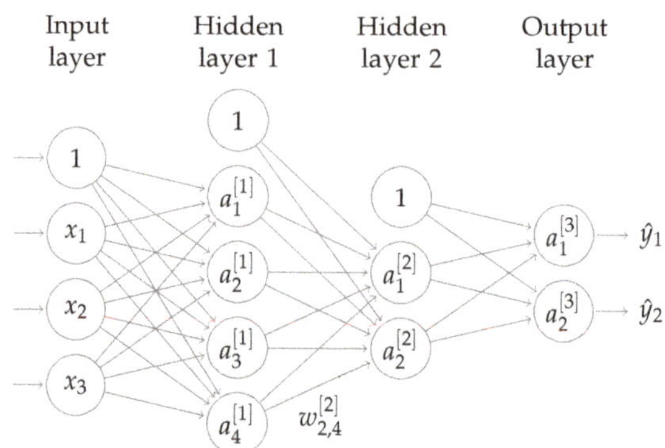

Figure 3. Basic neural network architecture with 3 input features, $\vec{x}$, 4 and 2 nodes in two hidden layers and 2 output features, $\hat{\vec{y}}$. The additional bias nodes are shown at the top of the first three layers. The weights are represented by the lines connecting the nodes, and weight $w_{2,4}^{[2]}$ is shown to highlight the notation.

We then performed backpropagation, where the error was propagated back (i.e., right to left) through the neural network, using the chain rule, to find the derivatives of the cost function with respect to each weight and bias term. The derivatives were used in a minimising algorithm, such as the Adam [32] or RMSProp optimisers, to update the values of the weights such that the prediction error was decreased.

The forward/backward propagation process was repeated until a set number of iterations was executed, or a specific accuracy level was reached. The learning phase was then stopped and the results tested. To ensure the test was unbiased, a *test set* was used, which is a reserved portion of the dataset on which the neural network has not been trained. Using the test set, we can identify if the model is overfitting, which is when the neural network performs significantly better on the training data than the test data, i.e., it fails to generalise. In such a case, we can use *regularization*, which penalises the neural network when it fits too well, to reduce overfitting. The test set is usually smaller than the training set, as the training phase requires as much data, with as much variation, as possible.

The performance of the neural network largely depends on its architecture; therefore, choosing the following parameters is important:

- number of layers;
- number of nodes per layer;
- activation function;
- number of iterations;
- minimising algorithm.

There is no particular rule on how to choose the optimal architecture, it is often a case of investigating the different combinations and finding one that is within a tolerance. A trade-off between time and performance may arise, and the user then has to decide what is more important. A genetic algorithm can be used to search more intelligently through a large set of combinations and is, in fact, used in this work; see the work of Mitchell [33] for an introduction to the subject.

In this work, the goal is to train a neural network (implemented using the TensorFlow framework for Python 3 (https://www.tensorflow.org/)) that can map from radar Doppler spectra to the corresponding ocean spectra. Although it may be possible for a neural network to learn to resolve the directional ambiguity that exists for a single radar, this is yet to be tested. Therefore, to resolve the directional ambiguity, two radars will be used to provide two spectra for each location. Consequently, the neural network will have two radar Doppler spectra as the input and the corresponding directional ocean spectrum as the output.

Depending on where the radars are situated, we define ϕ_1 and ϕ_2 as the beam angles, measured clockwise from north, from each radar to a measurement point. As the beam angles are different for each measurement location in the radar coverage area, the resulting Doppler spectra are also different. Therefore, either a different neural network must be trained for every desired location or a single neural network must be able to understand the relationship between ϕ_1, ϕ_2 and the related Doppler spectra. In this work, both methods will be tested:

1. the single-location neural network, or *SLNN*, trained in one location;
2. and the multi-location neural network, or *MLNN*, trained to understand the relationship between the radar beam angles ϕ_1 and ϕ_2 and the Doppler spectra.

For both neural network experiments, the training/test data were generated in the same way. Beginning with a large dataset of directional ocean spectra, from both the wave buoy and WW3 datasets described in Section 2.1, we used the radar cross-section expression, from Equations (2) and (4), to simulate the corresponding Doppler spectra at a particular location, for both radars. The radar frequencies, the depth of the ocean and the beam angles (ϕ_1 and ϕ_2) were fixed, and two 512-point Doppler spectra ranging from −1.27 Hz–1.27 Hz were simulated. These simulations should mimic the real data as closely as possible to make the neural network's task less complicated, and therefore, appropriate noise floor and general noise levels were included in the simulations, after analysing the available measured radar data.

Then, any Doppler spectrum input to the neural network (so either the simulated data or the measured validation data) was filtered and processed. An SNR of 10 dB for the second order peaks was set as a minimum requirement for a Doppler spectrum to be used in the neural network, and anything below was discarded, as high noise levels may complicate the neural network model. Any current-induced Doppler shifts were removed (which in this case only applies to the measured radar data) before the resulting Doppler spectra were normalised, so that the first order peaks were set to 1 and −1, and the highest Bragg peak was set to a fixed value, here 70 dB. Values ±0.6 around each peak were used as input to the neural network, which for this experiment gave a 174 point array for each radar, say $\sigma(\eta)_1$ and $\sigma(\eta)_2$ (where $\eta = \omega/\omega_B$). This smaller subset of points was used so that the neural network did not waste resources trying to understand the less important outer parts of the Doppler spectrum, whilst fully enclosing the second order continuum.

The output values, y_i, were the directional spectra, and they were interpolated onto a grid of 36×98, where there were 36 directional values in the range $[0, 2\pi]$ and 98 wavenumber values in the range $[0.004, 0.986]$. The values were then each multiplied by k, for scaling purposes, and flattened, to form a vector of a size of 3528, for processing.

In the neural network, an *elu* function, i.e.,

$$g(x) = \begin{cases} x, & x \geq 0 \\ \alpha(\exp(x) - 1), & x < 0 \end{cases},$$

for constant α, was added in the final layer to encourage the output spectrum values to be positive. The value of α was set to the TensorFlow-default value of 1, to avoid the cost of tuning another parameter. The cost function was the mean squared error of the predicted and actual values of $S(\vec{k})$; namely:

$$C(\vec{w}, \vec{y}) = \frac{1}{2M} \sum_{i=1}^{M} \left| \hat{\vec{y}}(\vec{w})_i - \vec{y}_i \right|^2, \tag{7}$$

where $\hat{\bar{y}}_i$ and $\bar{y}_i$ are the ith predicted and actual values, respectively, in a total of M values. To find the optimal neural network architecture, a genetic algorithm was used, whose criteria were to minimise the differences in the predicted and actual values in the test set.

Single-Location Neural Network

In training the SLNN, ϕ_1 and ϕ_2 were fixed and therefore unnecessary for the training process. Thus, we define $\vec{x}_i$ as the two radar Doppler spectra subsets joined together, $[\sigma(\eta)_1, \sigma(\eta)_2]$, which forms a vector of size 348, and $\vec{y}_i$ as the related directional ocean spectrum $S(\vec{k})$. To validate the method, the location of the simulations was set to the position nearest to the wave buoy where operational radar data were available. At this position, we found that $\phi_1 = 15°$, $\phi_2 = 264°$, and the ocean depth was 53 m; the rest of the radar information used in the simulation process is given in Section 2.1.1.

In this experiment, we used both the WW3-modelled and wave buoy-measured directional spectra in the simulations, as described in Section 2.1. After filtering by the SNR limit, a total of 4804 training examples were available, which were then split into 66%/33% training/test sets for the training phase. After some experimenting, the values shown in Table 2 were decided for the genetic algorithm through which to search.

Table 2. The sets of parameters defined for the genetic algorithm to search through, for the SLNN.

Number of Nodes	[179, 256, 512, 880]
Number of Layers	[6, 7, 8, 9]
Minimising Algorithm	[RMSProp, Adam]
Activation Function	[elu, relu]
Number of Iterations	[2500, 3000, 3500]

Multi-Location Neural Network

For a neural network being trained to understand the location during the inversion, namely the MLNN, the beam angles ϕ_1 and ϕ_2 (scaled to be proportional to the range of the Doppler spectra) must also be passed into the algorithm. Thus, we define $\vec{x}_i$ as the 350-point vector $[\phi_1, \sigma(\eta)_1, \phi_2, \sigma(\eta)_2]$ and, again, $\vec{y}_i$ as the related $S(\vec{k})$.

The test was carried out on a relatively small scale, where 9 locations, including the location of the wave buoy, were chosen, as shown in Figure 2. For the 9 positions, ϕ_1 ranged between 9° and 42°, and ϕ_2 ranged between 261° and 288°.

In this experiment, like the SLNN, both the wave buoy-measured and WW3-modelled values of $S(\vec{k})$ were used to simulate the training set. Post filtering, the dataset contained 109,971 data pairs, which were divided into the training/test sets by a 67%/33% split.

Similarly to the SLNN, some experimenting was carried out to decide on the parameters for the genetic algorithm. The resulting parameters are shown in Table 3.

Table 3. The sets of parameters for the genetic algorithm to search through for the MLNN.

Number of Nodes	[444, 497, 512, 666, 701, 768, 800, 999]
Number of Layers	[4, 5, 6, 7, 8, 9]
Minimising Algorithm	[RMSProp, Adam]
Activation Function	[elu, relu]
Number of Iterations	[2000, 2500, 3000, 3500]

3. Results

To measure the accuracy of the inversion, the values of mean wave direction, θ_m, peak wave direction, θ_p, peak period, t_p, energy period, t_E, and significant wave height, h_s, defined in Appendix A, were calculated from the neural network algorithm-predicted directional wave spectra and compared with those measured by the wave buoy and the Seaview method. To compare the results quantitatively, the root mean squared error (RMSE) and correlation coefficient (CC) of each parameter were calculated and presented for both the neural network and Seaview inversion methods. Scatter and time series plots are also given, to present the results qualitatively.

For both the SLNN and MLNN, each genetic algorithm was run for a total of 96 h, on a GPU in a high performance computing (HPC) cluster at the University of Sheffield, in its search for the optimal neural network parameters. In training the SLNN, the algorithm tested 173 combinations before the parameters were chosen, as summarised in Table 4. In the case of the MLNN, 67 combinations were tested before the parameters (also shown in Table 4) were chosen, noting that a lower number was tested within the time frame due to the larger training data size and the larger neural network required.

Table 4. The hyperparameters chosen by genetic algorithms for the single-location neural network (SLNN) and multi-location neural network (MLNN) inversion methods.

	SLNN	MLNN
Number of Layers	8	9
Number of Nodes	880	999
Minimising Algorithm	Adam	Adam
Activation Function	elu	relu
Number of Iterations	3000	3000

3.1. SLNN Results

The numerical results for the SLNN training and test sets are given in Table 5, and scatter plots are shown in Figure 4. The results showed a strong correlation between the simulated and predicted values of each ocean parameter, showing that SLNN can accurately invert simulated HF radar data.

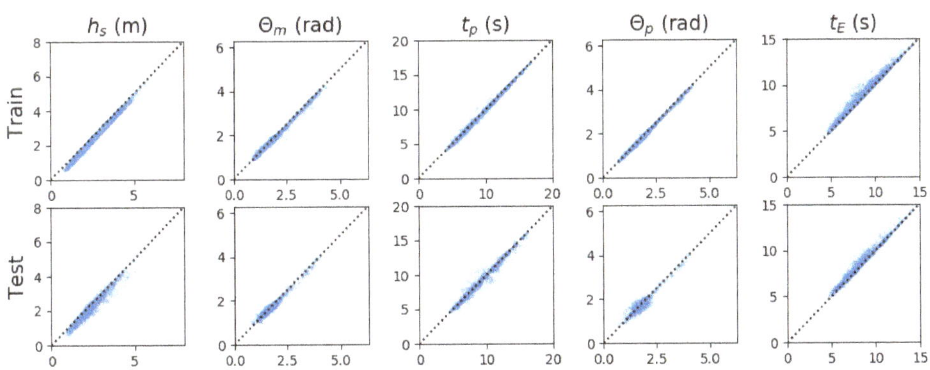

Figure 4. Scatter plots showing the results of the SLNN on the training (**top**) and test (**bottom**) datasets. The simulated values are shown on the x axes and the predicted values on the y axes.

The results of the trained neural network, when applied to the Wave Hub HF radar data, are shown in Table 6. The monthly comparisons of the ocean spectrum parameters are shown in Figures 5 and 6. Note that for this inversion method, there were 644 Doppler spectra available for inversion; this is more than the Seaview method due to a difference in the SNR restriction levels, however, a comparison of the same dataset is given in Table 6.

The results are similar to those of the test set, where both directional parameters, θ_m and θ_p, were predicted accurately, with small error and strong correlation. In comparison to the Seaview inversion, the inferred directional values appeared to be more accurate, with fewer outliers. On the other hand, the value of t_E was generally overestimated, an attribute not shared with the Seaview inversion, signalling that the neural network inversion was less accurate at predicting the power in longer ocean waves. The derived values of h_s were similar to the wave buoy and Seaview values at lower wave heights, but were underestimated at the higher wave heights, approximately when $h_s > 4$; a discussion about this difference is presented in Section 4. The comparison of t_p shows that the Seaview and SLNN methods performed similarly, where the SLNN results showed more variance, but fewer significant outliers.

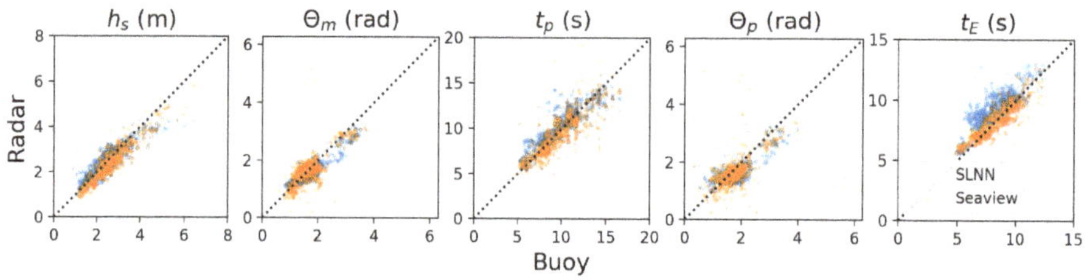

Figure 5. Scatter plots showing the results of the SLNN inversion on the Wave Hub HF radar data in November 2012. The wave buoy data, for the same period, are shown on the x axis of each plot, and the inverted radar data are on the y axis.

Table 5. Numerical results of the SLNN and MLNN inversion methods, on the training and test datasets. The correlation coefficients (CC) and root mean squared errors (RMSE) are given. The mean and standard deviation of each parameter, for both the training and test datasets, are also given to provide further context for the RMSE values.

	SLNN				MLNN				Data Statistics			
	Train		Test		Train		Test		Train		Test	
	CC	RMSE	CC	RMSE	CC	RMSE	CC	RMSE	Mean	SD	Mean	SD
h_s (m)	1.0	0.18	0.96	0.31	0.9	0.44	0.9	0.44	2.47	0.96	2.45	0.96
θ_m (rad)	0.96	0.15	0.91	0.19	0.86	0.28	0.87	0.28	1.89	0.65	1.85	0.61
t_p (s)	0.98	0.52	0.94	1.03	0.87	1.69	0.86	1.69	9.36	2.81	9.60	2.89
θ_p (rad)	0.95	0.16	0.88	0.26	0.82	0.32	0.82	0.33	1.90	0.64	1.85	0.61
t_E (s)	0.99	0.54	0.97	0.63	0.91	1.12	0.91	1.12	08.14	2.07	8.28	2.12

Table 6. Numerical results of the Seaview, SLNN and MLNN inversion methods, tested on the Wave Hub HF radar data. The Seaview results are for the inversion of 501 Doppler spectra, whereas the SLNN and MLNN results are for the inversion of 644 Doppler spectra. The correlation coefficients (CC) and root mean square errors (RMSE) are given. The mean and standard deviation of each parameter, for the wave buoy dataset, are also given to provide further context for the RMSE values.

	Seaview		SLNN		MLNN		Wave Buoy Statistics	
	CC	RMSE	CC	RMSE	CC	RMSE	Mean	SD
h_s (s)	0.93	0.4	0.87	0.47	0.8	1.51	2.50	0.94
θ_m (rad)	0.75	0.36	0.91	0.26	0.83	0.37	1.78	0.60
t_p (s)	0.83	1.47	0.82	1.56	0.75	3.18	9.77	2.55
θ_p (rad)	0.59	0.45	0.70	0.42	0.62	0.47	1.85	0.59
t_E (s)	0.85	0.85	0.80	1.64	0.82	2.53	8.18	1.62

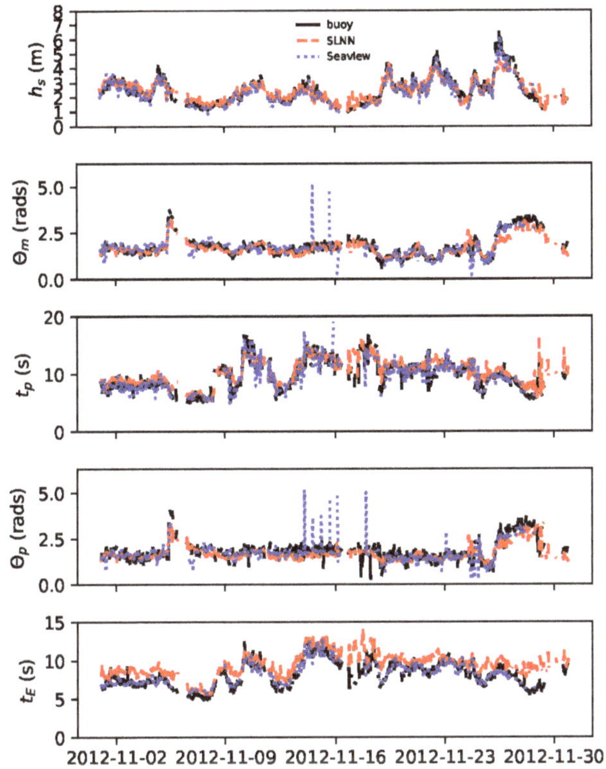

Figure 6. Time series plots, for November 2012, showing the results of the Seaview inversion on the Wave Hub HF radar data against the values measured by the wave buoy.

3.2. MLNN Results

The results of the inversion, when applied to the training and test sets, are shown numerically in Table 5 and visually in Figure 7. The results were similar for both the training and test sets, but showed that the neural network was not performing particularly well on either dataset.

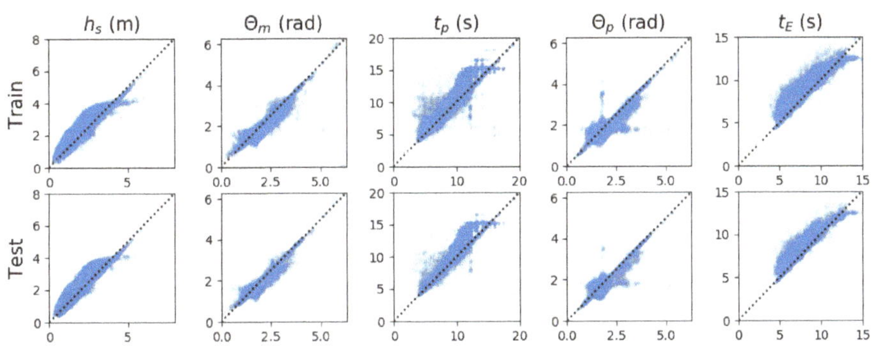

Figure 7. Scatter plots showing the results of the MLNN on the training (**top**) and test (**bottom**) datasets. The simulated values are shown on the x axis and the predicted values on the y axis.

The results of the trained neural network, when applied to the Wave Hub HF radar data, are shown in Table 6 and Figures 8 and 9. These results showed levels of under-performance similar to the training/test sets, with only θ_m inferred reasonably well. Possible reasons and remedies for this are discussed in Section 4, and solving this issue was imperative to this inversion method being a viable inversion option.

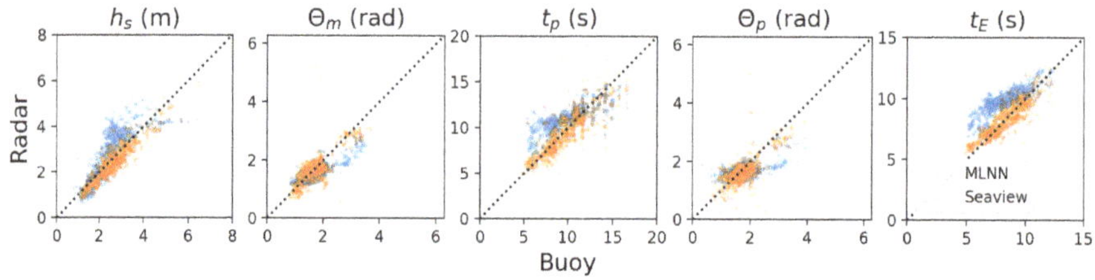

Figure 8. Scatter plots showing the results of the MLNN inversion on the Wave Hub HF radar data in November 2012. The wave buoy data, for the same period, are shown on the x axis of each plot, and the inverted radar data are on the y axis.

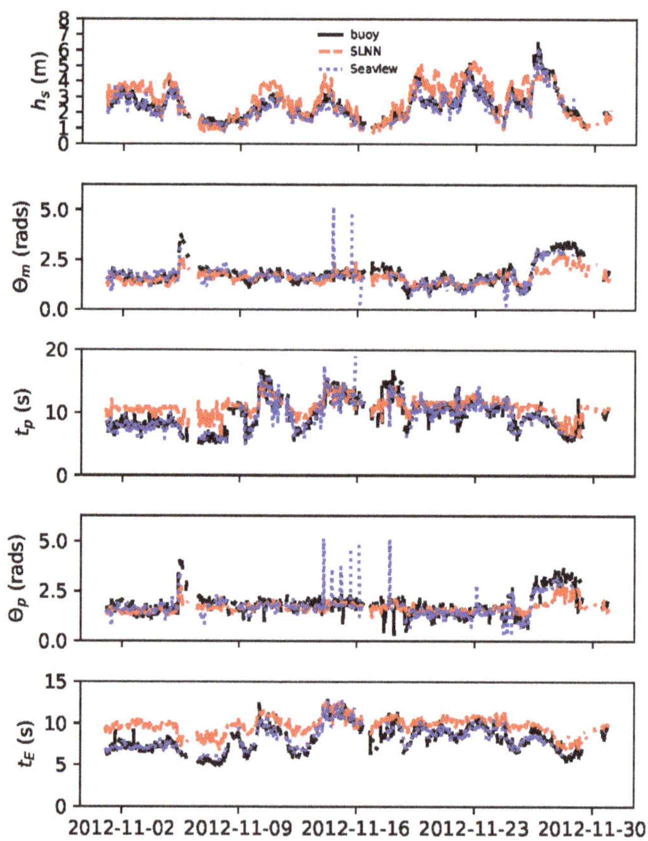

Figure 9. Time series plots, for November 2012, showing the results of the MLNN inversion on the Wave Hub HF radar data. The wave buoy data and Seaview inversion results are shown, for comparison.

An example of derived wave heights and mean wave direction for the nine trained locations is shown in Figure 10, and an inversion of Doppler spectra from a larger expanse of the ocean, for the same time, is shown in Figure 11. A Seaview inversion, for the same time and date of these examples, is shown in Figure 12. The Seaview map shows more directional variation than the MLNN map, which seems inclined to predict the mean value of θ_m. A discussion of possible ways to rectify this issue is given in Section 4, although the MLNN-derived values of h_s are encouraging.

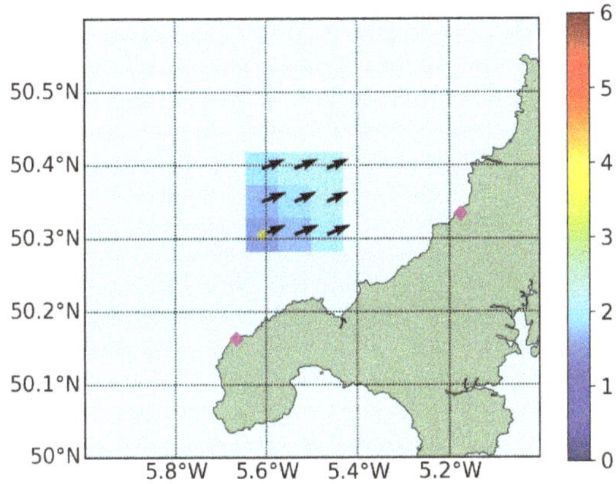

Figure 10. An example of MLNN-derived values at each of the nine locations that the MLNN was trained at, on 20 November 2012 at 03:00. Significant wave height, measured in metres, is shown by the background colour of each cell, and the mean wave direction is shown by the black arrows.

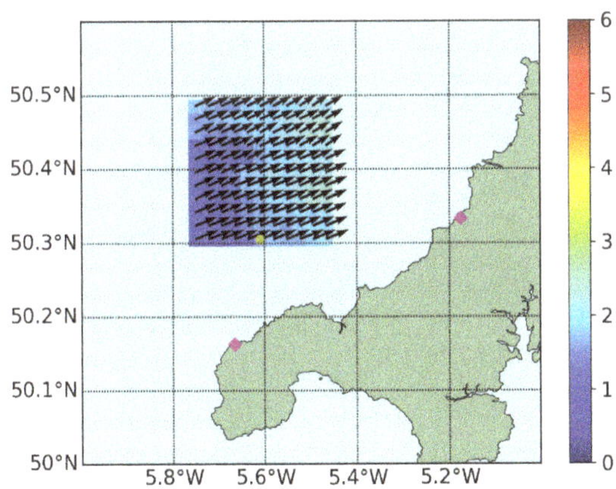

Figure 11. An example of a whole field inversion, for 20 November 2012 at 03:00, using the MLNN inversion method, where Doppler spectra from untrained locations were also inverted. Significant wave height, measured in metres, is shown by the background colour of each cell, and the mean wave direction is shown by the black arrows.

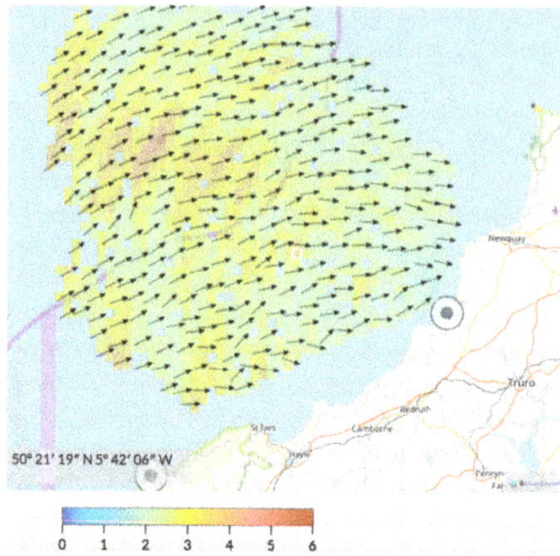

Figure 12. An example of the whole field inversion, for 20 November 2012 at 03:00, using the Seaview inversion method on the Wave Hub HF radar data (taken from https://phillips.shef.ac.uk/pub/svdv/whc/). Significant wave height, measured in metres, is shown by the background colour of each cell, and the mean wave direction is shown by the black arrows.

4. Discussion

The SLNN inversion showed generally good agreement with the wave buoy and the Seaview inversion method and showed the potential of machine learning algorithms in the field of remote sensing using HF radar. The derived values of θ_m were particularly encouraging; however, the value of t_E seemed to be overestimated. At sea states above approximately 4 m, the neural network was underestimating h_s, which was perhaps due to a limitation in the perturbation analysis used in the derivation of the radar cross-section. In the derivation, small slopes in comparison to the radar wavelength were assumed. Therefore, for high seas with large significant wave height, the theory broke down, and the simulations that the neural network was trained on did not represent the real data well. Wyatt et al. [34] explained how, in the Seaview method, the first order Bragg peaks were scaled, such that the second order continuum was increased in comparison to the first, for a given wave height. The scaling was based on the work of Creamer et al. [35], who showed that a second order contribution (due to the combination of first and third order ocean waves) was missed in the derivation of the radar cross-section by Barrick [2,3] and was significant at larger wave heights. The improvement in the results when the scaling was included was shown in the work of Wyatt et al. [34] and is something that could be included in future work, by modifying the simulations appropriately, to improve the accuracy in predicting h_s.

Interestingly, the numerical comparison of the SLNN and Seaview inversions for the same dates (shown in Table 7) did not affect the SLNN results much, indicating that the neural network inversion was performing well on the data with a lower SNR.

The MLNN inversion, although providing encouraging results, did not perform as well as the SLNN inversion. From looking at the results of the training and test datasets, the introduction of the radar beam angles clearly affected the model accuracy and hence why the results were not as good as the SLNN. In Figure 11, the MLNN-derived values of significant wave height and mean wave direction are shown at locations the neural network was not trained. The values of θ_m were all very similar, in contrast to the more varied values derived by the Seaview inversion, shown in Figure 12. A possible reason for this was that when the neural network was trained, the algorithm got stuck in a local minimum and learned to predict the average values for θ_m and θ_p. This was further suggested by the results of the inversion of the entire dataset, shown in Figure 8, where θ_p in particular seemed inclined to predict the average value of the dataset.

Table 7. Numerical results of a direct comparison of 501 inversions, of the Seaview, SLNN and MLNN inversion methods, tested on the Wave Hub HF radar data. The correlation coefficients (CC) and root mean squared errors (RMSE) are given. The results are for a subset of the Wave Hub data, on which the Seaview inversion results are based. The mean and standard deviation of each parameter, for the wave buoy dataset, are also given to provide further context for the RMSE values.

	Seaview		SLNN		MLNN		Wave Buoy Statistics	
	CC	RMSE	CC	RMSE	CC	RMSE	Mean	SD
h_s (s)	0.93	0.4	0.88	0.46	0.77	1.49	2.66	0.93
θ_m (rad)	0.75	0.36	0.88	0.27	0.68	0.4	1.71	0.52
t_p (s)	z 0.83	1.47	0.84	1.46	0.77	3.15	9.82	2.52
θ_p (rad)	0.59	0.45	0.69	0.4	0.7	0.43	1.80	0.56
t_E (s)	0.85	0.85	0.83	1.48	0.83	2.52	8.26	1.61

Therefore, more work must be done on successfully including the radar beam angle in the neural network algorithm. The main advantage of HF radar is its ability to provide measurements across a larger expanse of the ocean, as opposed to the fixed location of a wave buoy. As the neural network can work in a single location, there is no reason why it should not also be able to learn the relationship between ϕ_1 and ϕ_2 and the Doppler spectra to the same level of accuracy. This is seemingly a case of getting the correct neural network architecture and having a big enough dataset for the training process. A training dataset with the locations spaced further apart could be used, and perhaps a larger neural network would also be beneficial; additionally, a convolutional neural network may be more appropriate for this task.

There were a number of variables not yet experimented within the training phase, which may impact the accuracy of the neural network. The SNR minimum was fixed at 10 dB in this experiment; however, in the Seaview inversion, for example, the SNR minimum was 15 dB. Similarly, the points of the Doppler spectrum of 0.6 on either side of each Bragg peak was input into the neural network. Perhaps valuable information was being missed in the training phase or useless information was being passed in and wasting resources. The issue with neural networks is their "black box" nature, whereby it can be difficult to interpret the results. There are a vast number of other machine learning methods that could be used for this task, such as a random forest regression model, which may be more interpretable, and a comparison study, discovering which method is most appropriate, would be interesting.

The expense of the inversion process in both time and computing resources was fairly high. Other methods, for instance the Seaview inversion method, require no prior knowledge of the radar data. One just has to set variables, such as radar frequency and location, and then, the algorithm is ready to be implemented. As the neural network procedure is dependent on factors such as the locations of the radars, new simulations and hence another training process must be carried out for each new radar site. If a radar were to be moved, the whole process must be carried out again, meaning a few days of operational downtime would occur. Training a neural network to recognise all different radar frequencies and angles would be theoretically possible, although an exceptional amount of computing power and a much larger neural network and dataset would be necessary.

The neural network approach has proven itself to be capable of inverting HF radar Doppler spectra, in a single location, when the radar quality is good. The next step is to test the method in a more challenging situation, where the existing methods perform poorly, for instance for a bistatic phased array radar with low spatial resolution. Grosdidier et al. [22] showed that for their particular bistatic radar, the data were not accurately modelled by the radar cross-section, due to the limitations of the radar configuration at their site. A simulation method that includes these radar effects would better represent this type of data. Then, training a neural network on a large dataset of Doppler spectra, simulated in this way, should theoretically enable us to invert these more complicated Doppler spectra.

5. Conclusions

In this work, we showed that a neural network can accurately invert HF radar Doppler spectra to obtain directional wave spectra. Key ocean statistics, namely mean wave direction, peak wave direction, energy period, peak wave period and significant wave height, were all obtained from measured radar data and the results compared with the values obtained by the Seaview inversion method and those measured by a wave buoy.

However, a number of issues with this new approach were identified: (1) the radar beam angle was not successfully included in the neural network inversion method, meaning that Doppler spectra measured over a large area of the ocean cannot be accurately inverted; and (2) the expense of training the algorithm was high in both resources and time. Nonetheless, the method can potentially be used more easily with radar data affected by sidelobes or poor spatial resolution since these effects can be included in the neural network training process.

Author Contributions: Conceptualization, R.L.H. and L.R.W.; methodology, R.L.H. and L.R.W.; software, R.L.H. and L.R.W.; validation, R.L.H. and L.R.W.; formal analysis, R.L.H. and L.R.W.; investigation, R.L.H. and L.R.W.; resources, R.L.H. and L.R.W.; data curation, R.L.H. and L.R.W.; writing, original draft preparation, R.L.H.; writing, review and editing, L.R.W; visualization, R.L.H. and L.R.W.; supervision, L.R.W.; project administration, L.R.W.; funding acquisition, R.L.H. and L.R.W.

Funding: This research was funded by the Engineering and Physical Sciences Research Council (Grant Number EP/M506618/1). The APC for this work has been waived.

Acknowledgments: We are grateful to Daniel Conley and Guiomar Lopez at the University of Plymouth for the Wave Hub HF radar and wave buoy datasets.

Conflicts of Interest: R.L.H. declares no conflict of interest. L.R.W. is the Technical Director of Seaview Sensing Ltd, but her role in this paper is as a University of Sheffield Professor, and as such, she has only made scientific inputs and judgements on the work. The funders had no role in the design of the study; in the collection, analyses, or interpretation of data; in the writing of the manuscript; nor in the decision to publish the results.

Appendix A. Ocean Spectrum Statistics

Firstly, the non-directional spectrum, $E(f)$, is calculated from $S(f,\theta)$ by:

$$E(f) = \int_0^{2\pi} S(f,\theta)\, d\theta \tag{A1}$$

and the first Fourier coefficients a_1 and b_1 (see Kuik et al. [36]) are:

$$a_1 = \int_0^{2\pi} S(f,\theta)\cos\theta\, d\theta \qquad b_1 = \int_0^{2\pi} S(f,\theta)\sin\theta\, d\theta. \tag{A2}$$

From Tucker [37], the significant wave height, h_s, is defined as:

$$h_s = 4\sqrt{\int E(f)\, df}. \tag{A3}$$

To measure where the majority of the energy is in the spectrum, peak period, t_p, which is the inverse of the spectral peak frequency, f_p, and the energy period, t_E, which measures roughly where the power is based in frequency, as defined by Wyatt [20], are used. They are defined as:

$$t_p = \frac{\int_{f_1}^{f_2} E(f)\, df}{\int_{f_1}^{f_2} fE(f)\, df}, \tag{A4}$$

where $E(f_1) = E(f_2) = 0.8E_{max}$ and:

$$t_E = \frac{\int f^{-1}E(f)\, df}{\int E(f)\, df}. \tag{A5}$$

To measure directional information, the peak direction, θ_p, which is the direction of the most energetic waves in the spectrum, is defined as:

$$\theta_p = \arctan\left(\frac{b_1(f_p)}{a_1(f_p)}\right), \tag{A6}$$

for $f_p = 1/t_p$ and a_1 and b_1 as in Equation (A2). The mean wave direction is defined as:

$$\theta_m = \arctan\left(\frac{\int b_1(f)\,df}{\int a_1(f)\,df}\right). \tag{A7}$$

References

1. Rice, S.O. Reflection of electromagnetic waves from slightly rough surfaces. *Commun. Pure Appl. Math.* **1951**, *4*, 351–378. [CrossRef]
2. Barrick, D.E. First-order theory and analysis of MF/HF/VHF scatter from the sea. *IEEE Trans. Antennas Propag.* **1972**, *20*, 2–10. [CrossRef]
3. Barrick, D.E. *Remote Sensing of the Troposphere*; NOAA: Boulder, CO, USA, 1972; Chapter 12.
4. Paduan, J.D.; Washburn, L. High-frequency radar observations of ocean surface currents. *Annu. Rev. Mar. Sci.* **2013**, *5*, 115–136. [CrossRef] [PubMed]
5. Abascal, A.J.; Castanedo, S.; Medina, R.; Losada, I.J.; Alvarez-Fanjul, E. Application of HF radar currents to oil spill modelling. *Mar. Pollut. Bull.* **2009**, *58*, 238–248. [CrossRef] [PubMed]
6. Lipa, B.J.; Barrick, D.E. Extraction of sea state from HF radar sea echo: Mathematical theory and modeling. *Radio Sci.* **1986**, *21*, 81–100. [CrossRef]
7. Holden, G.J.; Wyatt, L.R. Extraction of sea state in shallow water using HF radar. *IEE Proc. F (Radar Signal Process.)* **1992**, *139*, 175–181. [CrossRef]
8. Lipa, B.J.; Barrick, D.E. *Analysis Methods for Narrow-Beam High-Frequency Radar Sea Echo*; Technical Report ERL 420-WPL 56; NOAA: Boulder, CO, USA, 1972.
9. Lipa, B.J. Derivation of directional ocean-wave spectra by integral inversion of second-order radar echoes. *Radio Sci.* **1977**, *12*, 425–434. [CrossRef]
10. Lipa, B.J. Inversion of second-order radar echoes from the sea. *J. Geophys. Res. Ocean.* **1978**, *83*, 959–962. [CrossRef]
11. Wyatt, L.R. A relaxation method for integral inversion applied to HF radar measurement of the ocean wave directional spectrum. *Int. J. Remote Sens.* **1990**, *11*, 1481–1494. [CrossRef]
12. Green, J.J.; Wyatt, L.R. Row-action inversion of the Barrick–Weber equations. *J. Atmos. Ocean. Technol.* **2006**, *23*, 501–510. [CrossRef]
13. Howell, R.; Walsh, J. Measurement of ocean wave spectra using narrow-beam HE radar. *IEEE J. Ocean. Eng.* **1993**, *18*, 296–305. [CrossRef]
14. Hisaki, Y. Nonlinear inversion of the integral equation to estimate ocean wave spectra from HF radar. *Radio Sci.* **1996**, *31*, 25–39. [CrossRef]
15. Gurgel, K.W.; Essen, H.H.; Schlick, T. An empirical method to derive ocean waves from second-order Bragg scattering: Prospects and limitations. *IEEE J. Ocean. Eng.* **2006**, *31*, 804–811. [CrossRef]
16. Hardman, R.L. Remote Sensing of Ocean Winds and Waves with Bistatic HF Radar. Ph.D. Thesis, University of Sheffield, Sheffield, UK, March 2019.
17. Tolman, H.L. User manual and system documentation of WAVEWATCH III TM version 3.14. *Tech. Note MMAB Contrib.* **2009**, *276*, 220.
18. Wyatt, L.R.; Green, J.; Gurgel, K.W.; Borge, J.N.; Reichert, K.; Hessner, K.; Günther, H.; Rosenthal, W.; Saetra, O.; Reistad, M. Validation and intercomparisons of wave measurements and models during the EuroROSE experiments. *Coast. Eng.* **2003**, *48*, 1–28. [CrossRef]
19. Wyatt, L.R.; Green, J.J.; Middleditch, A.; Moorhead, M.D.; Howarth, J.; Holt, M.; Keogh, S. Operational wave, current, and wind measurements with the Pisces HF radar. *IEEE J. Ocean. Eng.* **2006**, *31*, 819–834. [CrossRef]
20. Wyatt, L.R. Measuring the ocean wave directional spectrum 'First Five' with HF radar. *Ocean. Dyn.* **2019**, *69*, 123–144. [CrossRef]

21. Wyatt, L.R.; Liakhovetski, G.; Graber, H.C.; Haus, B.K. Factors affecting the accuracy of SHOWEX HF radar wave measurements. *J. Atmos. Ocean. Technol.* **2005**, *22*, 847–859. [CrossRef]
22. Grosdidier, S.; Forget, P.; Barbin, Y.; Guérin, C.A. HF bistatic ocean Doppler spectra: Simulation versus experimentation. *IEEE Trans. Geosci. Remote Sens.* **2014**, *52*, 2138–2148. [CrossRef]
23. Mathew, T.E.; Deo, M.C. Inverse estimation of wind from the waves measured by high-frequency radar. *Int. J. Remote Sens.* **2012**, *33*, 2985–3003. [CrossRef]
24. Zeng, Y.; Zhou, H.; Roarty, H.; Wen, B. Wind speed inversion in high frequency radar based on neural network. *Int. J. Antennas Propag.* **2016**, *2016*, 2706521. [CrossRef]
25. Shen, W.; Gurgel, K.W.; Voulgaris, G.; Schlick, T.; Stammer, D. Wind-speed inversion from HF radar first-order backscatter signal. *Ocean Dyn.* **2012**, *62*, 105–121. [CrossRef]
26. Wahle, K.; Stanev, E. Consistency and complementarity of different coastal ocean observations: A neural network-based analysis for the German Bight. *Geophys. Res. Lett.* **2011**, *38*, 1–4. [CrossRef]
27. Gurgel, K.W.; Antonischki, G.; Essen, H.H.; Schlick, T. Wellen Radar (WERA): A new ground-wave HF radar for ocean remote sensing. *Coast. Eng.* **1999**, *37*, 219–234. [CrossRef]
28. Lopez, G.; Conley, D.C. Comparisons of directional wave spectra and derived parameters from in situ observations and HF radar. **2019**, in press.
29. Lopez, G.; Conley, D.C.; Greaves, D. Calibration, Validation, and Analysis of an Empirical Algorithm for the Retrieval of Wave Spectra from HF Radar Sea Echo. *J. Atmos. Ocean. Technol.* **2016**, *33*, 245–261. [CrossRef]
30. IOWAGA Database. 2018. Available online: https://wwz.ifremer.fr/iowaga/ (accessed on 14 December 2018)
31. Clevert, D.A.; Unterthiner, T.; Hochreiter, S. Fast and accurate deep network learning by exponential linear units (elus). *arXiv* **2015**, arXiv:1511.07289.
32. Kingma, D.P.; Ba, J. Adam: A method for stochastic optimization. *arXiv* **2014**, arXiv:1412.6980.
33. Mitchell, M. *An Introduction to Genetic Algorithms*; MIT Press: Cambridge, MA, USA, 1998.
34. Wyatt, L.R.; Green, J.J.; Middleditch, A. HF radar data quality requirements for wave measurement. *Coast. Eng.* **2011**, *58*, 327–336. [CrossRef]
35. Creamer, D.B.; Henyey, F.; Schult, R.; Wright, J. Improved linear representation of ocean surface waves. *J. Fluid Mech.* **1989**, *205*, 135–161. [CrossRef]
36. Kuik, A.; Van Vledder, G.P.; Holthuijsen, L. A method for the routine analysis of pitch-and-roll buoy wave data. *J. Phys. Oceanogr.* **1988**, *18*, 1020–1034. [CrossRef]
37. Tucker, M.J.; Pitt, E.G. *Waves in Ocean Engineering*; Elsevier: Oxford, UK, 2001; Volume 5.

Journal of
Marine Science and Engineering

MDPI

Article

Comparison of HF Radar Fields of Directional Wave Spectra Against In Situ Measurements at Multiple Locations

Guiomar Lopez [1,*] and Daniel C. Conley [2]

1 Morphodynamique Continentale et Côtière, CNRS UMR 6143, Caen University, 14000 Caen, France
2 Coastal Processes Research Group, School of Biological and Marine Sciences, Plymouth University, Plymouth PL4 8AA, UK
* Correspondence: guiomar.lopez-fernandez@unicaen.fr

Received: 19 June 2019; Accepted: 12 August 2019; Published: 14 August 2019

Abstract: The coastal zone hosts a great number of activities that require knowledge of the spatial characteristics of the wave field, which in coastal seas can be highly heterogeneous. Information of this type can be obtained from HF radars, which offer attractive performance characteristics in terms of temporal and spatial resolution. This paper presents the validation of radar-derived fields of directional wave spectra. These were retrieved from measurements collected with an HF radar system specifically deployed for wave measurement, using an established inversion algorithm. Overall, the algorithm reported accurate estimates of directional spectra, whose main distinctive characteristic was that the spectral energy was typically spread over a slightly broader range of frequencies and directions than in their in situ-measured counterparts. Two errors commonly reported in previous studies, namely the overestimation of wave heights and noise related to short measurement periods, were not observed in our results. The maximum wave height recorded by two in situ devices differed by 30 cm on average from the radar-measured values, and with the exception of the wave spreading, the standard deviations of the radar wave parameters were between 3% and 20% of those obtained with the in situ datasets, indicating the two were similarly grouped around their means. At present, the main drawback of the method is the high quality signal required to perform the inversion. This is in part responsible for a reduced data return, which did not exceed 55% at any grid cell over the eight-month period studied here.

Keywords: HF radar; wave directional spectra; spatial wave fields; ADCP; wave buoy

1. Introduction

Knowledge of the offshore and coastal wave climate is crucial for ensuring the successful outcome of several human activities such as marine operations, coastal defence, or marine energy extraction. Most of the projects associated to these activities require high quality data, with good temporal and spatial resolution, at several stages of their development. Among the tools able to retrieve spatial data, HF radar is a remote sensing technique that has the capability of providing a relatively large coverage with high temporal and spatial resolution. Nowadays, this technology plays an essential role in monitoring surface currents ([1–3]), whereas the suitability of its wave products for different applications has only recently begun to be explored ([4–7]).

The wave directional spectrum retrieved by HF radars is derived from backscattered signals reflected primarily off waves satisfying the Bragg resonance condition. Standard spectral techniques are then applied to transform these signals into Doppler spectra [8] that can be subsequently inverted into wave directional spectra ([9–14]), or some of its summary parameters ([15–22]).

The focus of this work is the inversion method presented in [23], which is distributed as a software package by Seaview Sensing Ltd., Sheffield, UK (Seaview hereafter). Its results have been previously examined using datasets collected during different experiments (e.g., [7,24–26]), which have revealed the accuracy and limitations of the method and the HF radar technique itself. Reported inaccuracies have been found to originate predominantly from two different sources: limitations in the theory describing the relation between Doppler and ocean wave spectra [27], and the use of short averaging periods to derive wave directional spectra [28]. The main consequence of the former is an overestimation of high sea states ([24,25]), while the latter has been identified as the cause of noise in the measurements ([26,29,30]).

In this study, we used data collected with two operational phased-array Wellen Radars (WERA) to validate the Seaview algorithm. Results obtained with this method were derived from backscattered time series integrated over periods similar to those used by the in situ devices (17 min). With this approach, we intended to diminish the effect of short averaging on the results, to be able to clearly identify the contributions of other limitations as well as the ranges for which reliable data can be obtained.

The next section provides an overview of the study area and the devices that collected the data used here. This is followed by the intercomparison of radar and in situ wave integrated parameters and directional spectra, in Section 3. Finally, we provide a discussion of the results in Section 4, and the main conclusions derived from them are summarised in Section 5.

2. Materials and Methods

The measurements analysed in this work were collected over an area covering the Wave Hub; a grid connected wave energy site for testing offshore renewable energy technologies located in the north coast of Cornwall (UK). In this region, the wave climate is a combination of the long period swell that propagates across the North Atlantic and locally generated wind waves, which are both modified by a current field that is largely dominated by tidal streams reaching 1 m s^{-1} [31]. The area is well exposed to waves arriving from the west and southwest, and to some extent shielded by Ireland from waves coming from the northwest [31]. The prevailing wind directions are between south-southwest and northwest, with northeasterly winds increasing their frequency in late winter and spring. Winds of 8 m s^{-1} or greater occur about half of the time in autumn and winter, and only 25% and 15% during spring and summer, respectively [32]. Typical wave conditions are represented by waves of 1.6 m significant wave height and 6 s mean period, while the conditions with one-year return period can reach the 10 m significant wave height and 12 s period [31].

Two WERA radars were installed in 2011 to study the feasibility of using such devices to monitor wave conditions and provide useful data for Marine Renewable Energy (MRE) projects. The individual stations are located at Pendeen and Perranporth (Figure 1), approximately 40 km apart. Each site consists of a 16-element phased-array receiver and a 4-element transmitter. Independent measurements are synchronously acquired at the two stations for 17 min 45 s, every hour, at approximately 600 m range resolution and 7° angular resolution. The two radars operate around a center frequency of 12.3 MHz in a "listen-before-talk" mode [33], determining the cleanest frequency band within 250-kHz. At the center transmitting frequency, the backscatter is dominated by Bragg reflection off 12.5 m waves, and the theoretically measurable sea states range from wave heights between 0.4 and 8 m, and have an upper frequency limit of approximately 0.28 Hz, although the latter will change depending on the intersection angle between wind and radar beam. These limits are dictated by the theory behind the Seaview method, and the reader is referred to [27] for more information about them. Finally, with this configuration, the maximum range for wave measurements is about 50 km.

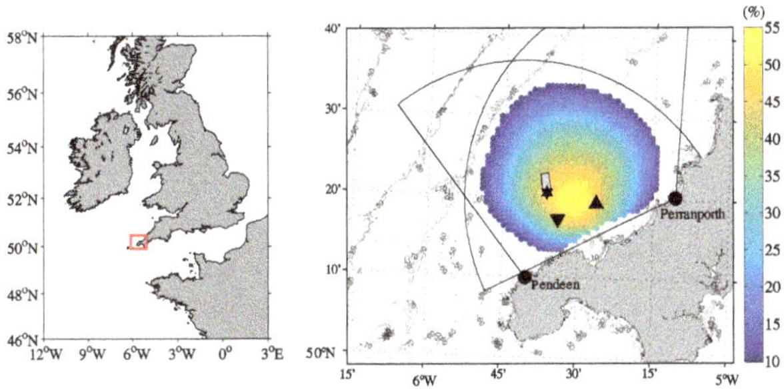

Figure 1. Study area: (**right**) the red rectangle demarcates the area shown in left figure, (**left**) radar stations (black circles), their coverage (50 km radius fans) and measuring grid. Depth contours are also shown and expressed in meters. The temporal data coverage of each radar cell is shown as a percentage of the time each cell had inverted wave data. The in situ mooring devices are also depicted: wave buoy (star), ADCP-W (inverted triangle) and ADCP-E (triangle). The grey rectangle delimits the Wave Hub test site for marine renewable energy devices.

2.1. Datasets

2.1.1. HF Radar

The dataset used in this study, chosen to coincide with a period when in situ measurements were available, was collected between April and November 2012. During this period, the two radars were simultaneously operational 98% of the time, while the maximum wave data return of the inversion algorithm at any one cell within the radar coverage was 55% (Figure 1). Low data return is generally associated with low signal-to-noise ratios (SNR), which are usually linked to low wind speeds. Under wind speeds between 0 and 4 m s^{-1}, the data availability at the three cells where the in situ devices were deployed (see Figure 1) did not exceed 12%. Wind direction seems to also have an effect on data return, and the highest rate of success on the inversion is attained when the wind is blowing from the west (see Figure 2). In that situation, data were inverted 60% of the time, while this percentage decreased to about 40% when the wind was blowing from the three remaining fetch-limited quadrants.

Radar directional ocean wave spectra were computed from the recorded backscatter using the Seaview algorithm. The method is an iterative scheme initialized with a Pierson–Moskovitz spectrum and a cardioid model for the directional distribution, both of which are defined using information directly extracted from the radar-measured Doppler spectrum [23]. This initial guess of the ocean wave spectrum is then used to derive a Doppler spectrum by integration of the equation relating the two [34]. The integrated result is then compared to that measured, and if the spectra differ from each other, the initial estimate of the ocean wave spectrum is adjusted in the following iteration until a certain convergence criterion is met [23]. The above process is performed as long as the SNR of the second order Doppler spectrum is higher than 15 dB, and the amplitude of the local minima around the strongest Bragg peak is 3 dB below the latter. The result is a wave number directional spectrum calculated over 30 directions and a variable band of wave numbers that depends on factors such as the transmitting frequency, the intersection angle between the radar beams, and the wind direction [35]. The wave number spectra retrieved by the algorithm were subsequently converted into directional frequency spectra using the wave dispersion relation as follows:

$$S(f,\theta) = S(k,\theta)\frac{4\pi k\sqrt{gk \tanh kd}}{g \tanh kd + gkd \operatorname{sech}^2(kd)} \tag{1}$$

where,

$$f = \frac{sqrtgk \tanh kd}{2\pi} \tag{2}$$

Frequency spectra were then obtained integrating the result of Equation (1) over the available range of directions. Finally, these spectra were interpolated to a 0.01 Hz resolution over the frequency range 0.05–0.25 Hz.

Wave parameters were then obtained from the frequency spectra and the first two directional Fourier coefficients. The time series derived from the radar and in situ measurements are shown in Figures 2 and 3, respectively. The calculations were performed for the whole frequency range (0.05–0.25 Hz) and three additional sub-ranges; 4–6 s, 6–10 s, and 10–20 s as follows,

$$H_s = 4\sqrt{\int_{f_1}^{f_2} S(f)df} \tag{3}$$

$$T_s = \frac{\int_{f_1}^{f_2} f^{-1}S(f)df}{\int_{f_1}^{f_2} S(f)df} \tag{4}$$

$$\theta_m = \int_{f_1}^{f_2} S(f)e^{i\theta(f)}df \tag{5}$$

where $\theta(f)$ is calculated from the first two Fourier coefficients as,

$$\theta(f) = tan^{-1}(b_1(f), a_1(f)) \tag{6}$$

$$\sigma_m = \int_{f_1}^{f_2} S(f)e^{i\sigma(f)df} \tag{7}$$

where σ_f is the frequency dependent directional spreading, calculated as,

$$\sigma(f) = (2(1 - \sqrt{a_1(f)^2 + b_1(f)^2}))^{1/2} \tag{8}$$

In all equations, S(f) is the frequency spectrum, while f_1 and f_2 represent the lower and upper integration limits. The coefficients a_1 and b_1 are the first two Fourier coefficients calculated from the directional distribution [36].

2.1.2. Mooring Data

During 2012, two Acoustic Doppler Current Profilers (ADCPs) and a wave buoy were deployed in the area to provide data with which to validate the radar measurements against. From 13 March to 20 December 2012, a Seawatch Mini II directional wave buoy was deployed at 53 m depth, 20 km offshore from Pendeen and 30 km from Perranporth. The ADCPs, two upward looking 600-kHz Teledyne RDI WorkHorse Broadband, were deployed on the same day as the wave buoy, recovered on 16 May, and redeployed at the same locations on 18 May. Finally, after a second recovery on 11 August, the last deployment began three days later, on 14 August. The eastern-most ADCP, referred to as ADCP-E here, was located at 24 km from Pendeen and 19 km from Perranporth at a depth of 33 m, while the other (ADCP-W) was moored at 16 km from Pendeen and 29 km from Perranporth, at 35 m depth.

The ADCP orbital velocities were processed into directional spectra using RDI's Wavesmon software. The iterative maximum likelihood method (IMLM [37]) was applied to compute directional spectra at 0.0078 Hz frequency resolution, and 36 directions at 15° resolution. In the processing, the effect of background currents was taken into account in order to avoid overestimation of wave heights due to the subsurface nature of the ADCP measurement [38]. The wave buoy time series were analyzed using WAFO [39]. Directional spectra were calculated using the extended maximum entropy method (EMEM [40]) at 128 frequency bins over 30 directions. Directional spectra were left unmodified, while the frequency spectra derived from them were interpolated to a 0.01 Hz frequency resolution and high- and low-pass filtered with cut-off frequencies of 0.05 and 0.25 Hz, respectively.

Integrated parameters were then calculated using Equations (3)–(8). Further details about the devices and data processing can be found in [41].

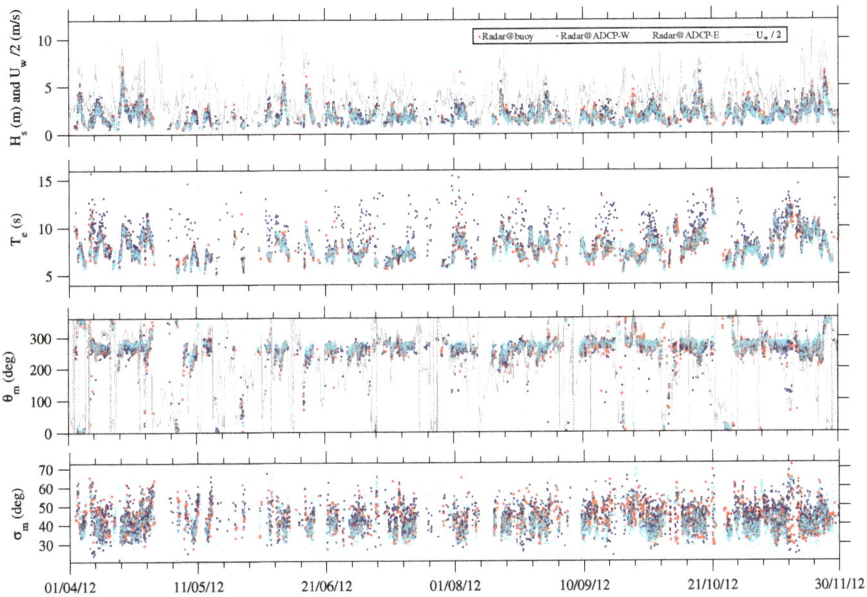

Figure 2. Radar wave parameters co-located with the wave buoy (red star), the ADCP-W (dark blue dots) and ADCP-E (light blue crosses) from April to November 2012 (**a**) significant wave height and wind speed (grey line); (**b**) energy period; (**c**) wave direction (direction of approach) and wind direction (grey line); and (**d**) mean directional spreading.

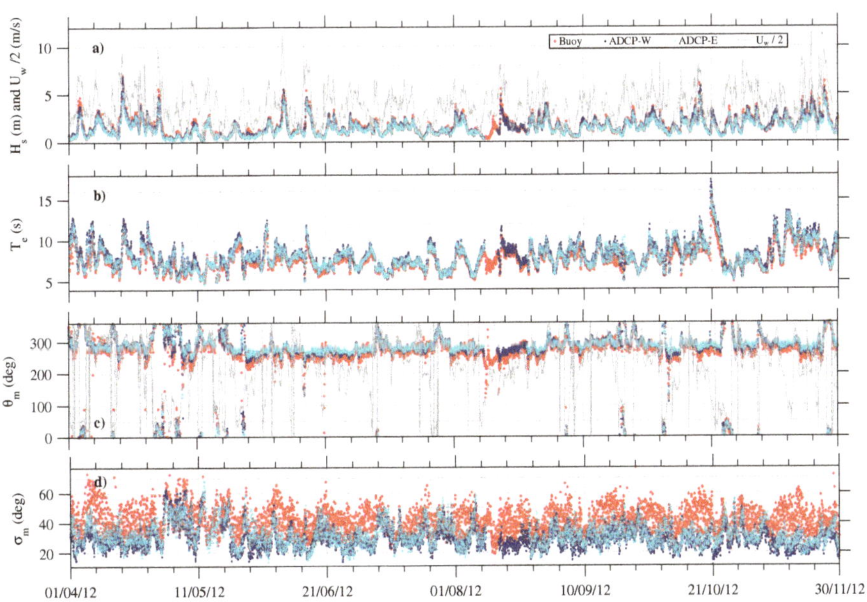

Figure 3. Same as previous figure, but showing the in situ-measured wave parameters.

2.2. Methodology

Common error metrics were used to evaluate the agreement between radar- and in situ-measured wave parameters. The root-mean-squared error (RMSE), the linear correlation (R) between two datasets, and the standard deviation of the data (σ) were calculated as,

$$\mathrm{RMSE} = \sqrt{\frac{1}{n}\sum_{i=1}^{n}[y_i - x_i]^2} \tag{9}$$

$$\mathrm{R} = \frac{\sum_{i=1}^{n}(y_i - \bar{y})(x_i - \bar{x})}{\sqrt{\sum_{i=1}^{n}(y_i - \bar{y})^2 \sum_{i=1}^{n}(x_i - \bar{x})^2}} \tag{10}$$

$$\sigma = \sqrt{\frac{\sum(x - \bar{x})^2}{N}} \tag{11}$$

In Equations (9) and (10), x corresponds to the reference value and y is the radar measurement. The variables with an over-bar indicate mean values.

Whenever required, directional variables were compared using expressions from circular statistics. Mean values of the angular data were calculated by vector addition as described in [42],

$$\bar{x} = \pm \tan^{-1}\left(\frac{S}{C}\right) \tag{12}$$

where S and C are given by,

$$S = \sum_{i=1}^{n}\sin(x_i) \quad \text{and} \quad C = \sum_{i=1}^{n}\cos(x_i) \tag{13}$$

Then, the circular standard deviation can be calculated as,

$$\sigma = \sqrt{2(1 - \rho)} \tag{14}$$

where $\rho = \sqrt{C^2 + S^2}/N$, with N the number of records. The RMSE for circular values was calculated as,

$$\mathrm{RMSE} = \sqrt{-2\ln\left[\frac{1}{N}\sum_{i=1}^{n}(\cos(x_i - x_j))\right]} \tag{15}$$

Finally, the linear correlation coefficient (Equation (10)) was replaced by its circular counterpart [42],

$$R = \frac{\sum\limits_{1\le i\lesssim j\le n}\sin(x_i - x_j)\,\sin(y_i - y_j)}{\sqrt{\sum\limits_{1\le i\lesssim j\le n}\sin^2(x_i - x_j)\sum\limits_{1\le i\lesssim j\le n}\sin^2(y_i - y_j)}} \tag{16}$$

3. Results

3.1. Integrated Parameters

Figure 4 shows the values of RMSE and correlation derived from the comparison of the radar and in situ datasets. Results are shown for the frequency bands mentioned in the previous section (Figure 4a–c) and for the full range of available frequencies (Figure 4d).

Significant wave height is consistently well estimated across all frequency bands. Nonetheless, the performance of the algorithm for this parameter is optimal for 6.7–10 s period waves. Standard deviations (not shown) are close to those of the in situ devices for the energy containing waves (Figure 4b) and the full range of frequencies (Figure 4d), correlations (R) are above 0.8, and RMS differences are between 24 and 50 cm. The locally generated wind waves exhibit the lowest correlation to the in situ measurements (between 0.83 and 0.87), as well as a greater variability, represented by a standard deviation almost 1.5 times higher than the in situ values. Spatially, the poorest agreement appears on the ADCP-W comparisons. This device was deployed at the edge of Perranporth's radar field of view, where the quality of the beamforming technique used by phased array radars to get azimuthal resolution and digitally steer the beam to the desired direction loses accuracy. In addition, the area is characterised by a slightly different surface current pattern than that of the rest of the radar's field of view, with generally stronger current magnitudes. The combination of these two factors appears to result in wave spectra with spurious energy on their low frequencies.

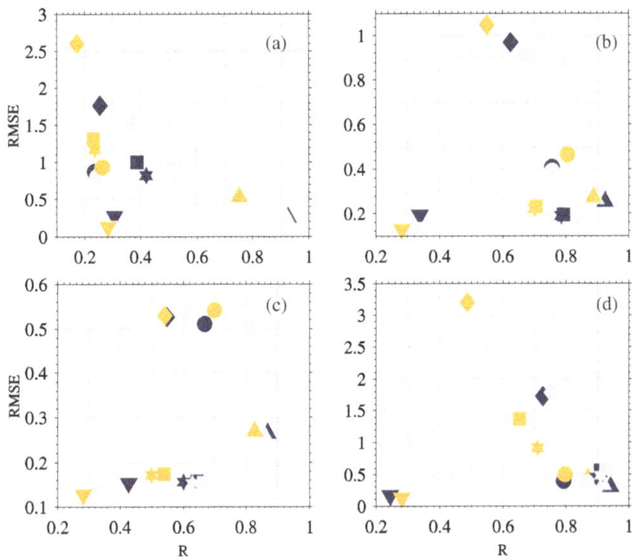

Figure 4. Correlation coefficient and RMSE obtained from the comparison between radar- and in situ-measured wave parameters. The three in situ devices are represented by a different color: buoy (blue), ADCP-E (grey) and ADCP-W (orange). The symbols denote the different wave parameters evaluated: energy period (T_e; star), mean period (T_m; square), peak period (T_p; diamond), significant wave height (H_s; triangle), mean wave direction (circle), and directional wave spread (spr; inverted triangle). The units of the RMSE values are meters, seconds, and radians for the wave height, period, and direction and wave spreading, respectively. Results are shown for three frequency bands and the full range of available frequencies: (**a**) 0.05–0.1 Hz; (**b**) 0.1–0.15 Hz; (**c**) 0.15–0.25; and (**d**) 0.05–0.25 Hz.

The results for the energy and mean periods are very similar between them and show the same pattern seen for the significant wave height, with correlations around 0.9 and 0.8 for the bulk of the energy spectrum and waves between 6.7 and 10 s, respectively. The results degrade at the lowest frequency band, where the radar shows lower standard deviation than in situ measurements, correlations decrease to values between 0.25 and 0.5, and RMS errors are between 0.9 and 1.3 s.

Among the non-directional parameters, peak period is the least correlated to the in situ measurements. At the lowest frequency band, the results obtained for the buoy and ADCP-E comparisons show lower standard deviation than the in situ measurements, and correlations below 0.4. The agreement improves when comparing shorter swell and the bulk of the spectrum, both of which show similar results. Correlations in these cases are between 0.6 and 0.75, and radar standard deviations are equivalent to the wave buoy and lower than the ADCP-E. As found for the other parameters, the radar estimate at the ADCP-W site shows the poorest agreement to the in situ measurements at the lower frequency bands due to the peculiarities previously described, while the results for the short waves are very close to those obtained at the other two sites.

The agreement on the mean wave directions estimated by the radar and the in situ devices is generally satisfactory, and as already observed for the other parameters, the poorest performance was found at the low frequency band, where the results are marginally correlated to the in situ values, and RMS errors reach 27°. In addition, the radar results show what it seems an unusual variability for this frequency band, which is represented by standard deviations two times higher than those of the ADCP results. The difference is smaller when comparing the results against the wave buoy, whose directions show higher variability than the ADCPs. Above 0.1 Hz, correlations increase to around 0.8, RMS errors lie between 11° and 19°, and both radar and in situ devices show similar standard deviations.

Finally, with correlations below 0.5, the radar directional spread shows marginal agreement to the in situ measurements at all frequency bands. It is worth mentioning the opposite behavior of the radar's standard deviation when compared to the wave buoy and the ADCPs, respectively. While the results show a higher variability on the radar's wave spread when compared to the ADCPs, the result is reversed when the comparison is made against the wave buoy (Table 1).

Table 1. Summary statistics of the radar and in situ H_s, T_e, D_m and directional spread comparisons calculated over the full range of frequencies (0.05–0.25 Hz), as shown in Figure 4d. The third and fourth columns are the in situ- and radar-measured mean values, respectively. The fifth column is the radar standard deviation normalized by the in situ value, the sixth is the RMSE (meters, seconds and degrees for H_s, T_e, D_m, and spr, respectively), the seventh is the correlation coefficient, and the last column is the number of records.

Device	Parameter	$\bar{x_{is}}$	$\bar{x_{rad}}$	std	RMSE	R	N
Buoy	H_s	2.18	2.20	0.91	0.30	0.94	2617
	T_e	7.94	7.95	1.27	0.56	0.91	
	T_p	8.92	8.81	0.99	1.72	0.73	
	D_m	276	265	1.06	22.3	0.79	
	spr	40.4	42.75	0.76	9.22	0.25	
ADCP-E	H_s	1.91	2.04	1.16	0.37	0.92	2366
	T_e	8.07	7.77	0.83	0.59	0.93	
	T_p	9.25	8.54	0.80	1.78	0.76	
	D_m	290	276	1.14	21.2	0.87	
	spr	29.4	40.9	1.11	7.93	0.32	
ADCP-W	H_s	2.11	2.19	1.04	0.45	0.87	2320
	T_e	8.48	8.51	1.02	1.38	0.66	
	T_p	9.78	9.99	1.25	3.20	0.49	
	D_m	284	265	1.21	28.5	0.80	
	spr	26.9	43.3	1.29	7.22	0.28	

3.2. Directional Wave Spectra

Direct comparison of directional spectra is not a straightforward task [43] and a qualitative analysis, based on the description of the graphical representation of the two types of results, is the common approach to this exercise. Here, we provide comparisons of three representative case studies.

The longest swell recorded over the period analyzed here arrived to the study area on the first hours of 21 October. At 1000 UTC, a moderate wind was blowing from the east at 7 m s^{-1} and the wave field was dominated by this long period swell, whose peak direction was determined to be about 280 ± 10° by both radar and in situ devices. At the ADCP-E site (Figure 5a,d), the radar shows a secondary mode close to 0.1 Hz and 120° that was not observed by the in situ device. The same had been observed by the radar at the other two locations the previous two hours, and there is some energy on the buoy spectrum around the same direction, which seems to indicate that is not an error on the radar estimate.

Figure 6 shows the integrated one-dimensional representations of spectral energy, direction and spread derived from the directional spectra of Figure 5. The spectra match very well across the three sites, and the biggest disagreement was observed on the ADCP-E comparison, which reveals a much broader radar spectrum, and a difference of 0.01 Hz on the estimated peak frequency that was also seen on the ADCP-W comparison. The directional spread measured by the different techniques shows what was seen on the previous section: radar measurements higher than the ADCPs over the low frequencies and slightly lower at the higher frequency end, and the opposite behavior on the radar-buoy comparison. Despite these differences, both radar and in situ devices show narrow distributions at the peak of the spectra, which broaden far from it.

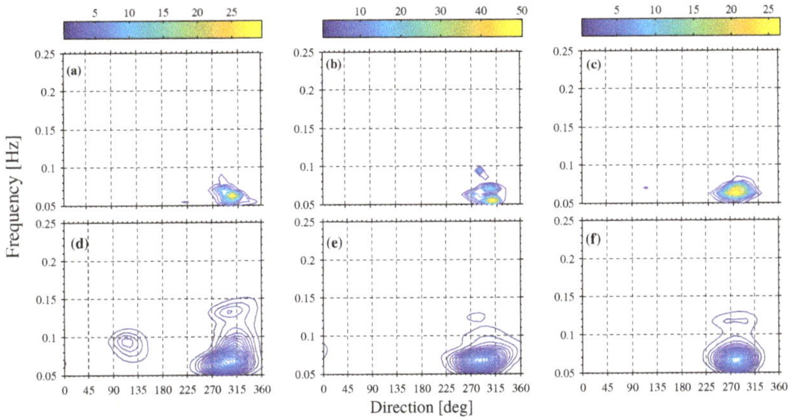

Figure 5. Radar and in situ estimates of wave directional spectra on 21 October 2012 at 1000 UTC. The upper panels show the results obtained from the in situ measurements, and the bottom panels show the radar estimates: ADCP-E spectra (**a**,**d**); and ADCP-W (**b**,**e**). Wave buoy (**c**,**f**). The colorbar represents the directional energy density ($m^2/Hz/deg$).

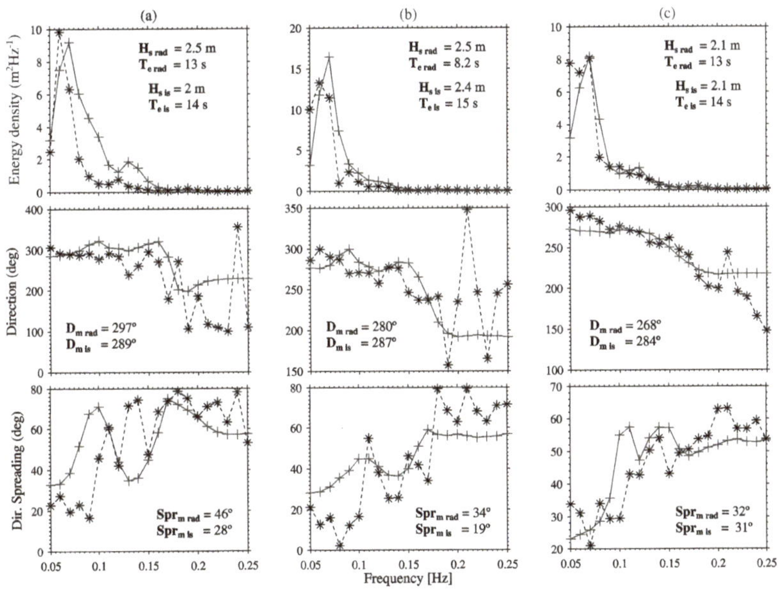

Figure 6. Radar and in situ estimates of one-dimensional spectra, mean direction and directional spreading as a function of frequency on 21 October 2012 at 1000 UTC: (**a**) ADCP-E (dashed black line and *) and radar (grey line and +); (**b**) ADCP-W (dashed black line and *) and radar (grey line and +); and (**c**) buoy (dashed black line and *) and radar (grey line and +).

On 24 September at 1100 UTC, the wave field was formed by a low frequency swell and a wind sea both very close in direction (Figure 7). On the afternoon of the day before, the wind speed had grown very rapidly from 5 to 11 m s^{-1} while turning north from a more easterly direction. At the time of the observations described here, the wind had been steadily blowing for about 14 h and had a westerly direction. The radar-derived two-dimensional spectra show broader contours than the in situ spectra, especially when compared to the ADCPs, whose results are sharper than those of the buoy, possibly because the greater number of degrees of freedom in the measurements [44]. The swell measured by the radar at the buoy site exhibits two modes propagating from 220° and 280°, and very close in frequency. Although the same is reported by the in situ device, the direction of arrival is about 10° further north and at slightly higher frequencies. The results of the wave directional spreading (Figure 8) show higher values on the low frequencies of the radar estimates as compared to the ADCPs, and lower above 0.19 Hz. This is reversed on the buoy comparison, where the radar spread is lower at the low frequency and higher above 0.16 Hz.

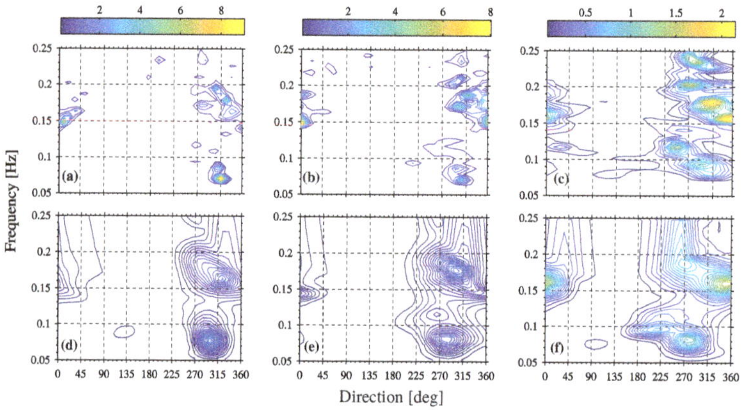

Figure 7. As in Figure 5, but for 24 September 2012 at 1100 UTC.

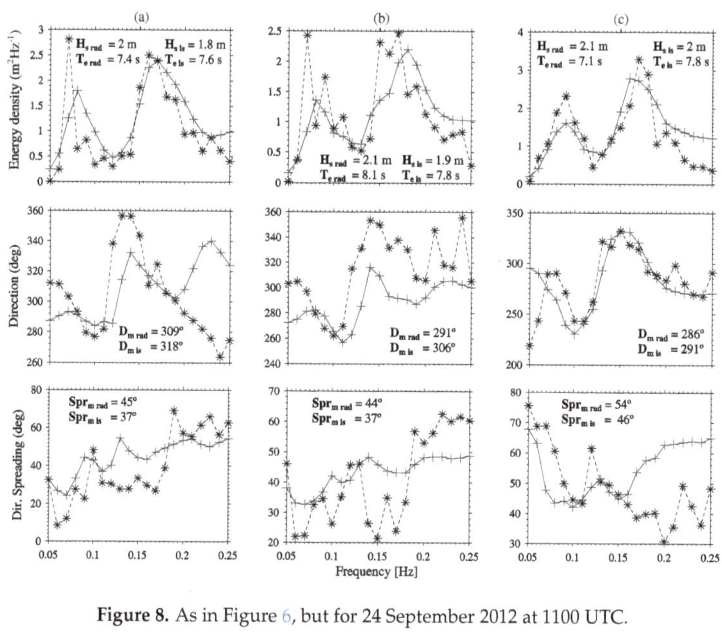

Figure 8. As in Figure 6, but for 24 September 2012 at 1100 UTC.

On 26 October at 0100 UTC, the directional spectrum measured at the buoy (Figure 9b,e) and ADCP-W (Figure 9c,f) sites shows a mixed sea formed by a weak south-westerly swell component and a wind sea generated by a wind blowing from the northeast with a speed of 10 m s^{-1}. The wind sea was also observed at the ADCP-E (Figure 9a,d), but there is no evidence of the low frequency component on the radar estimate and very little energy appears at about 0.09 Hz on the in situ spectrum. The swell measured by the radar at the ADCP-W location is more spread in direction than the in situ measurement, while at the wave buoy site the radar measurement is displaced toward the south with respect to the in situ result. The direction of the wind sea compares well at all locations and the different measuring techniques, but, as in the other cases, the contours are broader in the radar estimates. There are three modes on the radar measurement at the ADCP-E site, which are slightly stronger than the in situ measurement and result in a higher wave height (Figure 10). The direction of the spectral components above 0.13 Hz agree well, with differences up to 20°. The same can be observed at the low frequencies and the greater disagreement was found between 0.1 and 0.13 Hz. The directional spread estimated by the radar follows the same pattern as in the in situ results, showing a better agreement to the buoy measurement and higher values than the ADCPs up to 0.2 Hz.

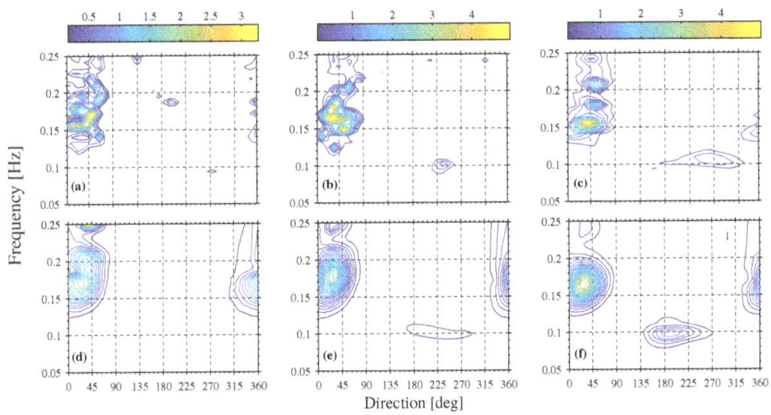

Figure 9. As in Figure 5, but for 26 October 2012 at 0100 UTC.

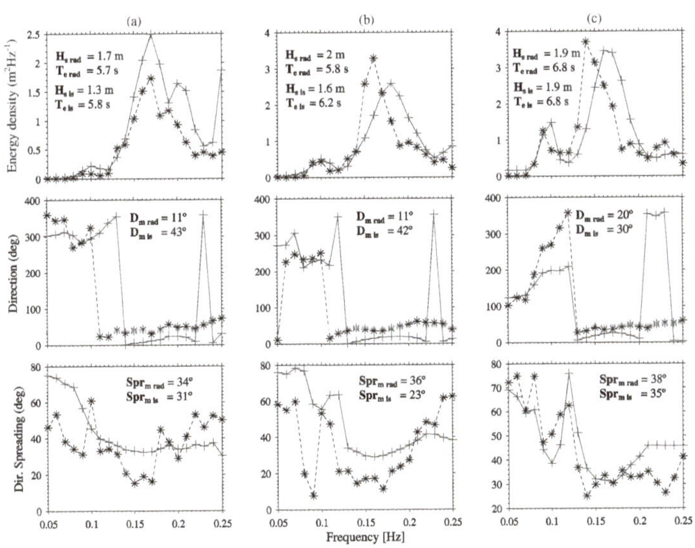

Figure 10. As in Figure 6, on 26 October 2012 at 0100 UTC.

4. Discussion

As referred to in the Introduction, commonly reported errors on the algorithm's results have been related to a limitation of the theory underpinning the inversion in high sea states, and short averaging periods. The latter result in an increased statistical sampling variability, which is inherent to deriving estimates of a theoretically infinite stochastic process from finite samples. Such source of variability was not found to significantly affect the wave parameters evaluated here, whose standard deviations were between 3% and 20% of those obtained with the in situ datasets. In addition, the maximum wave height of 7 m measured during our experiment was accurately estimated by the radar. The difference between this study and some of the most exhaustive available validations of the method (e.g., [24,25]) is the frequency at which the radars operate. While our system transmits at 12.3 MHz, a frequency which until now remained relatively untested in the literature for the Seaview algorithm in combination with WERA radars, the aforementioned works used radars operating at 27.75 MHz, a frequency limited in its description of wave heights higher than 6 m [27], which has been previously found to produce overestimates in the high frequency part of the spectrum for wave heights higher than 4 m [25].

In this work, the main origin of the differences found between radar and in situ wave spectra appears to be related to the lower spectral resolution of the former, which has been attributed to a smoothing procedure performed during the inversion to stabilize the solution ([45,46]). As a result, the spectral energy was spread over a wider range of frequencies than in the situ spectra. In terms of parameter estimation, this is generally translated in an underestimation of wave periods, which were nonetheless still strongly correlated to the in situ measurements. Comparison of the energy period against the wave buoy yielded values of 0.91 for the correlation coefficient, and a RMSE of 0.56 s. The peak period resulted in a correlation of 0.73, and RMSE of 1.72 s, statistics that compare well with the correlation of 0.8 and 1.3 s RMSE found for a WERA system operating at the low end of the HF band in South Australia [7]. Despite the higher spread of energy over frequencies, the radar spectra generally had a slightly weaker spectral peak, as previously observed in [24]. As a result, the overall energy content was equivalent to the in situ measurements. Consequently, significant wave height was accurately estimated by the radar, with a correlation of 0.94, and a RMSE of 0.3 m at the center of the radar's coverage. As found for the wave period, these values compare well with the correlation of 0.9 and 0.42 m RMSE reported in [7].

The inversion algorithm has shown good skill estimating mean wave direction, which was well correlated to the in situ measurements at all the frequency bands studied, with the only exception of the low end of the spectrum, below 0.1 Hz. While the poor agreement of the radar–ADCP-W comparison is certainly due to the presence of spurious energy in the radar spectra, we did not find clear evidence of the same issue at the other two locations examined. However, there was a strong variability on the radar-derived directions, which does not appear consistent with what it might be expected on this frequency band, where the measured directions should be mainly those of the prevailing swell, which enters the area from the west and southwest. Nevertheless, part of the differences might also be derived from the general difficulty to obtain accurate directional measurements at low frequencies, where the signal-to-noise ratio is usually low. There is evidence showing that the Fourier coefficients obtained from wave buoy measurements have greater variability and less accuracy at these frequencies [47], and it is not uncommon to find inconsistencies between measuring platforms at this frequency band [48].

The wave directional spread calculated from the radar and in situ measurements resulted in similar spectral curves that showed low values close to peak, and a widening of the distribution at the frequencies both sides of it. However, despite the overall resemblance of the frequency dependent curves, the radar integrated results showed a poor correlation to the in situ values. The differences were higher when contrasting the radar against the ADCPs, the spread of which was narrower than the radar's below the peak, broader above it, and resulted in lower integrated values than those obtained from both the radar and the wave buoy.

The different results obtained when comparing the radar directional spread to the ADCPs or the wave buoy appear to be related to a stronger tidal modulation on the estimates obtained with the latter instrument, which showed variations of up to 20° in 6-h cycles. Although the wave spread obtained from the radar and the ADCP measurements also showed signs of tidal modulation, the variations were not as significant as in the buoy's results. This is evidenced by the results of the autocorrelation function of the differences between the wave buoy and the ADCP-E directional spread, plotted in Figure 11a together with the autocorrelation of the differences between the radar spread at equivalent positions (Figure 11b). The residuals obtained from the in situ measurements show a clear tidal signal, with peaks in correlation every 6 h, which is not observed when comparing the radar's spread measured at the same locations. Using the ADCP-W measurements produced the same outcome (not shown). This suggests that the differences between the wave buoy and the ADCPs are in fact caused by the tidal current.

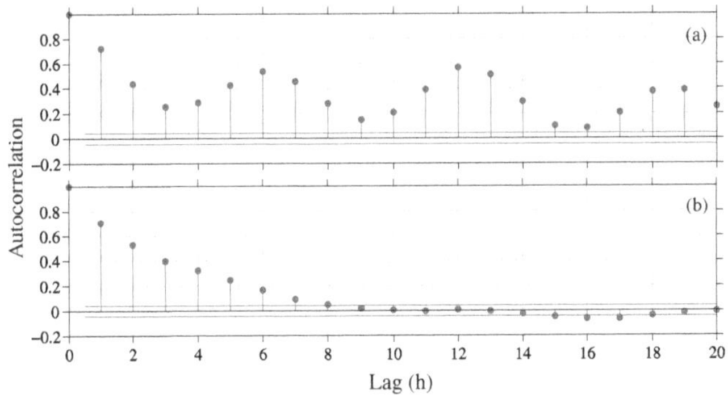

Figure 11. Autocorrelation function of the time series of the differences between directional spread measured at two locations: (**a**) autocorrelation of differences between the wave buoy and the ADCP-E directional spread; and (**b**) autocorrelation of the differences between the radar directional spread estimates obtained at the buoy and ADCP-E sites. The confidence bounds are shown in blue.

Six-hour modulation cycles in buoy-measured mean directional spreading were also reported in [47] for the same study area, where they observed a correlation between wave spread and current intensity, characterised by minimum values during slack water, which then increased during the flood and ebb phases of the tide. Both our wave buoy, and the devices used in [47] were deployed in deeper water than our ADCPs, located at sites where the water column was about 20 m shallower than at the buoy's location. Therefore, the effects of refraction in reducing directional spread might be responsible for the lower values measured by the ADCPs. However, differences in the measuring techniques themselves, as well as in their processing approaches might also be responsible of the differing results. In the presence of background currents, the gain applied to transfer the ADCP subsurface measurements to surface displacement will be overestimated if the value of the current is not taken into account in the calculations. To avoid this issue, the measured velocity data were used in the wave dispersion equation when calculating the ADCP directional spectra. This avoids the energy overestimation, but might also mean the calculated values are not strictly comparable to the buoy's or the radar's results.

Finally, irrespective of the parameter studied, the least satisfactory results were generally found when comparing the ADCP-W and radar measurements at that location. This was found to be related to the presence of spurious energy at the low frequencies of the radar spectra. This is due to the modification of the beam pattern as it is steered away from its boresight. Both the width of the main lobe and its sidelobe levels increase, degrading azimuthal resolution and increasing the sensitivity to unwanted signal returns from areas outside the main lobe, respectively. The latter becomes particularly significant when the currents vary across the radar coverage, and can completely hinder wave estimation.

Figure 12 a shows an example of the above. The figure depicts two Doppler spectra measured at 28 km range from Perranporth and different steering angles. In it, it is possible to see how the Bragg peak of the spectrum obtained at 14.6° (cell 3757) from Perranporth's normal appears as part of the signal obtained at 45° (cell 3682), where the ADCP-W was deployed. As explained in [49], when the differences in first-order Doppler shifts along the same range are large, the Bragg peaks originating from directions other than that of the main lobe can be incorrectly identified as second order structure, and result in artificially large low frequency components in the estimated ocean wave spectrum, as observed in our results, or in [50].

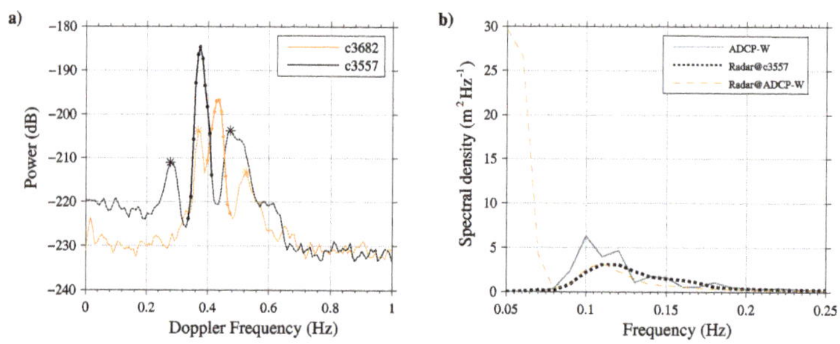

Figure 12. Radar measurement affected by antenna sidelobes: (**a**) The Doppler spectra measured at 14° (black line) and 45° (orange line) from Perranporth's normal. The first-order region of each spectrum is emphasized with a full line. The stars indicate the identified second order peaks. (**b**) The frequency spectra obtained at the same two cells, and measured by the ADCP-W, located at cell 3682.

5. Conclusions

HF radars allow measuring the wave directional spectrum from the coast and at relatively high spatial and temporal resolutions. Unlike traditional methods such as wave buoys, they provide a spatial snapshot of the wave field every few minutes, which is very useful in studying the spatiotemporally heterogeneous waters of the coastal seas. Although the most important factor to get good wave estimates from HF radar seems to be the quality of the signal, the algorithm used to convert such signal into ocean wave spectra is also relevant. In this study, we presented an evaluation of the Seaview algorithm [23], which enables the measurement of the wave directional spectrum from HF radar Doppler spectra. The radar results were compared against measurements acquired by three in situ devices deployed at depths that ranged between 33 and 53 m. The qualitative analysis of the directional spectra produced by the inversion algorithm evidenced high concordance between the radar and in situ results for different sea states. Comparisons of radar and in situ-derived wave parameters extracted from the bulk of the spectra resulted in correlations above 0.9 for the significant wave height and energy period, and above 0.79 for the mean direction, at two of the three locations examined. The remaining location, which presented poorer statistics when compared to the in situ measurements, was found to be affected from spurious energy at the low frequencies of the spectra, resulting from contamination of other steering angles. Increased levels of antenna sidelobes at high angles from the radar's boresight, combined with surface current heterogeneity along a range cell were identified as the source of this problem. A careful analysis of the surface current at the site, along with the examination of antenna patterns are therefore essential to delimit the area where accurate wave results can be expected.

The main drivers of previously reported inaccuracies in the algorithm's results, usually related to short measuring periods and a limitation of the theory underpinning the inversion in high sea states, have not been found to have a significant effect on the results. This suggests that the integration time used to calculate the radar's directional spectra, and the chosen transmitting frequency are well suited to measure the wave climate of the area. Nonetheless, adding a higher transmitting frequency would allow measuring low sea states, which with the current configuration resulted in data gaps.

Having established its accuracy, it seems that at present one of the main drawbacks of the method is the data return, which did not exceed 55% over the eight-month period studied here. Together with the above-mentioned possibility of adding a higher transmitting frequency to measure low sea states, improvement of the SNR, noise cancelling techniques, and a dedicated maintenance of the radars are possibly the essential requirements to increase the data return and enable the use of HF radar wave spectra at operational level.

Author Contributions: Conceptualization, G.L. and D.C.C.; Methodology, G.L. and D.C.C.; Writing—Original Draft Preparation, G.L.; and Writing—Review and Editing, D.C.C. and G.L.

Funding: This research was supported by the School of Marine Science and Engineering at Plymouth University, the Natural Environment Research Council (Grant NE/J004219/1) and MARINET, a European Community—Research Infrastructure Action under the FP7 Programme.

Acknowledgments: The authors thank Megan Sheridan, Davide Magagna and Peter Ganderton for their involvement in the deployment of the devices which provided the in situ data used in this work. We are also grateful to all of the Plymouth University Coastal Processes Research Group members, who at some point helped with the maintenance of the radar sites. We also express our gratitude to the anonymous reviewers for their helpful comments and suggestions.

Conflicts of Interest: The authors declare no conflict of interest.

References

1. Fujii, S.; Heron, M.L.; Kim, K.; Lai, J.W.; Lee, S.H.; Wu, X.; Wu, X.; Wyatt, L.R.; Yang, W.C. An overview of developments and applications of oceanographic radar networks in Asia and Oceania countries. *Ocean Sci. J.* **2013**, *48*, 69–97. [CrossRef]
2. Roarty, H.; Glenn, S.; Kohut, J.; Gong, D.; Handel, E.; Rivera, E.; Garner, T.; Atkinson, L.; Brown, W.; Jakubiak, C. Operation and application of a regional high-drequency radar network in the mid-atlantic bight. *Marine Technol. Soc. J.* **2010**, *44*, 133–145. [CrossRef]
3. Rubio, A.; Mader, J.; Corgnati, L.; Mantovani, C.; Griffa, A.; Novellino, A.; Quentin, C.; Wyatt, L.; Schulz-Stellenfleth, J.; Horstmann, J.; et al. HF radar activity in European coastal seas: Next steps toward a pan-European HF radar network. *Front. Mar. Sci.* **2017**, *4*, 8. [CrossRef]
4. Lorente, P.; Sotillo, M.G.; Aouf, L.; Amo-Baladrón, A.; Barrera, E.; Dalphinet, A.; Toledano, C.; Rainaud, R.; De Alfonso, M.; Piedracoba, S.; et al. Extreme Wave Height Events in NW Spain: A Combined Multi-Sensor and Model Approach. *Remote Sens.* **2018**, *10*. [CrossRef]
5. Lorente, P.; Basañez Mercader, A.; Piedracoba, S.; Perez-Muñuzuri, V.; Montero, P.; Alvarez-Fanjul, E. Long-term skill assessment of SeaSonde radar-derived wave parameters in the Galician coast (NW Spain). *Int. J. Remote Sens.* **2019**. [CrossRef]
6. Saviano, S.; Kalampokis, A.; Zambianchi, E.; Uttieri, M. A year-long assessment of wave measurements retrieved from an HF radar network in the Gulf of Naples (Tyrrhenian Sea, Western Mediterranean Sea). *J. Oper. Oceanogr.* **2019**, *12*, 1–15. [CrossRef]
7. James, C.; Collopy, M.; Wyatt, L.R.; Middleditch, A.; James, C.; Collopy, M.; Wyatt, L.R.; Middleditch, A.; James, C. Suitability of the Southern Australia integrated marine observing system's (SA-IMOS) HF-radar for operational forecasting. *J. Oper. Oceanogr.* **2019**, *12*, 1–13. [CrossRef]
8. Gurgel, K.W.; Essen, H.H.; Kingsley, S.P. High-frequency radars: Physical limitations and recent developments. *Coast. Eng.* **1999**, *37*, 201–218. [CrossRef]
9. Lipa, B. Derivation of directional ocean-wave spectra by integral inversion of second-order radar echoes. *Radio Sci.* **1977**, *12*, 425–434. [CrossRef]
10. Wyatt, L.R. A relaxation method for integral inversion applied to HF radar measurement of the ocean wave directional spectrum. *Int. J. Remote Sens.* **1990**, *11*, 1481–1494. [CrossRef]
11. Howell, R.; Walsh, J. Measurement of ocean wave spectra using narrow-beam HF radar. *IEEE J. Ocean. Eng.* **1993**, *18*, 296–305. [CrossRef]

12. Hisaki, Y. Nonlinear inversion of the integral equation to estimate ocean wave spectra from HF radar. *Radio Sci.* **1996**, *31*, 25–39. [CrossRef]
13. Hashimoto, N.; Tokuda, M. A Bayesian approach for estimation of directional wave spectra with HF radar. *Coast. Eng. J.* **2003**, *41*, 137–149. [CrossRef]
14. Hisaki, Y. Development of HF radar inversion algorithm for spectrum estimation (HIAS). *J. Geophys. Res. Oceans* **2015**, *120*, 1725–1740. [CrossRef]
15. Barrick, D.E. Extraction of wave parameters from measured HF radar sea-echo Doppler spectra . *Radio Sci.* **1977**, *12*, 415–424. [CrossRef]
16. Wyatt, L.R. Significant waveheight measurement with HF radar. *Int. J. Remote Sens.* **1988**, *9*, 1087–1095. [CrossRef]
17. Wyatt, L.R. An evaluation of wave parameters measured using a single HF radar system. *Can. J. Remote Sens.* **2002**, *28*, 205–218. [CrossRef]
18. Maresca, J.W.; Evans, M.W.; Georges, T.M. Measuring rms wave spectra using narrow-beam HF radar. *IEEE J. Ocean. Eng.* **1980**, *18*, 295–305.
19. Heron, S.F.; Heron, M.L. A comparison of algorithms for extracting significant wave height from HF radar ocean backscatter spectra. *J. Atmos. Ocean. Technol.* **1998**, *15*, 1157–1163. [CrossRef]
20. Essen, H.H.; Gurgel, K.W.; Schlick, T. Measurement of ocean wave height and direction by means of HF radar: An empirical approach. *Deutsche Hydrographische Zeitschrift* **2007**, *51*, 369–383. [CrossRef]
21. Gurgel, K.W.; Essen, H.H.; Schlick, T. An empirical method to derive ocean waves from second-order Bragg scattering: Prospects and limitations. *IEEE J. Ocean. Eng.* **2006**, *31*, 804–811. [CrossRef]
22. Alattabi, Z.; Cahl, D.; Voulgaris, G. Swell and wind wave inversion using a single Very High Frequency (VHF) radar. *J. Atmos. Ocean Technol.* **2019**, [CrossRef]
23. Wyatt, L.R. Limits to the inversion of HF radar backscatter for ocean wave measurement. *J. Atmos. Ocean. Technol.* **2000**, *17*, 1651–1665. [CrossRef]
24. Wyatt, L.R.; Thompson, S.P.; Burton, R.R. Evaluation of high frequency radar wave measurement. *Coast. Eng.* **1999**, *37*, 259–282. [CrossRef]
25. Wyatt, L.R.; Green, J.J.; Gurgel, K.W.; Nieto Borge, J.C.; Reichert, K.; Hessner, K.; Günther, H.; Rosenthal, W.; Saetra, O.; Reistad, M. Validation and intercomparisons of wave measurements and models during the EuroROSE experiments. *Coast. Eng.* **2003**, *48*, 1–28. [CrossRef]
26. Wyatt, L.R.; Green, J.J.; Middleditch, A. Directional spectra comparisons between HF radar and a wave model. In Proceedings of the IEEE Working Conference on Current Measurement Technology, Charleston, SC, USA, 17–19 March 2008; pp. 211–216. [CrossRef]
27. Wyatt, L.R.; Green, J.J.; Middleditch, A. HF radar data quality requirements for wave measurement. *Coast. Eng.* **2011**, *58*, 327–336. [CrossRef]
28. Wyatt, L.R.; Green, J.J.; Middleditch, A. Signal sampling impacts on HF radar wave measurement. *J. Atmos. Ocean. Technol.* **2009**, *26*, 793–805. [CrossRef]
29. Voulgaris, G.; Haus, B.K.; Work, P.; Shay, L.K.; Seim, H.E.; Weisberg, R.H.; Nelson, J.R. Waves initiative within SEACOOS. *Mar. Technol. Soc. J.* **2009**, *42*, 68–80. [CrossRef]
30. Savidge, D.; Amft, J.; Gargett, A.; Archer, M.; Conley, D.; Voulgaris, G.; Wyatt, L.; Gurgel, K.W. Assessment of WERA long-range HF-radar performance from the user's perspective. In Proceedings of the 2011 IEEE/OES/CWTM 10th Working Conference on Current, Waves and Turbulence Measurement, CWTM 2011, Monterey, CA, USA, 20–23 March 2011; pp. 31–38. [CrossRef]
31. Halcrow Group Limited *Wave Hub Development Phase Coastal Processes Study Report*; Technical Report; South West of England Regional Development Agency: Exeter, UK, 2006.
32. Department of Energy & Climate Change. *UK Offshore Energy Strategic Environmental Assessment OESEA3 Non-Technical Summary*; Technical Report; Department of Energy and Climate Change: London, UK, 2016.
33. Gurgel, K.W.; Schlick, T. Compatibility of FMCW modulated HF surface wave radars with radio services. In Proceedings of the International Radar Symposium, IRS 2007, Cologne, Germany, 5–7 September 2007; pp. 255–258.
34. Weber, B.L.; Barrick, D.E. On the nonlinear theory for gravity waves on the ocean's surface. part I: Derivations. *J. Phys. Oceanogr.* **2002**, *7*, 3–10. [CrossRef]

35. Middleditch, A.; Cosoli, S. The Australian coastal ocean radar network: Temporal and spatial scales of HF radar wave data. In Proceedings of the IEEE OCEANS 2016, Shanghai, China, 10–13 April 2016; pp. 1–8, [CrossRef]
36. Kuik, A.J.; van Vledder, G.P.; Holthuijsen, L.H. A Method for the routine analysis of pitch-and-roll buoy wave data. *J. Phys. Oceanogr.* **2002**, *18*, 1020–1034. [CrossRef]
37. Krogstad, H.E. Maximum likelihood estimation of ocean wave spectra from general arrays of wave gauges. *Model. Identif. Control* **1988**, *9*, 81–97. [CrossRef]
38. Teledyne RD Instruments Inc. *WAVES PRIMER: Wave Measurements and the RDI ADCP Waves Array Technique*; Technical Report; Teledyne RD Instruments Inc.: Poway CA, USA, 2013.
39. Brodtkorb, P.A.; Johannesson, P.; Lindgren, G.; Rychlik, I.; Ryden, J.; Sjo, E. WAFO—A MATLAB toolbox for random waves and loads. In Proceedings of the Tenth (2000) International Offshore and Polar Engineering Conference, Seattle, WA, USA, 28 May–2 June 2000; Volume 3, pp. 343–350.
40. Hashimoto, N. *Analysis of the Directional Wave Spectrum From Field Data*; World Scientific: Singapore, 2010; Volume 3, pp. 103–143.
41. Lopez, G.; Conley, D.C.; Greaves, D. Calibration, validation, and analysis of an empirical algorithm for the retrieval of wave spectra from HF radar sea echo. *J. Atmos. Ocean. Technol.* **2016**, *33*, 245–261. [CrossRef]
42. Bowers, J.A.; Morton, I.D.; Mould, G.I. Directional statistics of the wind and waves. *Appl. Ocean Res.* **2000**, *22*, 13–30. [CrossRef]
43. Krogstad, H.E.; Wolf, J.; Thompson, S.P.; Wyatt, L.R. Methods for intercomparison of wave measurements. *Coast. Eng.* **1999**, *37*, 235–257. [CrossRef]
44. Strong, B.; Brumley, B.; Terray, E.; Stone, G. The performance of ADCP-derived directional wave spectra and comparison with other independent measurements. In Proceedings of the OCEANS 2000 MTS/IEEE, Providence, RI, USA, 11–14 September 2002; Volume 2, pp. 1195–1203. [CrossRef]
45. Green, J.J.; Wyatt, L.R. Row-action inversion of the Barrick-Weber equations. *J. Atmos. Ocean. Technol.* **2006**, *23*, 501–510. [CrossRef]
46. Wyatt, L.R. Measuring the ocean wave directional spectrum 'First Five' with HF radar. *Ocean Dyn.* **2019**, *69*, 123–144. [CrossRef]
47. Saulnier, J.B.; Maisondieu, C.; Ashton, I.; Smith, G.H. Refined sea state analysis from an array of four identical directional buoys deployed off the Northern Cornish coast (UK). *Appl. Ocean Res.* **2012**, *37*, 1–21. [CrossRef]
48. Work, P.A. Nearshore directional wave measurements by surface-following buoy and acoustic Doppler current profiler. *Ocean Eng.* **2008**, *35*, 727–737. [CrossRef]
49. Wyatt, L.R.; Liakhovetski, G.; Graber, H.C.; Haus, B.K. Factors affecting the accuracy of SHOWEX HF radar wave measurements. *J. Atmos. Ocean. Technol.* **2005**, *22*, 847–859. [CrossRef]
50. Middleditch, A. Spectral Analysis in High Frequency Radar Oceanography. PhD Thesis, University of Sheffield: Sheffield, UK, 2006.

Journal of
Marine Science and Engineering

MDPI

Article

Estimation of the Significant Wave Height from Marine Radar Images without External Reference

Giovanni Ludeno [1,*] and Francesco Serafino [2]

1 Institute for Electromagnetic Sensing of the Environment, National Research Council, Napoli I-80124, Italy
2 Institute of BioEconomy (IBE), National Research Council, 50100 Florence, Italy; francesco.serafino@cnr.it
* Correspondence: ludeno.g@irea.cnr.it

Received: 16 October 2019; Accepted: 26 November 2019; Published: 27 November 2019

Abstract: In the context of the sea state monitoring by means of the X-band marine radar, the estimation of a significant wave height (H_s) is, currently, one of the most challenging tasks. For its estimation, a calibration is usually required using an external reference, such as in situ sensors, and mainly buoys. In this paper, a method that allows us to avoid the need for an external reference for H_s estimation is presented. This strategy is, mainly, based on the correlation between a raw radar image and the corresponding non-calibrated wave elevation image to which varying its amplitude by using a scale factor creates a mathematical model for the radar imaging. The proposed strategy has been validated by considering a simulated waves field, generated at varying sea state conditions. The results show a good estimation of the significant wave height, confirmed by a squared correlation coefficient greater than 0.70 for each considered sea state.

Keywords: significant wave height; marine radar; sea state monitoring

---

1. Introduction

X-band marine radar is a useful and well-assessed technology for sea state monitoring for both offshore and coastal areas. Within the last three decades, several algorithms, mainly based on spectral analysis, have been developed in order to retrieve the sea state parameters starting from the raw radar images [1–6]. Currently, the wave radar system is employed in many applicative scenarios since these algorithms provide accurate measurements of the sea states in terms of the period, length, and direction of the dominant waves, the significant wave height, and the reconstruction of the sea surface current and bathymetry fields [7–13].

In this context, one of the most interesting and challenging tasks is the estimation of the significant wave height (H_s). In literature, there are two classes of approaches for the H_s estimation. The first class is based on the calibration of the wave spectra thanks to an external reference, such as the wave buoy. In this frame, the methods based on spectral analysis [14] or approaches based on time analysis [15–17] or other techniques, including an iterative least square approach [18] and a wavelet-based algorithm [19], are recalled.

The second class is about the possibility to calibrate H_s without additional reference sensors [20–23]. In particular, in Wijaya, A.P. and van Groesen, E. 2016 [20], the estimation of H_s is carried out by correlating the shadowing phenomenon to a visibility function stored in a database. In Gangeskar, R. 2014 [21], by assuming a geometric shadowing condition, shadowed areas are first extracted from the image by an edge detection algorithm. Afterward, by using the calculated illumination ratios in local areas, the Root Mean Square (RMS) surface slope is derived by curve fitting Smith's function [24]. Lastly, H_s is estimated from the RMS surface slope and the average zero-crossing wave period. In Liu, X. et al. 2016 [22], few improvements of the algorithm in Gangeskar, R. 2014 [21] have been presented. In particular, the use of the upwind information to fix the azimuthal location of the subarea is employed to calculate the RMS of the surface slope. The average zero-crossing wave period can be estimated from the sea spectra obtained by processing the X-band radar images. Recently, a shadowing mitigation method, which allows the estimation of wave elevation fields in coastal areas through the sea clutter data obtained from the X-band marine radar systems in extreme grazing incidence angles without calibration or the empirical Modulation Transfer Function (MTF) adjustments, has been proposed [23].

The purpose of this work is to describe a novel and alternative approach for estimating the H_s measurements from the radar image without the use of an external reference. In particular, the basic idea is to estimate this sea wave parameter by performing a correlation procedure between raw radar images and the corresponding non-calibrated wave elevation images to which, varying its amplitude by means of a scale factor, a mathematical model for the radar imaging is applied [2]. Herein, we show a preliminary study of the proposed strategy by considering a simulated wave field provided by a Fourier domain approach [25] with the spectral proprieties of the Pierson-Moskowitz model [26]. The simulation of sea wave images is completed by applying the modulation effects (shadowing and tilt modulation) introduced from the particular acquisition geometric of the radar antenna (grazing angle) [2]. In order to evaluate the performance of the proposed approach, the wave field for different sea states in terms of significant wave height and peak period have been generated.

Therefore, the paper is organized as follows. Section 2 describes the generation of synthetic data and the proposed strategy. The results are presented in Section 3. Discussion and conclusions are at the end of the paper.

2. Materials and Methods

This section is devoted to describe both the mathematical models used for generating the synthetic data and the proposed algorithm for estimating H_s.

2.1. Generation of the Synthetics Data

2.1.1. Generation of the Wave Field

Synthetic sea wave images have been generated with the Fourier domain approaches, which was introduced by Mastin 1987 by employing measured spectral proprieties of real ocean waves [25].

The sea wave field $h(\bar{r}, t)$ is modelled as a sum of sinusoids with time-dependent amplitudes, defined as [27]:

$$h(\bar{r}, t) = \sum_{k} \widetilde{h}(\bar{k}, t) e^{i\bar{k}\cdot\bar{r}} \tag{1}$$

where $\bar{k} = (k_x, k_y)$ is the 2D wave vector whose amplitude $k = \sqrt{k_x^2 + k_y^2}$ represents the wavenumber. $\bar{r} = (x, y)$ is the generic point and t is the time. The set of complex Fourier domain amplitudes and their initial phase values for the wave elevation field at time zero is defined as:

$$\widetilde{h_0}(\bar{k}) = \frac{1}{\sqrt{2}}(\xi_{Re} + i\xi_{Im}) \sqrt{S_{PM}(\bar{k})} \tag{2}$$

where ξ_{Re} and ξ_{Im} are ordinary independent draws from a Gaussian random number generator, with mean 0 and standard deviation 1 [27].

In the present paper, the spectral properties of the wave field are modelled by the Pierson-Moskowitz (PM) model, $S_{PM}(\bar{k})$, suitable for fully developed wind seas [26] and it is defined as follows:

$$S_{PM}(\bar{k}) = \frac{\alpha g^2}{k^4} e^{-\frac{5}{4}(k_p/k)} \tag{3}$$

where $\alpha = 0.0081$ is the Phillips constant, g is acceleration due to the gravity, and k_p is the peak wave number. This latter is directly related to the wind speed at a height of 10 m above the sea level, $\mathcal{W}_{10}$, by means of the expression $k_p = g/\mathcal{W}_{10}{}^2$.

Assuming linear wave theory, sea surface waves are dispersive and their propagation mechanism are governed by the dispersion relation given by

$$\omega(\overline{k}) = \sqrt{gk\tanh(kd)} + \overline{k}\cdot\overline{U} \tag{4}$$

where ω is the angular frequency, d is the water depth, and $\overline{U} = (U_x, U_y)$ is the surface current vector. Given this dispersion relation $\omega(\overline{k})$, the Fourier domain amplitudes of the wave elevation field at time t can be expressed as:

$$\widetilde{h}(\overline{k}, t) = \widetilde{h}_0(\overline{k})e^{i\omega(\overline{k})t} + \widetilde{h}_0{}^*(-\overline{k})e^{-i\omega(\overline{k})t} \tag{5}$$

This relation preserves the complex conjugation property $\widetilde{h}^*(\overline{k}, t) = \widetilde{h}(-\overline{k}, t)$ and is convenient because it will allows us to compute $\widetilde{h}(\overline{k}, t)$ and, hence, $h(\overline{r}, t)$ via the inverse 2D Fast Fourier Transform (FFT) on demand at any time t without having to compute it at any other time [27].

2.1.2. Radar Imaging

Simplified X-band radar imaging models were developed by References [2,28,29]. These works identify two main mechanisms, which are considered responsible for the modulation of the Radar Cross Section (RCS) for X-band marine radars including the shadowing and the tilt modulation. Herein, for the generation of the synthetic radar images, the model proposed by Nieto et al. [2] is adopted and briefly summarized below. Other sources of RCS modulation have been considered to be either negligible, as in the case of hydrodynamic modulation or pre-filtered from the received radar signal and in the case of the local wind velocity modulation.

A shadowing effect can be interpreted, to a first approximation, as a geometrical phenomenon. At low grazing angles, the wave free surface, in a few areas, prevents the illumination of some other parts of the wave field causing the shadowing of the nearby wave facets. As a result, the shadowing phenomenon can be summarized and modelled as if the radar antenna is not actually receiving any signal from the shadowed parts of the sea surface [2,20,30,31]. Moreover, this phenomenon will be more severe with increase of the distance from the radar antenna.

Since, in the real data, the shadowed area is not equal to zero, areas are modelled as Gaussian white noise. The simulated shadowed image $M_{sh}(\overline{r}, t)$ is given by the equation below.

$$M_{sh}(\overline{r}, t) = \begin{cases} 1 & if\ no\ shadowing\ occurs \\ noise & otherwise \end{cases} \tag{6}$$

Regarding the tilt modulation, the power received by the radar antenna depends on the local slope of the observed surface. This causes a modulation of the received radar signal, which depends on the angle between the radar illumination ray and the normal vector to the wave surface. The received radar signal is simulated by carrying out the scalar product between $\overline{u}(\overline{r}, t)$ and $\overline{n}(\overline{r}, t)$ defined as 3D vectors in a Cartesian axes system, which refers to the viewpoint of the radar and the normal vector to the surface, respectively. Therefore, the tilt simulated image $M_{tilt}(\overline{r}, t)$ is given by factor $\mathcal{T}(\overline{r}, t) = \overline{n}(\overline{r}, t)\cdot\overline{u}(\overline{r}, t)$.

$$M_{tilt}(\overline{r}, t) = \begin{cases} T(\overline{r}, t) & if\ \mathcal{T}(\overline{r}, t) > 0\ and\ M_{sh}(\overline{r}, t) \neq noise \\ noise & otherwise \end{cases} \tag{7}$$

A pictorial view of the tilt model and the shadowing effect is reported in Figure 1.

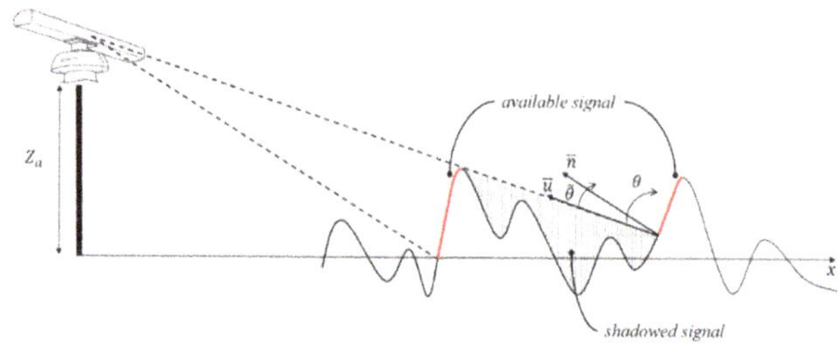

Figure 1. Geometrical scheme of the tilt model and shadowing effect.

2.2. Significant Wave Height Estimation Methodology

As previously mentioned, the shadowing phenomenon depends on both the H_s and the radar antenna height. Specifically, assuming given and known antenna height, the effect of shadowing on the radar image increases when the significant wave height grows. Consequently, the shadowing area contains intrinsic information of the H_s. Based on these assumptions, an algorithm that allows us to estimate H_s without an external reference was developed. In particular, the proposed algorithm is organized in three main steps whose flow diagram is shown in Figure 2.

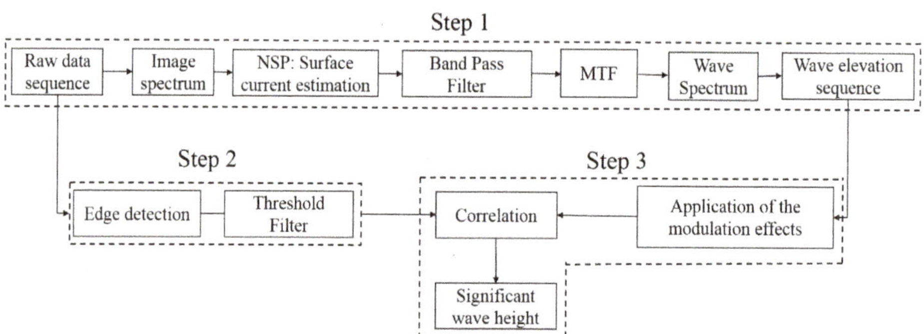

Figure 2. Data processing chain.

(1) Reconstruction of non-calibrated wave elevation by means of the inversion procedure [2].
(2) Estimation of the gray–level threshold in order to identify the areas affected by shadowing [21,22].
(3) Estimation of the significant wave height by means of the correlation between the raw radar image after the thresholding procedure (step 2) and the corresponding "reconstructed" radar image, which is achieved by applying the modulation effects to the non-calibrated wave elevation image retrieved during step 1.

2.2.1. Step 1: Inversion Procedure

The retrieval of the sea state parameters is carried out by exploiting the inversion procedure described by Nieto et al. 2004 [2].

Specifically, starting from the 3D Fourier Transform of the raw radar sequence, $I(\overline{r},t)$, the image spectrum $F(\overline{k},\omega)$ is achieved. In order to remove the non-stationary and non-homogenous trends in the image radar sequence, a High Pass filter is applied to the image spectrum and the spectrum $F_I(\overline{k},\omega)$ is obtained.

The second step of the procedure aims at extracting the linear components of the gravity waves from the image spectrum $F_I(\overline{k},\omega)$. To this aim, the dispersion relationship in Equation (4), which includes the parameters associated with the sea surface current and bathymetry, is used to build a Band Pass filter [32]. Therefore, the knowledge of the surface current $\overline{U} = (U_x, U_y)$ and bathymetry d

is a key point in the inversion procedure. In fact, an incorrect estimation of these parameters translates to an incorrect spectral filtering with a detrimental effect on estimating the sea state parameters. In this work, the normalized scalar product (NSP) [3] is used to estimate the bathymetry and the surface current. Once these parameters are estimated, it is possible to properly construct the band-pass filter (BP) $G(\overline{k},\omega,d,\overline{U})$ and to filter the image spectrum $F_I(\overline{k},\omega)$. The filtered image spectrum $\widetilde{F_I}(\overline{k},\omega)$ is not yet the desired sea wave spectrum because it is affected by the radar modulation effects. Thus, a modulation transfers function (MTF) is used to reduce these kinds of distortions. This function was empirically determined by Nieto et al. [2] and performed herein. From the wave spectrum $F_w(\overline{k},\omega)$, it is possible to determine the characteristic sea state parameters in terms of direction, length, period of the dominant waves, and the significant wave height. As far as H_s is concerned, this parameter is tied to the non-calibrated spectral moment of zero order m_0 by the following equation.

$$H_s = 4\sqrt{m_0} \tag{8}$$

The last step of the inversion algorithm is to establish the non-calibrated wave elevation sequence $\eta(\overline{r},t)$ as a function of the time and spatial variables by applying an inverse 3D FFT to the sea wave spectrum.

2.2.2. Step 2: Estimating Shadow Threshold

As mentioned in Section 2.1.2, the radar images are affected by the shadowing, for which no information can be recovered for the sea spots that are not in the line of sight (LOS). Moreover, at given and known radar antenna height Z_a, this phenomenon becomes more prominent when H_s increases and the distance from the radar antenna grows. Hence, in order to discriminate the shadowed areas from the wave pattern, an edge detection procedure is applied to the radar image $I(x,y)$. Such a technique involves the convolution of $I(x,y)$ with a pixel difference operator $D_n(x,y)$ for each of $n = 1, 2, \ldots, 8$ directions [21,22]. The kernel of the difference operator is given by

$$D_1(x,y) = \begin{matrix} 0 & 0 & 0 \\ 0 & 1 & -1 \\ 0 & 0 & 0 \end{matrix} \tag{9}$$

The operator takes a single kernel mask and rotates it in 45-degree increments through all eight compass directions: N, NW, W, SW, S, SE, E, and NE. Furthermore, the convolution outcome is given by:

$$I_{E_n}(x,y) = I(x,y) \bigotimes D_n(x,y) \tag{10}$$

To each edge image I_{E_n}, a threshold value equal to the upper N-percentile of the pixels is used and the thresholding of the edge image number n is given by

$$I_{E_n}(x,y) = \begin{cases} 1, & I_{E_n}(x,y) > \textit{upper N-percentile} \\ 0, & \textit{otherwise} \end{cases} \tag{11}$$

Subsequently, an overall edge image $I_T(x,y)$ is obtained by summing the eight edge images by the following two-step procedure.

$$I_T(x,y) = \sum_{n=1}^{8} I_{E_n}(x,y) \tag{12}$$

$$I_F(x,y) = \begin{cases} 1, & I_T(x,y) \in [1,\tau_F] \\ 0, & \textit{otherwise} \end{cases} \tag{13}$$

The filtering is implemented to remove the single pixel noise that has edges in more than τ_F directions. The pixels of intensity value of 1 in $I_F(x,y)$ can now be used to identify the pixels that are not in the shadow in the raw radar image.

The corresponding pixel values ρ from the radar image is used to create a histogram $F_D(\rho)$ of the statistics of image pixels located between the shadow and no shadow, from which the shadow threshold τ_s can be determined by means of the histogram mode as follows [14].

$$\tau_s = mode(F_D(\rho)) \tag{14}$$

The mode in Equation (14) is the data value that occurs the most in the considered radar image. The statistical distribution as well as the shadow threshold is determined for each radar image [22].

2.2.3. Step 3: Estimating Wave Height

The idea behind the estimation of the true (not scaled) significant wave height is to correlate the radar image after the thresholding procedure (step 2), $I_{th}(x,y)$, with a "reconstructed" radar image, $\widetilde{I}(x,y)$, achieved by applying the modulation effects, described in Section 2.1.2, to the corresponding reconstructed wave elevation image $\widetilde{\eta}(\overline{r},t)$. Additionally, $\widetilde{\eta}(\overline{r},t)$ is normalized through a dimensionless factor equal to the maximum of the considered sequence.

By assuming the radar antenna height Z_a is known, the correlation can be performed at a single time t of the considered $\widetilde{\eta}(\overline{r},t)$ by varying its amplitude through a scaling factor ε.

The best scaling factor $\hat{\varepsilon}$, which will allow us to calibrate $\widetilde{\eta}(\overline{r},t)$, can be estimated by looking for the maximum of the correlation function, i.e.,

$$\hat{\varepsilon} = \underset{\varepsilon}{\operatorname{argmax}} \frac{COV\left(I_{th}(x,y), \widetilde{I}(x,y)\right)^2}{Var(I_{th}(x,y)) \cdot Var\left(\widetilde{I}(x,y)\right)} \tag{15}$$

where $\widetilde{I}(x,y) = M_{sh,tilt}(\varepsilon \cdot \widetilde{\eta}(x,y))$ for each $\varepsilon = 0.2,\ 0.4, \ldots . \varepsilon^{max}$ and $I_{th}(x,y)$ is the thresholded radar image. Therefore, the scaling factor is used to calibrate the wave elevation sequence retrieved in step 1 and $\hat{H}_s$ is calculated by using Equation (8) from the corresponding calibrated spectral moment of zero order.

3. Results

In order to validate the proposed strategy, seven datasets were generated by considering a different peak period $\mathbf{T_p}$, peak wavelength $\boldsymbol{\lambda_p}$, significant wave height $\mathbf{H_s}$, and wind speed $\mathcal{W}_{10}$ (see Table 1 for the details). A deep-water condition and the absence of surface current are assumed (i.e., $\tanh(kd) = 1$ and $\overline{U} = 0\ m/s$ in Equation 4).

Table 1. Input parameters for the generation of simulated sea states.

Sea State	Sea State Parameters			
	H_s (m)	T_p (s)	λ_p (m)	$\mathcal{W}_{10}$ (m/s)
SS1	1.05	4.38	30.02	6.84
SS2	2.20	6.20	60.05	9.68
SS3	3.22	7.59	90.08	11.85
SS4	4.33	8.77	120.11	13.69
SS5	5.59	9.80	150.13	15.31
SS6	6.51	10.74	180.16	16.77
SS7	7.25	11.60	210.19	18.11

The time histories were generated considering a space-time grid corresponding to a typical configuration of a realistic wave radar device. Specifically, each dataset was constructed in a full circular area with an outer radius of 1004 m, and with the spatial spacing equal to $\Delta\bar{r} = 4m$ and $N_t = 128$ time instants with the interval between two successive images being equal to $\Delta t = 2s$. The main propagation direction for all considered sea states is from the north (0 degrees) to the south. The synthetic radar images have been generated by applying the shadowing and tilt modulation models reported in Equations (6) and (7), respectively. For all considered datasets, the radar antenna was located in the middle of the radar image with given height equal to $Z_a = 20\ m$, which corresponds to a typical real installation [11,12,20], above the mean sea level (see Figure 1). In the following, the images related to the sea state SS3 (see Table 1) at time instant $t_{64} = 128$ s are shown.

Figure 3a,b show the synthetic sea wave image, obtained by the Fourier domain approach described in Section 2.1.1, and the corresponding radar image, respectively.

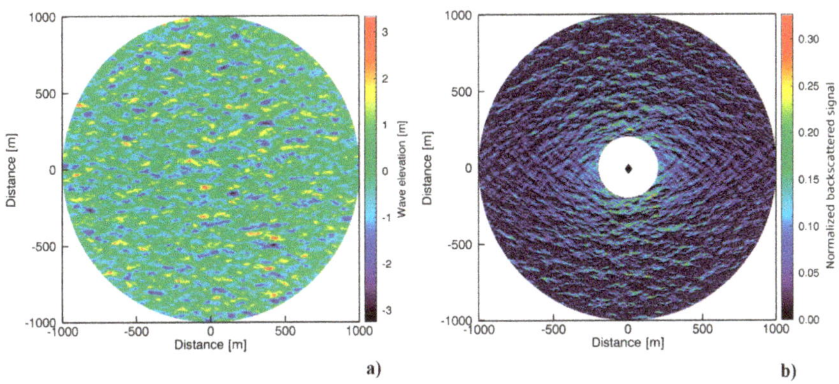

Figure 3. Images of the resulting synthetic data for sea state SS3. (**a**) The synthetized sea surface image. (**b**) The shadowed and tilt waves of the image at the same time instant ($t_{64} = 128s$).

For each considered dataset, the three steps of the proposed procedure, described in Section 2.2, are applied. In particular, step 1 provides the reconstruction of the non-calibrated wave elevation sequence (see Figure 4).

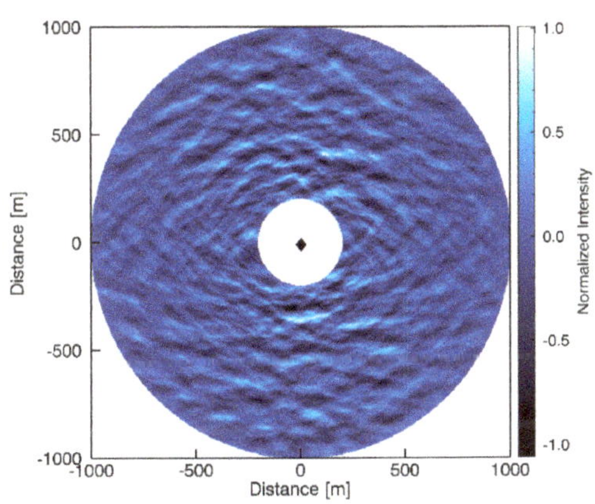

Figure 4. Non-calibrated reconstructed wave elevation image at the same instant of Figure 3a.

Afterward, in step 2, the edge detection technique is applied to the radar image to separate the shadow area from the remaining areas, according to the procedure described in Section 2.2.2.

Figure 5a shows the statistical distributions of the edge pixels (red circles) compared to the distributions of all image pixels (black circles). The shadow threshold is indicated by the dashed line.

The shadowed radar image, represented in binary scale, is obtained thresholding the radar image and is depicted in Figure 5b.

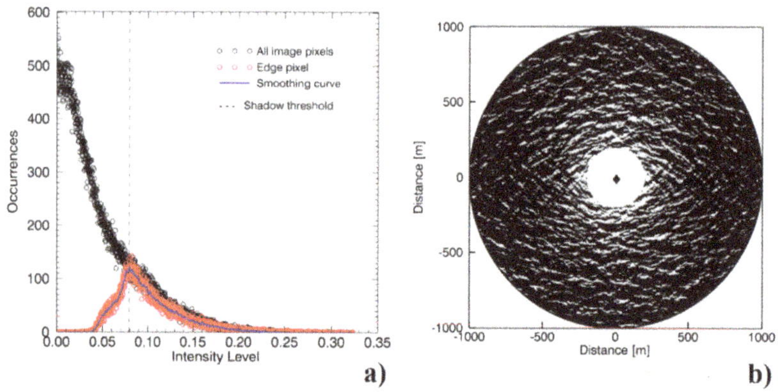

Figure 5. (**a**) Distribution of image pixels located between the shadow and no shadow (red circles) compared with distribution of all image pixels (black circles). The smoothing curve is indicated by a blue line. The shadow threshold is indicated by a dashed line. (**b**) Shadow image corresponding to the radar image in Figure 3b.

The modulation phenomena are then applied to the non-calibrated wave elevation image by varying its amplitude, which is correlated to the thresholded radar image (see Equation (14)). The behaviors of the R^2 of each considered dataset when varying the scaling factor ε are plotted in Figure 6a and $\hat{\varepsilon}$ is retrieved from their maximum value, which is used to calibrate the corresponding wave elevation sequence. Figure 6b shows the calibrated wave elevation image.

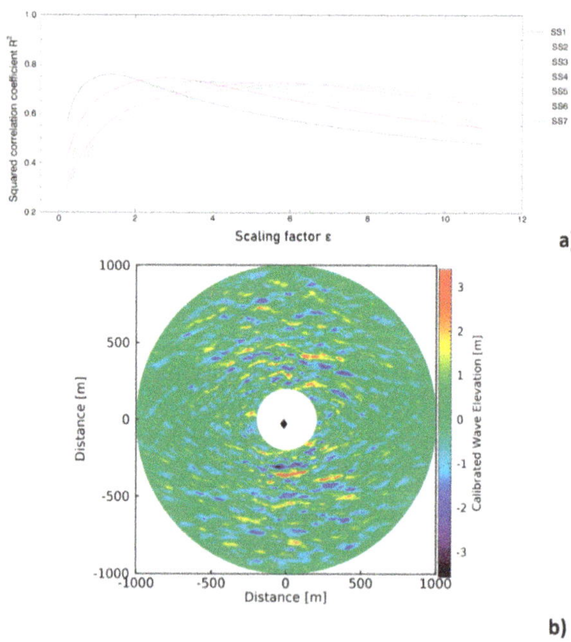

Figure 6. (**a**) Behaviours of the R^2 for each sea state when varying the amplitude of the normalized wave elevation image. (**b**) Calibrated wave elevation image at the same instant of Figure 3.

In order to evaluate the robustness of the proposed strategy, it is applied to 20-time instants of each considered sea state. Specifically, the radar and non-calibrated wave elevation images from time instant $t_{54} = 108$ s to $t_{73} = 146$ s are considered. The behavior of the estimated $\hat{H}_s$ for each time instant is depicted in Figure 7, from which it is possible to note that, with increasing significant wave height, there is more variability for the estimated values.

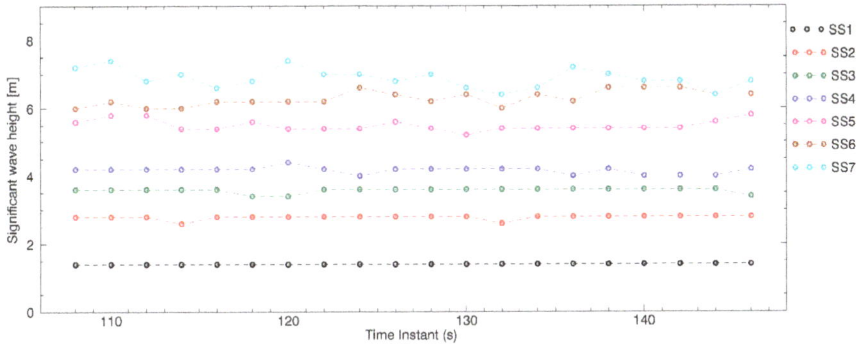

Figure 7. Estimated significant wave height for the considered 20 times instants for each sea state.

This trend can be explained in terms of the standard deviation and mean squared correlation coefficient that are depicted in Figure 8a,b, respectively. It is evident that the standard deviation increases and the squared correlation decreases with increasing $\hat{H}_s$.

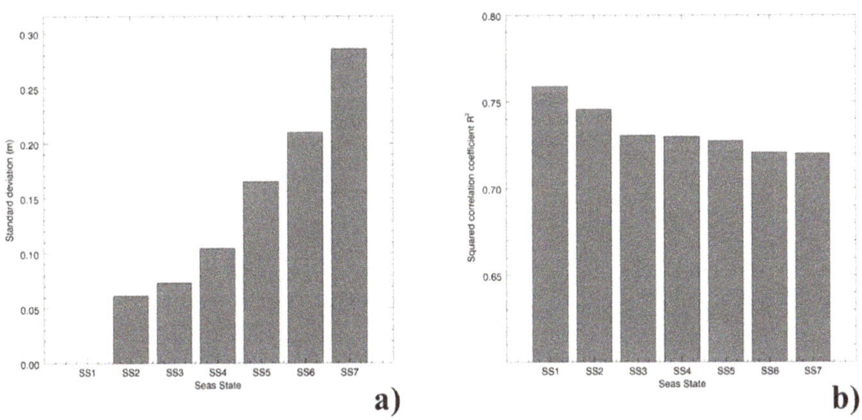

Figure 8. Statistical analysis. (**a**) Standard deviation of the considered 20 estimate significant wave height for each sea state. (**b**) Mean squared correlation coefficient of the considered 20-estimate significant wave height for each sea state.

At the end of the procedure, the significant wave height is calculated using Equation (8) and considering the calibrated moment of the zero order of the sea wave spectrum and each calibrated wave elevation sequence. In addition, in Figure 9, the generated and calibrated wave elevation profiles from the south to the north for each considered sea state are shown.

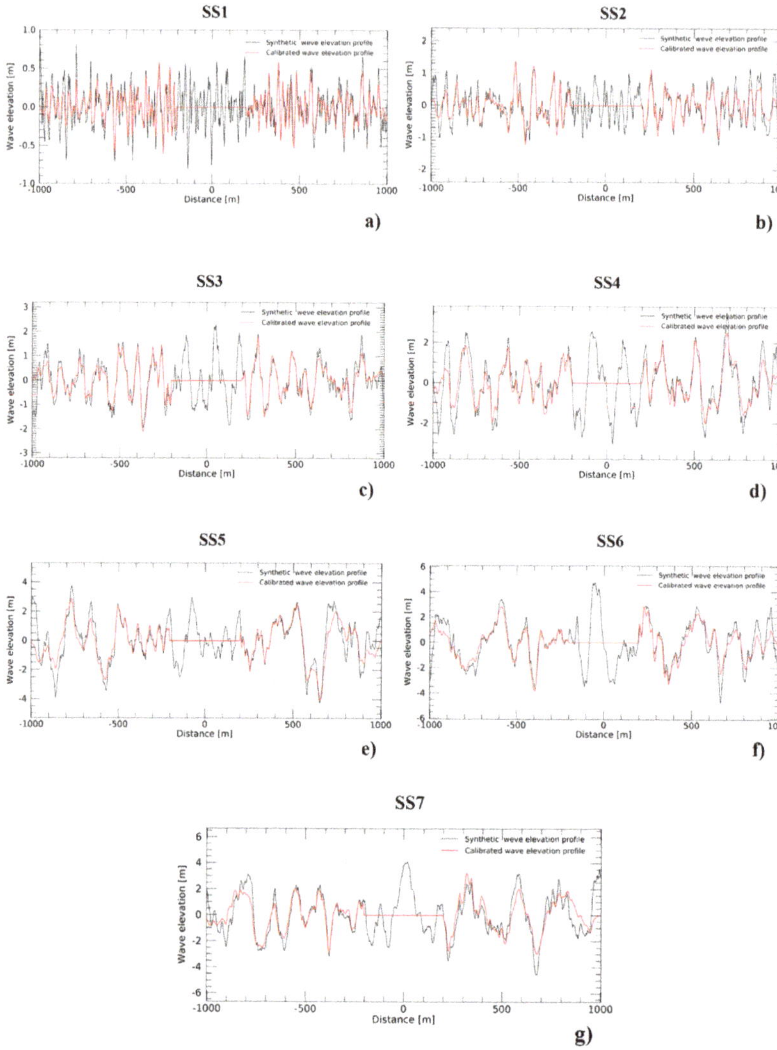

Figure 9. (**a–g**) Comparison between synthetic and calibrated wave elevation profiles from the south to the north for each considered sea state.

4. Discussion and Conclusions

The present work proposes a strategy for estimating the significant wave height from X-band radar images without the use of external sensors in order to calibrate the sea wave spectrum. Specifically, the strategy is based on the research of the maximum correlation, in time domain and at the same time instant, between a raw radar image and the "reconstructed" radar image.

To reach such a purpose, the wave field is generated through baseline simulations aimed at reproducing different wave conditions. In particular, herein, as a preliminary study of the proposed strategy, a fully developed sea state in the deep-water condition (PM spectrum) was adopted.

The radar imaging was performed by applying the modulation effects (i.e., shadowing and tilt modulation) to the wave images in order to recreate the radar data sequence. The inversion procedure described in Section 2.2.1 was applied and the non-calibrated wave elevation sequence was provided.

In order to perform the proposed strategy, two main steps were considered: i) a procedure based on the estimation of the gray-level threshold to discriminate the region of the image radar affected by the shadowing phenomenon, ii) application of the modulation effects to the corresponding non-calibrated wave elevation image when varying its amplitude by means of a scale factor. Such images are then exploited to estimate the $\hat{\varepsilon}$ value when the maximum of the correlation is reached. For each considered sea state, the proposed strategy provides a squared correlation coefficient value (see Figure 6a) greater than 0.70 and the estimated $\hat{\varepsilon}$ values were used to calibrate the corresponding wave elevation sequences $\widetilde{\eta}(\bar{r}, t)$.

Furthermore, it is noted that the behavior of R^2 decreases when increasing the $\hat{H}_s$ (see Figure 8b). This behavior is likely related to two phenomena that affect the reconstruction of the calibrated wave elevation sequence: i) the sea state monitoring by means of the wave radar system deals with a finite-length time series. By performing the Fast Fourier Transform, some errors, named edge effects, occur at the beginning and at the end of the calibrated wave elevation (Figure 9); ii) the shadowing phenomenon becomes more prominent when increasing the distance from the radar antenna [33].

Lastly, the present work can be considered as a proof of concept for estimating the significant wave height without the use of an external sensor, which allowed us, then, to avoid the operation of calibration during the installation of a wave radar system.

The application of the proposed approach on real-world data is expected in the near future. Lastly, the refinement of the method and testing it in various sites is required.

Author Contributions: Conceptualization, F.S. Methodology, F.S. Software, G.L. Validation, G.L. Formal analysis, G.L. Investigation, G.L. Writing—original draft preparation, G.L. Writing—review and editing, G.L. and F.S.

Funding: Please add: This research received no external funding.

Conflicts of Interest: The authors declare no conflict of interest.

References

1. Young, R.; Rosenthal, W.; Ziemer, F. Three-dimensional analysis of marine radar images for the determination of ocean wave directionality and surface currents. *J. Geophys. Res.* **1985**, *90*, 1049–1059. [CrossRef]
2. Nieto Borge, J.C.; Rodriguez, G.R.; Hessner, K.; Gonzales, P.I. Inversion of marine radar images for surface wave analysis. *J. Atmos. Ocean. Technol.* **2004**, *21*, 1291–1300. [CrossRef]
3. Serafino, F.; Lugni, C.; Soldovieri, F. A novel strategy for the surface current determination from marine X-Band radar data. *IEEE Geosci. Remote Sens. Lett.* **2010**, *7*, 231–235. [CrossRef]
4. Senet, C.M.; Seemann, J.; Ziemer, F. The near-surface current velocity determined from image sequences of the sea surface. *IEEE Trans. Geosci. Remote Sens.* **2001**, *39*, 492–505. [CrossRef]
5. Shen, C.; Huang, W.; Gill, E.W.; Carrasco, R.; Horstmann, J. An Algorithm for Surface Current Retrieval from X-band Marine Radar Images. *Remote Sens.* **2015**, *7*, 7753–7767. [CrossRef]
6. Gangeskar, R. Ocean current estimated from X-band radar sea surface, images. *IEEE Trans. Geosci. Remote Sens.* **2002**, *40*, 783–792. [CrossRef]
7. Bell, P.S.; Osler, J.C. Mapping bathymetry using X-band marine radar data recorded from a moving vessel. *Ocean. Dyn.* **2011**, *61*, 2141–2156. [CrossRef]
8. Ludeno, G.; Orlandi, A.; Lugni, C.; Brandini, C.; Soldovieri, F.; Serafino, F. X-band marine radar system for high-speed navigation purposes: A test case on a cruise ship. *IEEE Geosci. Remote Sens. Lett.* **2014**, *11*, 244–248. [CrossRef]
9. Lund, B.; Graber, H.C.; Hessner, K.; Williams, N.J. On shipboard marine X-band radar near-surface current "calibration". *J. Atmos. Ocean. Technol.* **2015**, *32*, 1928–1944. [CrossRef]
10. Lund, B.; Collins, C.O.; Graber, H.C.; Terrill, E.; Herbers, T.H. Marine radar ocean wave retrieval's dependency on range and azimuth. *Ocean. Dyn.* **2014**, *64*, 999–1018. [CrossRef]
11. Raffa, F.; Ludeno, G.; Patti, B.; Soldovieri, F.; Mazzola, S.; Serafino, F. X-band wave radar for coastal upwelling detection off the southern coast of Sicily. *J. Atmos. Ocean. Technol.* **2017**, *34*, 21–31. [CrossRef]
12. Ludeno, G.; Brandini, C.; Lugni, C.; Arturi, D.; Natale, A.; Soldovieri, F.; Gozzini, B.; Serafino, F. Remocean system for the detection of the reflected waves from the costa concordia ship wreck. *IEEE J. Sel. Top. Appl. Earth Obs. Remote Sens.* **2014**, *7*, 3011–3018. [CrossRef]

13. Senet, C.M.; Seemann, J.; Flampouris, S.; Ziemer, F. Determination of bathymetric and current maps by the method DiSC based on the analysis of nautical X-band radar image sequences of the sea surface (November 2007). *IEEE Trans. Geosci. Remote Sens.* **2008**, *46*, 2267–2279. [CrossRef]
14. Nieto-Borge, J.C.; Hessner, K.; Jarabo-Amores, P.; La Mata-moya, D.D. Signal-to-noise ratio analysis to estimate ocean wave heights from X-band marine radar image time series. *IET Radar Sonar Navig.* **2008**, *2*, 35–41. [CrossRef]
15. Gangeskar, R. Wave height derived by texture analysis of X-band radar sea surface images. In Proceedings of the IEEE 2000 IGARSS, Honolulu, HI, USA, 24–28 July 2000; Volume 7, pp. 2952–2959.
16. Wetzel, L.B. Electromagnetic scattering from the sea at low grazing angles. In *Surface Waves and Fluxes;* Kluwer: Boston, MA, USA, 1990; pp. 109–172.
17. Buckley, J.R.; Aler, J. Enhancements in the determination of ocean surface wave height from grazing incidence microwave backscatter. In Proceedings of the 1998 IEEE IGARSS, Seattle, WA, USA, 6–10 July 1998; Volume 5, pp. 2487–2489.
18. Huang, W.; Gill, E.W.; An, J. Iterative least-squares-based wave measurement using X-band nautical radar. *IET Radar Sonar Navig.* **2014**, *8*, 853–863. [CrossRef]
19. An, J.; Huang, W.; Gill, E.W. A Self-Adaptive Wavelet-Based Algorithm for Wave Measurement Using Nautical Radar. *IEEE Trans. Geosci. Remote Sens.* **2015**, *53*, 567–577.
20. Wijaya, A.P.; van Groesen, E. Determination of the significant wave height from shadowing in synthetic radar images. *Ocean Eng.* **2016**, *114*, 204–215. [CrossRef]
21. Gangeskar, R. An Algorithm for Estimation of Wave Height from Shadowing in X-Band Radar Sea Surface Images. *IEEE Trans. Geosci. Remote Sens.* **2014**, *52*, 3373–3381. [CrossRef]
22. Liu, X.; Huang, W.; Gill, E.W. Wave height estimation from ship-borne X-band nautical radar images. *J. Sens.* **2016**. [CrossRef]
23. Navarro, W.; Velez, J.C.; Orfila, A.; Lonin, S. A Shadowing Mitigation Approach for Sea State Parameters Estimation Using X-Band Remotely Sensing Radar Data in Coastal Areas. *IEEE Trans. Geosci. Remote Sens.* **2019**, *57*, 6292–6310. [CrossRef]
24. Smith, B.G. Geometric shadowing of a random rough surface. *IEEE Trans. Antennas Propag.* **1967**, *15*, 668–671. [CrossRef]
25. Mastin, G.; Watterberg, P.; Mareda, J. Fourier Synthesis of Ocean Scenes. *IEEE Comput. Graph. Appl.* **1987**, *7*, 16–23. [CrossRef]
26. Pierson, W.J., Jr.; Moskowitz, L. A proposed spectral from for fully developed wind seas based on the similarity theory of SA Kitaigorodskii. *J. Geophys. Res.* **1964**, *69*, 5181–5190. [CrossRef]
27. Tessendorf, J. Simulating Ocean Water. *SIGGRAPH* **2001**, *1*, 5.
28. Naaijen, P.; Wijaya, A.P. Phase Resolved Wave Prediction from Synthetic Radar Images. In Proceedings of the ASME 2014 33rd International Conference on Ocean, Offshore and Arctic Engineering, San Francisco, CA, USA, 8–13 June 2014; Volume 8A. Ocean Engineering.
29. Dankert, H.; Rosenthal, W. Ocean surface determination from X-band radar-image sequences. *J. Geophys. Res.* **2004**, *109*. [CrossRef]
30. Ludeno, G.; Postacchini, M.; Natale, A.; Brocchini, M.; Lugni, C.; Soldovieri, F.; Serafino, F. Normalized Scalar Product Approach for Nearshore Bathymetric Estimation from X-Band Radar Images: An Assessment Based on Simulated and Measured Data. *IEEE J. Ocean. Eng.* **2018**, *43*, 221–237. [CrossRef]
31. Postacchini, M.; Ludeno, G. Combining Numerical Simulations and Normalized Scalar Product Strategy: A New Tool for Predicting Beach Inundation. *J. Mar. Sci. Eng.* **2019**, *7*, 325. [CrossRef]
32. Nieto Borge, J.C.; Guedes Soares, C. Analysis of directional wave fields using X-band navigation radar. *Coast. Eng.* **2000**, *40*, 375–391. [CrossRef]
33. Fucile, F.; Ludeno, G.; Serafino, F.; Bulian, G.; Soldovieri, F.; Lugni, C. Some Challenges in Recovering Wave Features From a Wave Radar System". In Proceedings of the Annual International Offshore and Polar Engineering Conference (ISOPE), Rhodes (Rodos), Greece, 26 June–1 July 2016.

Journal of
Marine Science and Engineering

MDPI

Article

Synergistic Use of Synthetic Aperture Radar and Optical Imagery to Monitor Surface Accumulation of Cyanobacteria in the Curonian Lagoon

Francesca De Santi [1,*], Giulia Luciani [1,2], Mariano Bresciani [1], Claudia Giardino [1], Francesco Paolo Lovergine [3], Guido Pasquariello [3], Diana Vaiciute [4] and Giacomo De Carolis [1]

1 Institute for Electromagnetic Sensing of the Environment (IREA), National Research Council (CNR) of Italy, 20133 Milan, Italy; luciani.g@irea.cnr.it (G.L.); bresciani.m@irea.cnr.it (M.B.); giardino.c@irea.cnr.it (C.G.); giacomo.decarolis@cnr.it (G.D.C.)
2 Department of Civil and Environmental Engineering, Politecnico di Milano, 20133 Milan, Italy
3 Institute for Electromagnetic Sensing of the Environment (IREA), National Research Council (CNR) of Italy, 70126 Bari, Italy; lovergine.f@irea.cnr.it (F.P.L.); pasquariello.g@irea.cnr.it (G.P.)
4 Marine Research Institute, Klaipeda University, LT-92294 Klaipeda, Lithuania; diana.vaiciute@jmtc.ku.lt
* Correspondence: desanti.f@irea.cnr.it

Received: 30 October 2019; Accepted: 10 December 2019; Published: 14 December 2019

Abstract: Phytoplankton blooms in internal water bodies are an unpleasant sight that often emerges on top like a layer of foam containing high concentrations of toxins (scum event). Monitoring the concentration of algae and the occurrence of scum in lakes and lagoons has become a topic of interest for management and science. Optical remote sensing is a validated tool but unfortunately it is highly hindered by clouds. For regions with frequent cloud cover, such as the Baltic region, this means loss of data, which limits the purpose of sensing to spatially and temporally characterize any scum for a comprehensive ecological analysis. The aim of this paper is to investigate whether the use of synthetic aperture radar (SAR) images can compensate for the weaknesses of optical images for cyanobacteria bloom monitoring purposes in the event of cloudy skies. A "ready to use" approach to detect cyanobacteria bloom in the Curonian Lagoon based on the level 2 ocean product of Sentinel-1 images is proposed. This method is empirically validated for the images of summer/autumn 2018 of the Curonian Lagoon.

Keywords: scum; hypertrophic ecosystem; Sentinel-1; Sentinel-2; Sentinel-3; cloudiness

---

1. Introduction

Cyanobacteria, also known as blue-green algae, are aquatic bacteria affecting inland and coastal communities around the world and are some of the oldest living organelles on the Earth. Cyanobacteria can be found in many different environments such as freshwater and marine ecosystems. A mature blue-green algae bloom often surfaces atop as a layer of scum which is an unpleasant sight with potentially serious economic and health-related impacts [1].

In recent years, blooms of cyanobacteria have been increasing in magnitude, frequency and in geographical spread, due to climate change and human activity [2]. Warmer temperatures indeed favor cyanobacteria that generally grow better at higher temperatures than do other phytoplankton species [3]. Moreover, the warmer is the water surface, the steepest is the vertical stratification of inland waters, with a consequent reduction of vertical mixing layer. Many cyanobacteria take advantage of these stratified conditions by forming intra-cellular gas vesicles, which make the cells buoyant and able to accumulate in dense surface blooms [4]. Furthermore, global warming has extended the optimal growth period since water column stratifies earlier in spring and destratifies later in autumn [5].

Therefore, on top of water temperature, factors like turbulence levels, nitrogen fixation, long day-light and the buoyant ability of cyanobacterial cells play their part in the occurrence of scums [6].

While further understanding of biological and physical growth dynamics of various algae species is still ongoing, water managers are interested in monitoring the state of lakes and lagoons to be able to raise a red flag in time. In-situ data is the most widely used data-set for monitoring inland water bodies such as lakes and canals. These data-sets provide highly accurate, precise and in depth information on the state of a location at a point in time but are limited to few local points and their maintenance often reduces the in-situ coverage of a water body.

Satellite data have the advantage of taking frequent in time acquisitions and of providing a full spatial coverage at a coarse level, which allows trend detection in space and time. The Copernicus program, through Sentinel satellites constellations, ensures free, full and open access Earth Observation data availability for at least 20 years to come. Continuous, large scale studies of inland water ecosystems, biogeochemical processes and essential climate change variables can be performed for the very first time and with a spatial resolution suitable even for small lakes. These studies have the potential to tackle global scientific issues and to be a resource of information for management purposes [7]. Among the satellites of the Sentinels constellation, the recently launched Sentinel-1 (S1), Sentinel-2 (S2) and Sentinel-3 (S3) have the potential to provide information on bio-optical and physical properties of the waters.

Optical remote sensing is a validated tool for sensing, monitoring and developing better understanding of the state phytoplankton blooms from inland for example, References [8–10], to oceanic waters for example, Reference [11], successful applications in different study areas such as the Baltic Sea for example, References [12–15].

However, it is highly hindered by clouds. For regions with frequent cloud cover (such as the Baltic region) this means lack of data, which derails the purpose of sensing and makes difficult to spatially and temporally characterize any scum for a comprehensive ecological analysis. Moreover, cloudy conditions often favor the growth and aggregation of cyanobacteria on the surface since the consequent light reduction leads the algae to have excess buoyancy [16]. Therefore, the lack of data becomes even more serious with respect to the monitoring and management of aquatic environments and understanding of ecological phenomena in progress [17,18].

The weaknesses of optical images for cyanobacteria bloom monitoring purposes in the event of cloudy skies could be compensated by Synthetic-aperture radar (SAR) observations given the ability of the microwaves radiation to penetrate through cloud cover independent of weather conditions. It is indeed expected that cyanobacterial blooms lead to significant attenuation of SAR signal since they act as a viscous lid covering the water surface. To assess whether and to what extent microwaves can detect scum events is the aim of the present study.

A first step in this direction has been made by Svejkoky and Shandley [19], which integrated AVHRR and ERS-1 data to analyze and interpret a specific phytoplankton bloom phenomenon verified in the coastal area of California. In particular, the authors exploited the finer SAR spatial resolution to better identify the spatial distribution of the bloom quantified from the optical image.

Thereafter, Bresciani et al. [20] observed a significant decrease of the SAR signal for high values of chlorophyll-a cyanobacteria concentration for a wind speed between 2 and 6 m/s, by using data gathered during the Envisat mission from optical (MERIS) and microwave (ASAR) synchronous sensors. These findings support the use of microwave remote sensing as an additional tool to increase the temporal coverage of optical sensors. However, the use of radar alone to monitor algal blooms would be challenging. Indeed it is in general difficult to distinguish whether the decrease of the SAR signal is created by algal blooms or by other causes without additional information. On the other hand, optical/SAR integration might help differentiate natural slicks from those caused by oil or human activity.

Unfortunately, with the end of the Envisat mission, synchronous acquisition of optical and microwaves images are not longer available making difficult to rigorously improve the methodology to detect cyanobacteria bloom from SAR. Therefore, this paper aims to propose a ready to use approach to assign the presence of cyanobacteria blooms on empirical basis by using the Level 2 (L2) product of the SAR satellite Sentinel 1.

A good candidate as case study for the application of the proposed approach is the Curonian Lagoon, located in the South-Eastern Baltic Sea region. Indeed, while S1-L2 products are generally available only for open ocean, the closeness of the lagoon to the Baltic Sea makes that an exception. Moreover, cyanobacteria blooms occurring in the Curonian Lagoon have been widely investigated with dedicated field campaigns [21] and the optical water properties of the Curonian Lagoon in the presence of cyanobacteria have been ascertained [22,23].

2. The Curonian Lagoon

The Curonian Lagoon is the largest European lagoon. It is a shallow water body situated in the southern part of the Baltic Sea with a total area of 1584 km^2, a mean depth of 3.8 m and a total volume of water approximately of 6.2 km^3, see Figure 1.

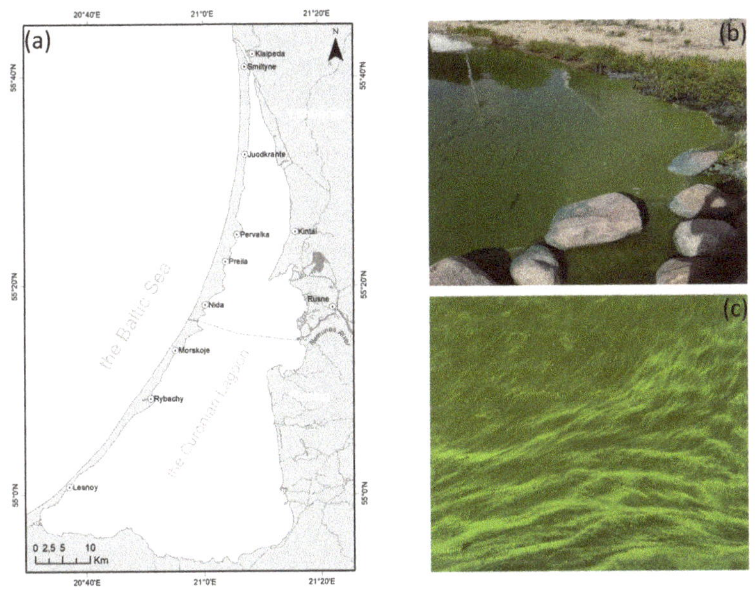

Figure 1. (**a**) Study area: the Curonian Lagoon; (**b**,**c**) examples of scum phenomena formed by cyanobacteria on the water surface of the Curonian Lagoon.

The Curonian Lagoon is a highly dynamic and complex ecosystem with many interacting processes. It is an open system, influenced by the exchange of the fresh Nemunas river water and brackish water of the Baltic sea. Indeed, it is separated from the Baltic by a narrow stretch of sandy beaches and woodland (0.5–4.0 km wide) called the Curonian Spit and is connected to Baltic's water thought the Klaipeda Strait. Water salinity in the northern part of the lagoon may fluctuate between 0.1–7 psu and is representatives of marine, brackish and fresh-water species live. The southern part is constantly freshwater and characterized by a relatively closed water circulation with lower current velocities, where the wind is the main driver [24,25].

Pollution from a combination of shipping, military and industrial sources is heavily affecting the Lagoon influencing local tourism and economics. Untreated sewage also have increased the pathogenic organisms loads [26]. Not surprisingly, fishing and bathing in the lagoon have declined significantly in modern times. Moreover, the Lagoon receives water from the River Nemunas, that together with agriculture and urban activities is a major sources of phosphorus (P) and nitrogen (N), contributing significantly to the eutrophication processes underway [27].

A peak of nutrient concentrations is generally observed in winter and early spring, in correspondence of diatom blooms (mainly *Asterionella formosa*, *Aulacoseira islandica*, *Stephanodiscus* spp., *Diatoma* spp.), which lead to a decreasing of nitrate concentrations. This phenomenon and the associated rapid regeneration of phosphorus compounds create a favorable environment for cyanobacteria blooms in terms of N/P ratio [28], mainly involving *Aphanizomenon flos-aquae*, *Planktothrix* sp. and *Microcystis* spp. The *A. flos-aquae*, the most dominant cyanobacteria, are one of those containing microscopic vesicles and therefore are able to regulate their buoyancy and to drive the accumulations on the water surface [18]. Starting from late June and early July and lasting until the end of October, cyanobacteria species regularly become prevalent during which time they can bloom intensively [29,30].

In addition to the exceptional availability of the second level products of Sentinel-1, the Curonian Lagoon is a perfect case study for this research because is highly hindered by clouds. The effective presence and impacts of clouds can be exploited from Mercury et al. [31], who built a composite of MODIS-Terra 2001–2011 providing the global percentage of cloudiness for the 10-year period. It is found that the Curonia Lagoon has an yearly cloud coverage of the 76%. This percentage reduce to the 64% in the summer time. These values are not too far from what observed in the present study, indeed among the 150 optical acquisitions in the time interval considered (summer-autumn 2018), 90 are cloud covered.

3. Materials and Methods

Satellite images of Sentinel missions acquired between July and October 2018 are considered for this study. The Sentinel missions are a group of five satellites part of the European Space Agency (ESA) Copernicus program, for the observation and monitoring of the Earth's surface, and the development of operational applications for environmental monitoring. All data gathered by the Sentinel satellites are provided for free, full and open access to users. More specifically, data from three of the Sentinel missions are used in this study.

- Sentinel-1 (S1), operating all-weather, day and night and performing C-band synthetic aperture radar imaging, enabling them to acquire imagery regardless of weather condition with a spatial resolution of 10 m.
- Sentinel-2 (S2) with the on-board Multispectral Instrument (MSI) provides high-resolution optical imaging over land and coastal waters. It measures the Earth's reflected radiance in 13 spectral bands, at visible and mid-infrared wavelengths and at various spatial resolutions (10, 20, 60 m).
- Sentinel-3 (S3) that makes use of multiple sensing instruments, of which data acquired by OLCI (Ocean and Land Colour Instrument) are used in this study. It is a medium-resolution imaging spectrometer with 21 spectral bands with wavelengths ranging from the optical to the near-infrared at approximately 300 m.

All of these missions consist of a constellation of two twin polar-orbiting satellites.

3.1. Optical Images (Sentinel-2/MSI and Sentinel-3/OLCI)

Within this study 33 S3/OLCI and 27 S2/MSI images acquired from July and October 2018 have been analyzed. The 60 optical images are analyzed to map chlorophyll a (Chl-a) concentrations using two different band ratio algorithms applied to atmospherically-corrected data. Given the very low transparency typical of the waters of the Curonian Lagoon [32] and the tendency of algae dispersed in the water column to accumulate near the surface, satellite-based estimates of Chl-a concentrations are presumably relevant only for the upper layer of water and thus can be considered as a proxy of the scum presence. Indeed the shallow, weakly stratified lagoon remains very turbid due to solids (mainly sand and silt) being resuspended as a result of water mixing caused by local winds and intensive primary production [33,34]. The Secchi disk depth varies from 0.3 to 2.2 m [29] and is always relatively low, so it is impossible to see the bottom at this shallow optical depth.

The atmospheric correction is carried out using the Second Simulation of the Satellite Signal in the Solar Spectrum-Vector code (6SV) [35] previously used in other satellite applications for the Curonian Lagoon [20]. The parametrization of 6SV code in case of the Curonian Lagoon is performed with Aerosol Optical Thickness (AOT) values gathered from AERONET station Gustav Dalen Tower located in the Baltic Sea and with water vapour values retrieved from daily MODIS products, via NASA Giovanni interface [36]. Chl-a concentration is produced after the application of a semi-empirical band-ratio model that uses the reflectance in the red and near-infrared (NIR) spectral regions [21,37]. The scum presence is calculated according to the approach by Bresciani et al. [20], with the ratio between reflectances in the near infrared (NIR) range. As indicated in INFORM D5.15 [38] (p. 140), Equation (1) is used for MSI

$$\text{Chl-a}\left[\frac{\text{mg}}{\text{m}^3}\right] = (76.36 \pm 2.29)\frac{\text{Ref}_{705}}{\text{Ref}_{665}} - (51.57 \pm 0.26), \quad (1)$$

and Equation (2) for OLCI

$$\text{Chl-a}\left[\frac{\text{mg}}{\text{m}^3}\right] = (52.19 \pm 1.81)\frac{\text{Ref}_{708}}{\text{Ref}_{665}} - (32.07 \pm 0.57), \quad (2)$$

where Ref_x indicates the reflectance of the band with central wavelength x.

3.2. SAR Images

The Synthetic Aperture Radar (SAR) imagery, equipped with dedicated models and analysis tools, can provides information related to the water-surface, such as wind waves, currents, bathymetry, biological activity and precipitation. To first order, ocean SAR backscatter depends on the wind-generated short waves. Therefore, to estimate the wind-related impact on the SAR image is mandatory to interpret features such as the occurrence of phytoplankton bloom, oil spill, current fronts, surface eddies and others. In this purpose, ESA processes and delivers a Level-2 (L2) ocean product (OCN) designed to provide geophysical parameters related to wind, waves and surface velocity to the users. All acquisitions over the Curonia Lagoon are processed up the Level-2. The ocean wind component of the L2OCN product is available for all acquisition modes. For this study, only the geophysical parameters relied on L2OCN products acquired in interferometric wide (IW) swath mode are considered.

In the L2OCN product, the normalised radar cross section (NRCS) is used to derive the 10 m ocean surface winds using the geophysical model (GMF) called CMOD-IFR2 [39]. Since SAR is a single fixed antenna instrument, it cannot estimate both wind speed and direction and some a priori wind information is needed. ESA takes the additional wind vector information from the ECMWF (European Centre for Medium-Range Weather Forecasts) atmospheric model. To retrieve the ocean surface wind, radar parameters (NRCS values, incidence angle and track angle) and the a priori wind information are combined into a Bayesian scheme, described by Kerbaol et al. [40]. Although SAR wind retrievals are available for VV and HH polarization, only VV related winds are used throughout the paper. Both the a priori ECMWF wind vector and the related SAR retrieved winds are stored in the L2OCN netcdf files.

This wind inversion procedure has been successfully validated and applied to open ocean images. However, less work has been carried out for closed or semi-enclosed water basins, like the Curonia Lagoon, where the geographical boundaries may affect air–sea processes and sea states and wind regimes are controlled by short temporal and spatial fetches.

Figure 2a shows the scatter-plot between the *a priori* wind velocity given by ECMWF, w_{ECMWF} and the related SAR retrieved wind velocity, w_{S1}, for 35 S1 images between 1 May 2019 and 30 June 2019, when no scum events were observed. As the w_{S1} is processed from w_{ECMWF}, these two velocities have to clearly be correlated by construction. However, in Figure 2a it is evident that other features can affect SAR backscatter to the first order. In this study we are not interested to investigate all these features but rather to discuss the impact of the presence of phytoplankton blooms on water surface.

In Bresciani et al. [20], it is shown that phytoplankton blooms at water surface acts as a viscous layer by reducing the gravity-capillarly waves generation and propagation. From the SAR sensor point of view, scum phenomena are characterized by a reduced backscattered energy. Indeed, a drastic decrease in the NRCS (with attenuation greater than 8–10 dB) for Chl-a concentrations higher than 50 mg/m^3 and wind speeds in the range of 2 to 6 m/s is observed. This NRCS attenuation in turn leads to a retrieved SAR wind speed significantly smaller in respect to the ECMWF a priori guess.

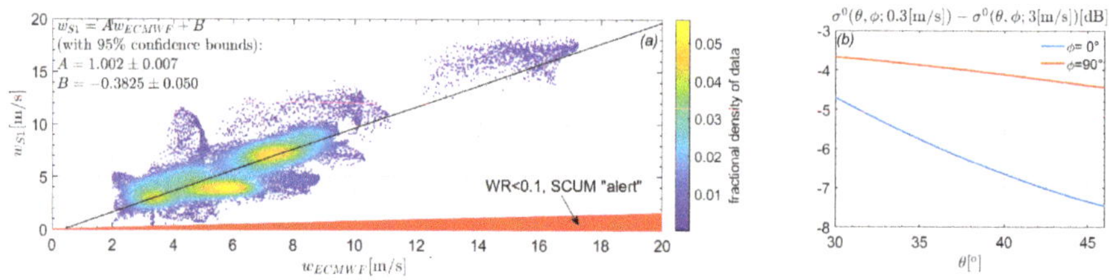

Figure 2. (**a**) Scatter-density plot between the a-priori European Centre for Medium-Range Weather Forecasts (ECMWF) wind velocity, w_{ECMWF} and the wind velocity estimated from Sentinel 1, w_{S1}, from L2OCN products of internal points of the Curonian lagoon. Wind data showed are from 1 of May 2019 and 30 of June 2019, when no scum events were observed. The condition WR < 0.1, which sets empirically a "scum" alert, is colored in red ($R^2 = 0.72$, rmse = 1.7589); (**b**) Attenuation of the backscattering for WR = 0.1 as a function of the incidence angle θ. For data showed in panel (**a**), $\theta \in [30°, 46°]$. Effect of wind direction ϕ is also showed, where $\phi = 0°$ corresponds to wind blowing towards the radar.

The L2OCN data displayed in Figure 2a refer to a data-set where no phytoplankton bloom (or ice) events have been registered and that is independent from the one considered in Section 4 for results interpretation.

Empirically a "scum alert" can be raised when the retrieved wind is lower than the worst case observed in a no-scum calibration data-set. This relation can be expressed as

$$\mathrm{w}_{S1} < \mathrm{w}_{\mathrm{fit}} - \mathrm{err}_{max},$$

where

$$\mathrm{w}_{\mathrm{fit}} = 1.002\,\mathrm{w}_{ECMWF} - 0.3825$$

is obtained by Gaussian fit as shown in Figure 2a and

$$\mathrm{err}_{max} = \max\left\{\frac{\mathrm{w}_{\mathrm{fit}} - \mathrm{w}_{S1}}{\mathrm{w}_{\mathrm{fit}}}\ \text{s. t. } \mathrm{w}_{S1} < \mathrm{w}_{\mathrm{fit}}\right\}.$$

Notice that condition $\mathrm{w}_{S1} < \mathrm{w}_{\mathrm{fit}}$ refines the range as only damping effects needs to be considered. For the data-set in Figure 2a, $\mathrm{err}_{max} = 0.8993$.

Let us define the ratio

$$\mathrm{WR} = \frac{\mathrm{w}_{S1}}{\mathrm{w}_{ECMWF}} \qquad (3)$$

to monitor phytoplankton bloom from SAR, where w_{S1} is the wind speed estimated from Sentinel 1 and w_{ECMWF} indicates the a-priori ECMWF wind speed. As mentioned before, there are different factors able to significantly damp the backscatter but the knowledge of the lagoon's characteristics and the seasonality of the events allow us to attribute the smallness of WR most likely to the presence of cyanobacteria's bloom at the water surface.

Defining an index that does not come directly from the amplitude of the SAR signal, but rather from a derived product such as the retrieved wind, is due to the strong dependence of the NRCS on the incidence angle θ. In Figure 2b it is shown how the condition WR $= 0.1$ is read in term of σ^0 for incidence angle values characteristic of S1 images for the Curonian Lagoon ($\theta \in [30^\circ, 46^\circ]$) and different wind blowing direction. The use of an ocean-backscatter model such as CMOD allows to simply remove the incidence-angle dependence of the σ^0 and therefore to enable the estimation of phytoplankton bloom statistics in SAR images.

The *a priori* wind velocity also sets the applicability limits of this approach: for ECMWF wind speed below 2 m/s a weak backscattering area is likely due to a lack of wind than to the presence of floating substance on water surface. Therefore, this method can not be applied when w_{ECMWF} falls in this range.

4. Results

As mentioned in Section 1, given the lack of a continuous and dedicated field campaign and the unavailability of synchronous acquisition of optical and radar images, a rigorous validation for the proposed method cannot be performed. Indeed, the dependencies of cyanobacteria on solar radiation and on air temperature and their ability to consequently perform vertical migration via buoyancy regulation makes their concentration highly variable within the time span of a daily cycle [16,21,41]. It is however viable to assess whether and to what extent the condition WR < 0.1 can be adopted as "SCUM alert" in terms of qualitative interpretation.

4.1. Discussion

Figure 3 collects results of two days in which all three satellites have done an acquisition, showing from left to right the S3 Chl-a product, the S2 Chl-a product (whose validation is documented in EOMORES D5.3 [42]), the S1 WR index map and the spatial and temporal evolution of ECMWF reanalysis wind vector. Areas affected by cloud cover are masked out in Chl-a maps as well as in S1 images areas where w_{ECMWF} is lower than 2 m/s. Times are expressed in UTC (Coordinated Universal Time).

Given the unavailability of synchronous SAR/optical images and the absence of in-field measurements in conjunction with the S1 acquisition, this research is focused to investigate whether the "anomaly" observed in SAR wind retrieval is compatible with a spatio-temporal evolution of scum observed in optical images.

First line of Figure 3 refers to the 29 August 2018 and is an example of slow bloom evolution typical of calm or gentle wind conditions [16].

The area with Chl-a $\gtrsim$ 100 mg/m^3, where visible, remains almost unvaried in the time between S3 and S2 acquisition.

S1 image is acquired six hours later the S2 ones. The area colored in red in Figure 3c is a little wider and slightly shifted to East in respect to the high Chl-a concentration area observed in Figure 3a,b. It is reasonable to envision that the associated wind speed ($\approx$1 m/s) is too low to transport the surface accumulation of cyanobacteria elsewhere too far in the lagoon but large enough to be responsible for a slight convection. Considering the wind direction during the afternoon, see orange arrows in panel Figure 3d, this convection should be eastward. Moreover, as shown in Bresciani et al. [41], Chl-a concentration tends to increase during the day, reaching a maximum at mid afternoon for calm weather condition. We can therefore conclude that the red area observed for the S1 afternoon image is overall compatible with the evolution of the Chl-a maps driven by vector spatio-temporal variation. These finding support the hypothesis that WR < 0.1 can indicates phytoplankton bloom.

Second line in Figure 3 refers to 10 October 2018 and is an example of medium wind strength—wind speed ($\approx$3.5 m/s) is high enough to be able to transport the bloom on the water surface but not enough to increase the mixing layer depth and the turbulence level in the water column.

Comparison between Chl-a products from S3/OLCI, panel (e) and S2/MSI, panel (f), highlights how the phenomenon can quickly evolve in less than two hours. WR map from S1, panel (g), shows an alert area consistent with a realistic spatio-temporal evolution of the high Chl-a concentration area observed from optical images and the whole day wind vector direction and intensity.

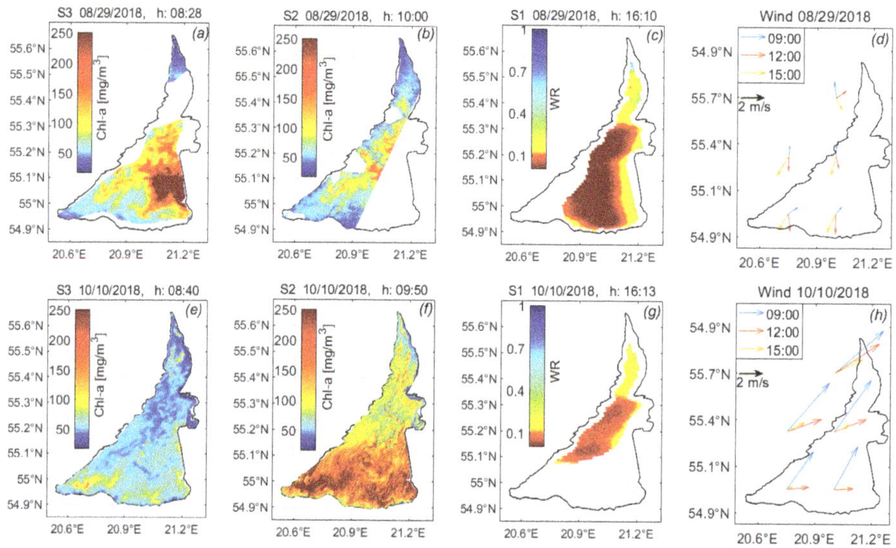

Figure 3. Qualitative comparison between optical and radar products for single day acquisitions: Chl-a maps concentration derived from S3 (**a**,**e**) and from S2 (**b**,**f**); WR index maps from S1 (**c**,**g**); Top panels show results of 28 August 2018, bottom panels show results of 10 October 2018. (**d**,**h**) report wind vector spatiotemporal evolution during the time interval between acquisitions. Wind vector come from ERA5 hourly data on single levels reanalysis database.

These results seem to overall support the usability of the proposed approach, that in turn would confirm the leading role of solar radiance and wind speed and direction for daily spatial distribution and accumulation of positively buoyant cyanobacteria as discussed in References [20,41].

In Figure 4, results from those days where acquisitions from S1 and S3 are available are shown. The first line refers to the 17 August 2018 and also presents a case where presumably bloom increases its area in the afternoon and moves accordingly with wind direction. Second line refer to the 18 October 2018 morning time. S3 indicates bloom area in the south-est region; WR map locate the alert area in a region that is almost congruent with that showed in panel (e) with no significant spatio-temporal evolution.

Figure 5 collects results of those days where acquisition from S1 and S2 are available. Behaviours similar to those showed above are observed. 9 September 2018 (first line) WR map suggests again a case of afternoon spreading of bloom covered area consistent with wind direction. 18 October 2018 (second line) S1 acquisition is in the early morning and identifies an alert area bigger than the bloom area observed from S2 image acquired 5 h later.

These results suggest that the proposed approach based on wind ratio point out areas where water surface is affected by some factor that during summer/autumn time, as long as wind speed is greater than 2 m/s, can reasonably be identified as cyanobacteria's bloom. S1-WR product can therefore contribute to the spatial and temporal characterization of the phenomena in order to improve the ecological analysis.

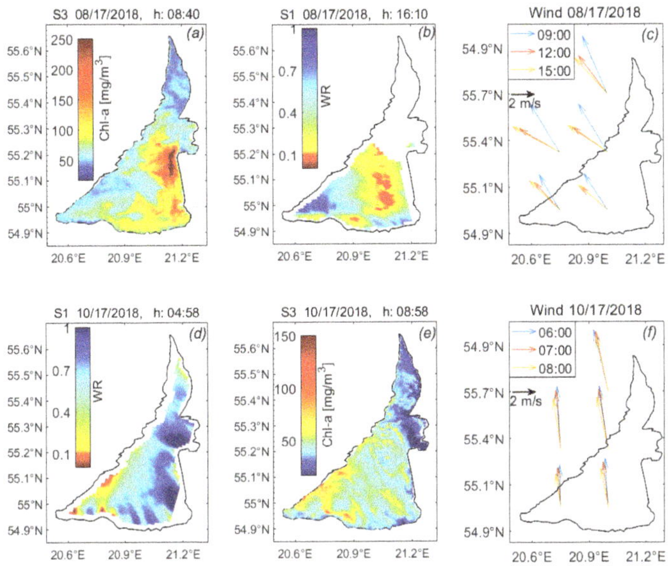

Figure 4. Qualitative comparison between S3 and S1 products for single day acquisitions: Chl-a maps concentration derived from S3 (**a**,**e**) and WR index maps from S1 (**b**,**d**); Top panels show results of the 17 August 2018, bottom panels show results of the 17 October 2018. Panels (**c**,**f**) report wind vector spatio-temporal evolution during the time interval between acquisitions. Wind vector come from ERA5 hourly data on single levels reanalysis database.

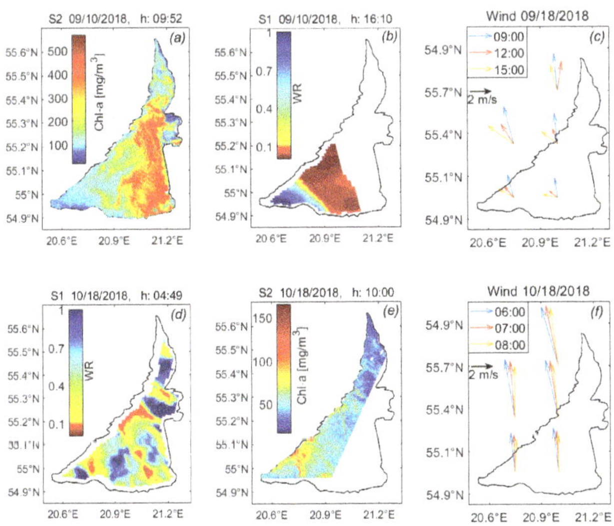

Figure 5. Qualitative comparison between S2 and S1 products for single day acquisitions: Chl-a maps concentration derived from S2 (**a**,**e**) and WR index maps from S1 (**b**,**d**); Top panels show results of the 9 September 2018, bottom panels show results of the 18 October 2018; Panels (**c**,**f**) report wind vector spatio-temporal evolution during the time interval between acquisitions. Wind vector come from ERA5 hourly data on single levels reanalysis database.

4.2. Application Example

An example of how the use of the proposed index could enrich a temporal characterization of the lagoon is shown in Figure 6. Let us recall that the main goal of the SAR based analysis here proposed consists in fill the gaps leaved by optical remote sensing techniques. This can be the case of no acquisition, of cloudy sky condition and of afternoon hours. As done in previous sub-section, we have thus to investigate whether wind ratio classification is compatible with the optical band ratio ones. Given the the large geographic extension of the lagoon and the wide window in which phytoplankton blooms can occurs, that is, almost 4 months, we propose hereafter the simplest and most synthetic approach for a temporal characterization of the lagoon.

Following the Water Framework Directive 2000/60/EC [43], the simplest temporal characterization of the lagoon can be performed classifying as "No BAD" images where Chl-a < 72 mg/m^3 in all the domain and as "BAD" any other. According to this paper's hypothesis and the results showed above, the equivalent classification for S1 images corresponds to "No BAD" if WR > 0.1 everywhere in the lagoon and "BAD" otherwise.

Considering that C band radar can be additionally influenced by rain, the total precipitations time series is also showed for completeness. ERA5 hourly data on single level reanalysis database provides the total precipitation time series, represented in Figure 6 as gray background when is greater than zero. It seems that WR index based classification never gives false alarm caused by rainfalls. Indeed most of the days with precipitation and S1 acquisition are classified as "NO BAD" and the remaining cases are nearby other "BAD" classified images. This confirms that the WR index approach is actually self-sustained and that other meteorologic parameters are not needed.

Time series in Figure 6 again highlights that there would be a a significant lack of information due to the cloud cover if considering only optical data. Indeed, in this case there will be no data for the 62.2% of the days considered, of which the 67% can be recovered by integration through S1 results. Moreover even when optical data are available, SAR results can be used to evaluate the bloom persistence, as it happens for example on 08/29 and on 10/17.

Overall there are 28 days in which optical and SAR images are both acquired. Among these, in 16 days optical and SAR classify in the same way the lagoon; for the remaining 12 days some hypothesis can be formulated by distinguishing three different cases.

- SAR acquisition around 05:00 UTC classified as "No BAD" and optical images classified as "BAD", this is the case of the 08/24, 09/18, 10/06 and 10/12.
  WR index seems still working reasonably since bloom typically forms later in the morning as the air temperature increases [41].
- Optical images classified as "BAD" and SAR acquisition around 16:00 classified as "No BAD".
  - for the days 07/24, 08/23, 09/03, 10/15, 10/21 and 10/28 we observe that $w_{\text{ECMWF}} > 6$ m/s. Since the threshold wind speeds required for vertical mixing in shallow inland lakes typically goes from 3.1 m/s [44] to 4 m/s [45], it is therefore reasonable to assume that wind can produce shear forces on the water surface able to destabilize cyanobacteria's buoyancy.
  - For the remaining two days (08/16 and 09/09) additional information is missing and it is difficult to understand if the WR index is failing or if bloom actually disappears for some unknown reason.

Summarizing, the SAR based approach returns a lagoon's classification consistent with optical images results and meteorological data (high wind speed) for 22 cases, giving an agreement rate of the 79%. The different classification can the explained by the temporal acquisition gap for the 14% of the cases and solely the remaining 7% show a possible disagreement.

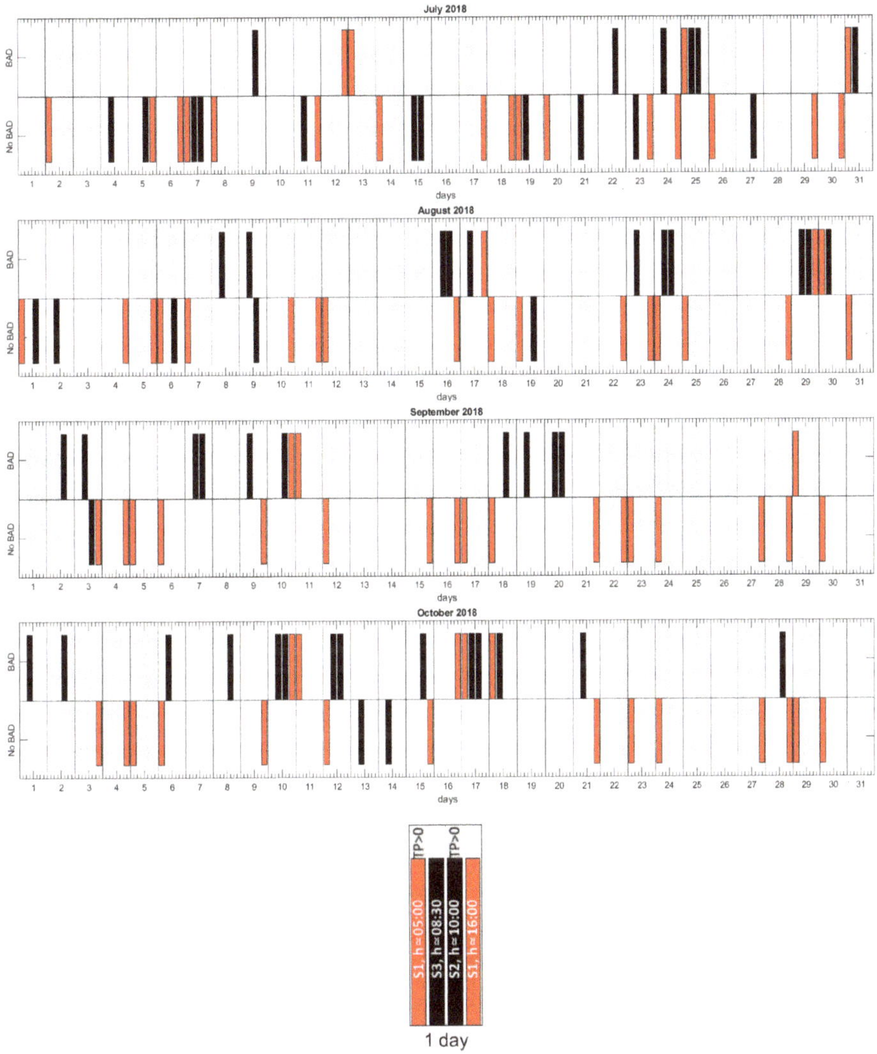

Figure 6. Temporal trends of WR (red bars), Chl-a concentrations (black bars) and Total Precipitation (TP) (gray bars) from July to October 2018. Images with Chl-a < 72 mg/m^3 or WR > 0.1 in all the domain are classified as "No BAD" [43]. Total Precipitation, indicated as a grey background when positive, come from ERA5 hourly data on single level reanalysis database and when is greater than zero is indicated as gray background. Temporal scheme is explained in the insert on the left.

5. Conclusions

This paper proposes a "ready to use" approach to detect cyanobacteria bloom in the Curonian Lagoon based on the dependency between wind vector and radar backscatter. The open source and self-contained database of the level 2 ocean product Sentinel-1 images is considered. This method has been applied for Sentinel-1 images acquired in summer/autumn 2018. Qualitative comparison with optical imagery gathered from Sentinel-2 and Sentinel-3 satellites combined with meteorological data reveals that the method shows reasonable results for most of the analysed cases (the 79%). It presents the advantage to be an easy and self-sustained approach able to provide observations independently on the presence of atmospheric haze or cloud cover.

The presented findings strongly support the use of L2 S1 products to improve the spatio-temporal detection of algal bloom in Curonian Lagoon and complete the observations from optical sensors. Nonetheless, due to the unavailability of coincident pairs of SAR and optical imagery, a dedicated field observations could however refine the wind ratio threshold empirically determined in this paper. Overall, the combination of the three Sentinels (1, 2 and 3) is providing unprecedented observations in frequency and availability to water ecosystems studies concerned with trends and episodic events. The possibility of having multiple spatial scales (from 10 of S2 to 1 km of S1 L2 products) with different sensing time (from the earlier S3 to the later S1) finally support the understanding of cascade effect processes (e.g., cyanobacterial growth with scum formation), characterized by high degree of change and across a range of scales.

Author Contributions: Conceptualization, M.B., G.D.C. and F.D.S. methodology, F.D.S., G.L., M.B., C.G., F.P.L., G.P., D.V. and G.D.C.; writing—original draft preparation, F.D.S.; writing—review and editing, F.D.S., G.L., M.B., C.G., F.P.L., G.P., D.V. and G.D.C.; supervision, M.B., G.D.C. and C.G.

Acknowledgments: This work was supported by the EU Horizon 2020 programme through the EOMORES project [grant number 730066].

Conflicts of Interest: The authors declare no conflict of interest.

Abbreviations

The following abbreviations are used in this manuscript:

S1	Sentinel-1
S2	Sentinel-2
S3	Sentinel-3
L2	Level 2 product for Sentinel-1
SAR	Synthetic Aperture Radar
AVHR	Advanced Very High Resolution Radiometer
ERS-1	European Remote-Sensing satellite
MERIS	MEdium Resolution Imaging Spectrometer
ASAR	Advanced Synthetic Aperture Radar (ASAR)
L2	Level 2 product for Sentinel-1
MODIS	Moderate Resolution Imaging Spectroradiometer
MSI	Multispectral Instrument
OLCI	Ocean and Land Colour Instrument
Chl-a	Chlorophyll- a concentration
6SV	Second Simulation of the Satellite Signal in the Solar Spectrum-Vector code
AOT	Aerosol Optical Thickness
NIR	Near-InfraRed reflectance
L2OCN	Level-2 (L2) Ocean product (for S1)
NRCS	Normalised Radar Cross Section
ECMWF	European Centre for medium-range weather forecasts

References

1. Paerl, H.W.; Paul, V.J. Climate change: Links to global expansion of harmful cyanobacteria. *Water Res.* **2012**, *46*, 1349–1363. [CrossRef]
2. Taranu, Z.E.; Gregory-Eaves, I.; Leavitt, P.R.; Bunting, L.; Buchaca, T.; Catalan, J.; Domaizon, I.; Guilizzoni, P.; Lami, A.; McGowan, S.; et al. Acceleration of cyanobacterial dominance in north temperate-subarctic lakes during the Anthropocene. *Ecol. Lett.* **2015**, *18*, 375–384. [CrossRef]
3. Reynolds, C.S. *The Ecology of Phytoplankton*; Cambridge University Press: Cambridge, UK, 2006.
4. Bartram, J.; Chorus, I. *Toxic Cyanobacteria in Water: A Guide to Their Public Health Consequences, Monitoring and Management*; CRC Press: Boca Raton, FL, USA, 1999.

5. PaeRL, H.; HussMann, J. Blooms like it hot. *Science* **2008**, *320*, 57–58. [CrossRef]
6. Walsby, A.E. Stratification by cyanobacteria in lakes: A dynamic buoyancy model indicates size limitations met by Planktothrix rubescens filaments. *New Phytol.* **2005**, *168*, 365–376. [CrossRef]
7. Tyler, A.N.; Hunter, P.D.; Spyrakos, E.; Groom, S.; Constantinescu, A.M.; Kitchen, J. Developments in Earth observation for the assessment and monitoring of inland, transitional, coastal and shelf-sea waters. *Sci. Total Environ.* **2016**, *572*, 1307–1321. [CrossRef]
8. Palmer, S.C.; Kutser, T.; Hunter, P.D. Remote sensing of inland waters: Challenges, progress and future directions. *Remote Sens. Environ.* **2015**, *157*, 1–8. [CrossRef]
9. Matthews, M.W. Eutrophication and cyanobacterial blooms in South African inland waters: 10 years of MERIS observations. *Remote Sens. Environ.* **2014**, *155*, 161–177. [CrossRef]
10. Bresciani, M.; Giardino, C.; Lauceri, R.; Matta, E.; Cazzaniga, I.; Pinardi, M.; Lami, A.; Austoni, M.; Viaggiu, E.; Congestri, R.; et al. Earth observation for monitoring and mapping of cyanobacteria blooms. Case studies on five Italian lakes. *J. Limnol.* **2017**, *76*. [CrossRef]
11. Bracher, A.; Vountas, M.; Dinter, T.; Burrows, J.; Röttgers, R.; Peeken, I. Observation of cyanobacteria and diatoms from space using Differential Optical Absorption Spectroscopy on SCIAMACHY data. In Proceedings of the 5th EGU General Assembly 2008, Vienna, Austria, 16 April 2008.
12. Kahru, M.; Elmgren, R. Multidecadal time series of satellite-detected accumulations of cyanobacteria in the Baltic Sea. *Biogeosciences* **2014**, *11*, 3619. [CrossRef]
13. Kutser, T.; Metsamaa, L.; Strömbeck, N.; Vahtmäe, E. Monitoring cyanobacterial blooms by satellite remote sensing. *Estuar. Coast. Shelf Sci.* **2006**, *67*, 303–312. [CrossRef]
14. Öberg, J. Cyanobacteria blooms in the Baltic Sea. In *HELCOM Baltic Sea Environment Fact Sheets 2017*; Helcom: Helsinki, Finland, 2016.
15. Reinart, A.; Kutser, T. Comparison of different satellite sensors in detecting cyanobacterial bloom events in the Baltic Sea. *Remote Sens. Environ.* **2006**, *102*, 74–85. [CrossRef]
16. Reynolds, C.; Walsby, A. Water-blooms. *Biol. Rev.* **1975**, *50*, 437–481. [CrossRef]
17. Bresciani, M.; Bolpagni, R.; Laini, A.; Matta, E.; Bartoli, M.; Giardino, C. Multitemporal analysis of algal blooms with MERIS images in a deep meromictic lake. *Eur. J. Remote Sens.* **2013**, *46*, 445–458. [CrossRef]
18. Walsby, A.E.; Ng, G.; Dunn, C.; Davis, P.A. Comparison of the depth where Planktothrix rubescens stratifies and the depth where the daily insolation supports its neutral buoyancy. *New Phytol.* **2004**, *162*, 133–145. [CrossRef]
19. Svejkovsky, J.; Shandley, J. Detection of offshore plankton blooms with AVHRR and SAR imagery. *Int. J. Remote Sens.* **2001**, *22*, 471–485. [CrossRef]
20. Bresciani, M.; Adamo, M.; De Carolis, G.; Matta, E.; Pasquariello, G.; Vaičiūtė, D.; Giardino, C. Monitoring blooms and surface accumulation of cyanobacteria in the Curonian Lagoon by combining MERIS and ASAR data. *Remote Sens. Environ.* **2014**, *146*, 124–135. [CrossRef]
21. Bartoli, M.; Zilius, M.; Bresciani, M.; Vaičiūtė, D.; Lubiene-Vybernaite, I.; Petkuviene, J.; Giordani, G.; Daunys, D.; Ruginis, T.; Benelli, S.; et al. Drivers of cyanobacterial blooms in a hypertrophic lagoon. *Front. Mar. Sci.* **2018**, *5*, 434. [CrossRef]
22. Bresciani, M.; Giardino, C.; Stroppiana, D.; Pilkaitytė, R.; Zilius, M.; Bartoli, M.; Razinkovas, A. Retrospective analysis of spatial and temporal variability of chlorophyll-a in the Curonian Lagoon. *J. Coast. Conserv.* **2012**, *16*, 511–519. [CrossRef]
23. Giardino, C.; Bresciani, M.; Pilkaityte, R.; Bartoli, M.; Razinkovas, A. In situ measurements and satellite remote sensing of case 2 waters: First results from the Curonian Lagoon. *Oceanologia* **2010**, *52*, 197–210. [CrossRef]
24. Ferrarin, C.; Razinkovas, A.; Gulbinskas, S.; Umgiesser, G.; Bliūdžiutė, L. Erratum to: Hydraulic regime-based zonation scheme of the Curonian Lagoon. *Hydrobiologia* **2010**, *652*, 397–397. [CrossRef]
25. Razinkovas, A.; Bliudziute, L.; Erturk, A.; Ferrarin, C.; Lindim, C.; Umgiesser, G.; Zemlys, P. Curonian lagoon: A modelling study-Lithuania. In *Modeling Nutrient Loads and Response in River and Estuary Systems*; Committee on the Challenges of Modern Society, North Atlantic Treaty Organization: Brussels, Belgium, 2005; pp. 194–222.
26. Katarżytė, M.; Mėžinė, J.; Vaičiūtė, D.; Liaugaudaitė, S.; Mukauskaitė, K.; Umgiesser, G.; Schernewski, G. Fecal contamination in shallow temperate estuarine lagoon: Source of the pollution and environmental factors. *Mar. Pollut. Bull.* **2018**, *133*, 762–772.[CrossRef] [PubMed]

27. Zilius, M.; Bartoli, M.; Bresciani, M.; Katarzyte, M.; Ruginis, T.; Petkuviene, J.; Lubiene, I.; Giardino, C.; Bukaveckas, P.A.; de Wit, R.; Razinkovas-Baziukas, A. Feedback Mechanisms Between Cyanobacterial Blooms, Transient Hypoxia, and Benthic Phosphorus Regeneration in Shallow Coastal Environments. *Estuaries Coasts* **2014**, *37*, 680–694. [CrossRef]
28. Pilkaitytė, R.; Razinkovas, A. Factors controlling phytoplankton blooms in a temperate estuary: Nutrient limitation and physical forcing. In *Marine Biodiversity*; Springer: Berlin/Heidelberg, Germany, 2006; pp. 41–48.
29. Gasiūnaitė, Z.R.; Daunys, D.; Olenin, S.; Razinkovas, A. The curonian lagoon. In *Ecology of Baltic Coastal Waters*; Springer: Berlin/Heidelberg, Germany, 2008; pp. 197–215.
30. Wasmund, N.; Nausch, G.; Gerth, M.; Busch, S.; Burmeister, C.; Hansen, R.; Sadkowiak, B. Extension of the growing season of phytoplankton in the western Baltic Sea in response to climate change. *Mar. Ecol. Prog. Ser.* **2019**, *622*, 1–16. [CrossRef]
31. Mercury, M.; Green, R.; Hook, S.; Oaida, B.; Wu, W.; Gunderson, A.; Chodas, M. Global cloud cover for assessment of optical satellite observation opportunities: A HyspIRI case study. *Remote Sens. Environ.* **2012**, *126*, 62–71. [CrossRef]
32. Vaičiūtė, D.; Bresciani, M.; Bartoli, M.; Giardino, C.; Bučas, M. Spatial and temporal distribution of coloured dissolved organic matter in a hypertrophic freshwater lagoon. *J. Limnol.* **2015**, *74*. [CrossRef]
33. Galkus, A. Vandens cirkuliacija ir erdvine drumstumo dinamika vasara Kuršių marių ir Baltijos jūros Lietuvos akvatorijose [Summer water circulation and spatial turbidity dynamics in the Lithuanian waters of Curonian lagoon and Baltic Sea]. *Geogr. Metrašt* **2003**, *36*, 3–16.
34. Mėžinė, J.; Ferrarin, C.; Vaičiūtė, D.; Idzelytė, R.; Zemlys, P.; Umgiesser, G. Sediment Transport Mechanisms in a Lagoon with High River Discharge and Sediment Loading. *Water* **2019**, *11*, 1970. [CrossRef]
35. Vermote, E.F.; Tanré, D.; Deuze, J.L.; Herman, M.; Morcette, J.J. Second simulation of the satellite signal in the solar spectrum, 6S: An overview. *IEEE Trans. Geosci. Remote Sens.* **1997**, *35*, 675–686. [CrossRef]
36. Available online: http://https://giovanni.gsfc.nasa.gov/giovanni/ (accessed on 12 December 2019).
37. Cazzaniga, I.; Bresciani, M.; Colombo, R.; Della Bella, V.; Padula, R.; Giardino, C. A comparison of Sentinel-3-OLCI and Sentinel-2-MSI-derived Chlorophyll-a maps for two large Italian lakes. *Remote Sens. Lett.* **2019**, *10*, 978–987. [CrossRef]
38. INFORM. INFORM Prototype/Algorithm Validation Report Update, D5.15; INFORM: 2016. Available online: https://www.google.com/url?sa=t&rct=j&q=&esrc=s&source=web&cd=1&ved=2ahUKEwjZnIq566_mAhW-UhUIHZEoBTUQFjAAegQIBBAC&url=http%3A%2F%2Finform.vgt.vito.be%2Ffiles%2Fdocuments%2FINFORM_D5.15_v1.0.pdf&usg=AOvVaw2S4Q_ofKQQBon-LYcm4a9_ (accessed on 12 December 2019).
39. Bentamy, A.; Queffeulou, P.; Quilfen, Y.; Katsaros, K. Ocean surface wind fields estimated from satellite active and passive microwave instruments. *IEEE Trans. Geosci. Remote Sens.* **1999**, *37*, 2469–2486. [CrossRef]
40. Kerbaol, V.; Wind, T.S.O.; Team, C. Improved Bayesian Wind Vector Retrieval Scheme Using ENVISAT ASAR Data: Principles and Validation Results. 2007. Available online: http://citeseerx.ist.psu.edu/viewdoc/summary?doi=10.1.1.434.4013 (accessed on 12 December 2019).
41. Bresciani, M.; Rossini, M.; Morabito, G.; Matta, E.; Pinardi, M.; Cogliati, S.; Julitta, T.; Colombo, R.; Braga, F.; Giardino, C. Analysis of within-and between-day chlorophyll-a dynamics in Mantua Superior Lake, with a continuous spectroradiometric measurement. *Mar. Freshw. Res.* **2013**, *64*, 303–316. [CrossRef]
42. EOMORES Consortium. D5.3 Final Validation Report; 2020; Under Review. Available online: https://eomores-h2020.eu/results/#deliverables (accessed on 12 December 2019).
43. Available online: https://eur-lex.europa.eu/eli/dir/2000/60/oj (accessed on 12 December 2019).
44. Cao, H.S.; Kong, F.X.; Luo, L.C.; Shi, X.L.; Yang, Z.; Zhang, X.F.; Tao, Y. Effects of wind and wind-induced waves on vertical phytoplankton distribution and surface blooms of Microcystis aeruginosa in Lake Taihu. *J. Freshw. Ecol.* **2006**, *21*, 231–238. [CrossRef]
45. Huang, C.; Li, Y.; Yang, H.; Sun, D.; Yu, Z.; Zhang, Z.; Chen, X.; Xu, L. Detection of algal bloom and factors influencing its formation in Taihu Lake from 2000 to 2011 by MODIS. *Environ. Earth Sci.* **2014**, *71*, 3705–3714. [CrossRef]

Journal of
Marine Science and Engineering

MDPI

Article

Implementation of the Listen-Before-Talk Mode for SeaSonde High-Frequency Ocean Radars

Simone Cosoli

Ocean Graduate School and the UWA Oceans Institute, The University of Western Australia, 35 Stirling Highway, Crawley, WA 6009, Australia; simone.cosoli@uwa.edu.au; Tel.: +61-8-6488-7314

Received: 27 November 2019; Accepted: 16 January 2020; Published: 19 January 2020

Abstract: The International Telecommunication Union (ITU) Resolution 612, in combination with Report ITU-R M2.234 (11/2011) and Recommendation ITU-R M.1874-1 (02/2013), regulates the use of the radiolocation services between 3 and 50 MHz to support high frequency oceanographic radar (HFR) operations. The operational frame for HFR systems include: band sharing capabilities, such as synchronization of the signal modulation; pulse shaping and multiple levels of filtering, to reduce out-of-band interferences; low radiated power; directional transmission antenna, to reduce emission over land. Resolution 612 also aims at reducing the use of spectral bands, either through the application of existing band-sharing capabilities, the reduction of the spectral leakage to neighboring frequency bands, or the development and implementation of listen-before-talk (LBT) capabilities. While the LBT mode is operational and commonly used at several phased-array HFR installations, the implementation to commercial direction-finding systems does not appear to be available yet. In this paper, a proof-of-concept is provided for the implementation of the LBT mode for commercial SeaSonde HFRs deployed in Australia, with potential for applications in other networks and installations elsewhere. Potential critical aspects for systems operated under this configuration are also pointed out. Both the receiver and the transmitter antennas may lose efficiency if the frequency offset from the resonant frequency or calibration pattern are too large. Radial resolution clearly degrades when a dynamical adaptation of the bandwidth is performed, which results in non-homogeneous spatial resolution and reduction of the quality of the data. A recommendation would be to perform the LBT-adapt scans after a full measurement cycle (1-h or 3-h, depending on the system configuration) is concluded. Mutual cross-interference from clock offsets between two HFR systems may bias the frequency scans when the site computers controlling data acquisitions are not properly time-synchronized.

Keywords: high-frequency ocean radar; interference mitigation; frequency band adaptation

1. Introduction

Shore-based High-frequency oceanographic radar (HFR) systems, operating in the frequency range between 3–30 MHz, are nowadays routinely used for a number of different scientific applications and operational purposes. Depending upon the operational parameters, HFRs can resolve spatial scales in the range of a few hundred meters to hundreds of kilometers, at time scales of minutes, hour, day, seasonal or longer. These capabilities make them very attractive, for instance, for oil spill mitigation purposes, search and rescue operations, and ingestion into ocean circulation numerical models [1]. Over the past decades, the overall number of operational HFR systems increased significantly to the point that regional and global networks have been established [1]. Along the coasts of the US, for instance, more than 100 HFR systems are presently in operation, and the number of deployments steadily increases both in the Asia-Pacific regions and Europe. Extensive validation analyses and deployments in a variety of environments have also proven their general reliability [1–4].

The most commonly deployed systems are supplied by two manufacturers, Codar Ocean Sensors (COS) and Helzel MessTechnik, who provide commercial direction-finding (SeaSonde) and phased-array (Wellen Radar, or WERA) HFR systems, respectively. Other HFR systems are also available, such as the PISCES or the University of Hawaii's HFRs. See, for instance, [1–4] and references herein for a more exhaustive literature review and for an assessment of their interoperability.

In 2012, the International Telecommunication Union (ITU) Resolution 612 [5] allocated specific frequency bands between 3 and 50 MHz to support high frequency oceanographic radar (HFR) operations. The allocation of specific frequency band to HFRs was a formal recognition of the benefits to society through environmental protection, disaster preparedness, public health protection, improved meteorological operations, increased coastal and maritime safety and enhancement of national economies. ITU Resolution 612 [5] was also an acknowledgement of the efforts to reduce interferences to services within the same or neighboring frequency bands.

In combination with Report ITU-R M2.234 (11/2011; [6]) and Recommendation ITU-R M.1874-1 (02/2013; [7]), Resolution 612 [5] set the operational frame for HFR systems. An oceanographic radar should operate in low radiated power mode, with peak effective isotropic radiated power (EIRP) below 25 dBW. Band sharing capabilities should be implemented in order to reduce to a minimum the spectral occupancy of a regional or global deployment of radars. These capabilities include, for instance, the synchronization of the signal modulation and the use of accurate GPS timing [8]. The reduction of spurious emissions and offband leakage of the transmitted signal to neighboring band, resulting in unwanted out-of-band interferences, is achieved through pulse shaping and multiple levels of filtering. Finally, the use of directional transmit antenna is also recommended, where applicable and as required by the regulatory bodies, so to reduce the EIRP in the direction of the transmit antenna backlobe, and limit as much as possible emission over land.

In Australia, the Integrated Marine Observing System (IMOS) Ocean Radar Facility operates commercial-type direction-finding SeaSonde and phased-array Wellen Radar (WERA). HFR systems operate within the ITU Region 3 frequency band with secondary-type licenses, meaning that operations must not cause harmful interference to primary users within or outside the allocated frequency band.

Unfortunately, this is not the case and several breach warning notices have been issued in the past. The most recent episodes, occurred in December 2017 and July 2018, and involved the freshly installed SeaSonde systems in New South Wales (NSW) operating in the 5.250–5.275 MHz ITU frequency band. These systems were reported as an interference source to primary services at various locations across the country, particularly to two 3 KHz fixed and mobile services at 5.264 MHz and 5.2715 MHz operating remote area travellers safety networks.

Following the complaint, systems were forced off for 12 months while mitigation measures were implemented as requested by the Australian Communications and Media Authority (ACMA). Operations could resume only upon successful implementation of the mitigation measures in compliance with relevant ITU regulation for HFR systems (Resolution 612 [5], Report ITU-R M2.234 (11/2011; [6]) and Recommendation ITU-R M.1874-1 (02/2013; [7]).

To reduce direct impact of the SeaSonde HFR systems on the affected users, transmit power was decreased to the minimum (even below 1 W), the center operating frequency was shifted to 5.257 MHz, the bandwidth was reduced to 11 kHz (resulting in a radial range resolution of approximately 13.5 km). As a further mitigation measure, prototype modified components were provided by the system manufacturers so to limit spurious emissions through pulse shaping with longer ramping-deramping times, longer pulse width, and modified weighting windows.

SeaSonde make use of the proprietary GPS synchronization band sharing techniques described in [8] to reduce band usage. However, it was also decided to operate in interrupted, short-burst acquisition mode in which the transmitter is on/off for 512 s/88 s and to observe radio silence periods in order to facilitate possible distress or emergency calls. This was possible thanks to the advanced programming features available in the SeaSonde hardware [9] and the Applescript ® programming language [10], which require no access to the proprietary source code.

Finally, a customized prototype non-resonant, wide-beam, travelling-wave directional transmit antenna providing up to 30 dB front-to-back lobe suppression was developed by the IMOS Ocean Radar Facility [11] and deployed on field. Operating frequency and bandwidth for the two SeaSonde systems deployed in NSW are, respectively, 5.257 MHz and 11.029 kHz, the latter resulting in a radial range resolution of approximately 13.5 km. No significant performance losses in terms of range can be observed in comparison to more conventional, uninterrupted transmit and operation modes, thanks to the long integration times.

In an attempt to further mitigate the band usage, a listen-before-talk operation mode for SeaSonde systems was developed and is described here, in which the SeaSonde perform automatic frequency scans (Figure 1) before each short-duration transmit cycle, the frequency scan results are analyzed and optimized operation parameters are provided in real-time mode. Although feasible in principle, an application of the bandwidth reduction to real-time data stream is however discouraged to avoid inconsistencies during the radial averaging in the final stage of the radial map processing chain.

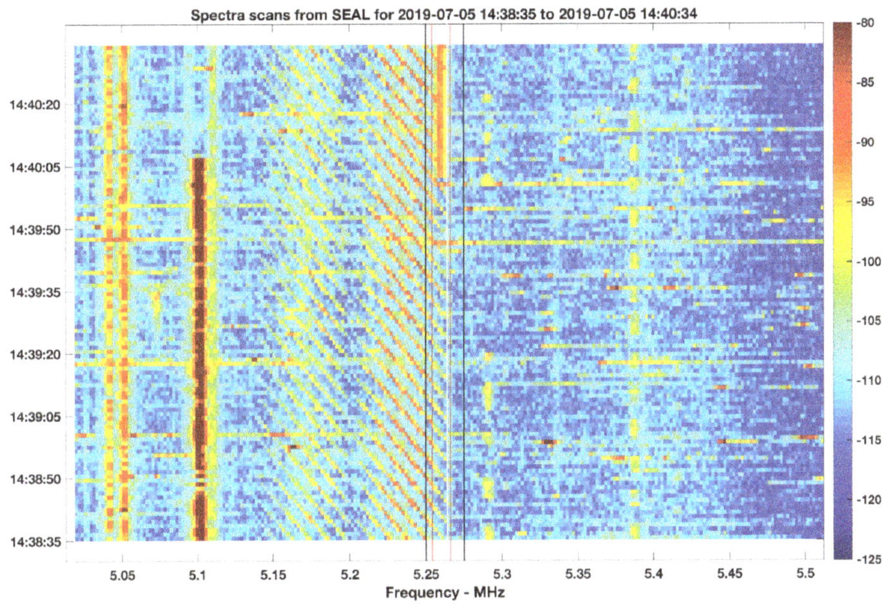

Figure 1. Example of a 2-min spectrum scan collected through receive channel 3 (the so called 'monopole' in the standard SeaSonde terminology) at the Seal Rocks SeaSonde HFR station (station code: SEAL). The station is located approximately 200 km N of Sydney (NSW). The scan was collected between 20190705T143835Z–20190705T144034Z and contains several recurrent features. Vertical black lines indicate the 5.250–5.275 MHz ITU band; red vertical lines show the 5.254–5.266 MHz band in use to the SeaSonde station located at Red Head, Newcastle (NSW), approximately 100 km apart from Seal Rocks; this station resumed transmission at 20190705T144000Z after being silent for approximately 90 s. Note the 50 kHz chirps (of unknown origin) contaminating the ITU frequency band allocated to HFR operations. Unit for signal power is dB.

The concepts of listen-before-talk, or listen-before-transmit (LBT), and dynamic bandwidth adaptation, or automatic frequency agility (AFA), are relatively straightforward: a technique used in radio communications in which a device first senses its radio environment before transmitting. Then, transmission will begin: unmodified, if the channel is free; modified, to adapt to the available portion of the channel otherwise; use a different channel, if available and allowed to operate on.

LBT is part of the so-called fair coexistence code of practice and is included the regulatory requirements worldwide in several applications, from medical to communications [12,13]. Within the HFR operator's community, only phased array systems routinely implement the LBT mode with dynamical adaptation at their installations [14]. Typically, a frequency pre-scan is made across the entire licensed band before each data acquisition. The pre-scan data are used to identify regions of lowest external noise, and operational settings are optimized for subsequent acquisition cycle. It is worth noting that prescan and adapt measures were also suggested in [5–7], however SeaSonde HFR operators involved in the preliminary feasibility study deemed the consequence of a reduced bandwidth unacceptable in terms of data quality.

The problem of band sharing, coexistence and off-band leakage is now becoming of primary importance within the HFR community as it involves all ITU frequency bands allocated to oceanographic HFR systems. The paper is organized as follows: Section 2 details the implementation of the proposed LBT methodology. Section 3 presents examples of its application to real time data at locations across Australia and discusses some of the implications and assumptions for the LBT mode. Concluding remarks are presented in Section 4.

2. Materials and Methods

The proposed implementation of the listen-before-talk (LBT) mode for SeaSonde HFRs is meant to deal with in-band interferences, that is both interfering transmitter and victim receiver operate in the same spectrum band. It is a multi-step process that can be summarized as follows (Figure 2):

(1) perform a scan of the radio spectrum before data acquisition (Spectrum scan-SSA; Figure 2);
(2) analyze the frequency scan, and identify interference sources (Signal Detected; Figure 2);
(3) adapt the acquisition parameters and resume operations (Adapt SSC; Figure 2);
(4) resume operations (Start-Stop DAQ; Figure 2);
(5) repeat the cycle.

Control of transmit and data acquisition cycles are possible using the advanced SeaSonde software features, documented in [9], and the AppleScript programming language [10]. Similarly, SeaSondeAcquisition can be programmed via AppleScript to operate in SpectrumAnalyzer mode, perform a frequency scan around the operating frequency, and save the results of the spectrum scans in time stamped resource indexed file format. Data collected during the frequency scan through the receive channel consists of a set of consecutive time sweeps containing received signal voltage (V) over frequency [15]. Once the spectrum scan is complete, the integrity of the file is checked, and the content decoded, following [15], and converted to signal power (dB) for further analyses.

An example of a frequency scan through the vertical omnidirectional whip (the so called 'monopole' in the standard SeaSonde terminology), is provided in Figure 1. In this example, collected at the Seal Rocks SeaSonde HFR station (station code: SEAL) between 20190705T143835Z–20190705T144034Z, the raw file is composed of 120 consecutive sweeps collected at 1 Hz sampling rate. Raw voltage data can be converted to signal power (SP) in dB, as follows Equation (1):

$$SP = 20 * \log_{10}(V) - 67, \tag{1}$$

where −67 is a correction factor for the processing gain.

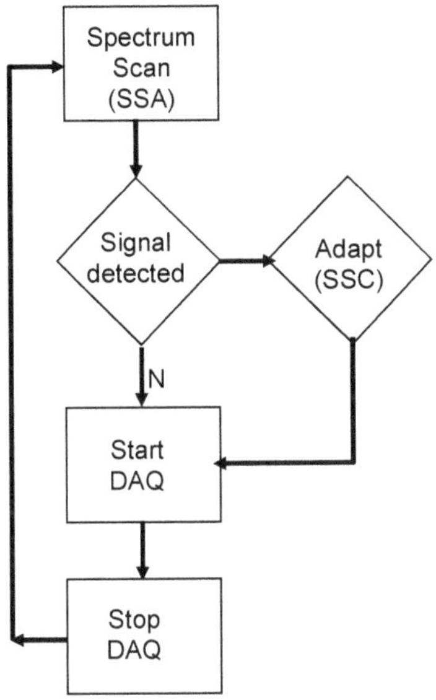

Figure 2. Flowchart for the proposed listen-before-talk mode: perform a spectrum scan using the built-in spectrum analyzer capability in the SeaSondeAcquisition (SSA) suite; adapt and pass acquisition parameters to SeaSondeController (SSC) suite; start and stop acquisition (DAQ) steps controlling the pulsed mode in the SeaSondeController suite.

In Figure 1, vertical black lines indicate the 5.250–5.275 MHz ITU band in use to the HFR systems; red vertical lines show the band in use to the paired SeaSonde station located at Red Head, Newcastle (NSW), approximately 100 km south of Seal Rocks. This station resumed transmission at 20190705T144000Z after being silent for approximately 90 s. The white line in the spectrum denotes the 3 KHz channel at 5.264 MHz that originally reported interferences to the ACMA. Note the 50 KHz chirps (of unknown origin) contaminating the ITU frequency band allocated to HFR operations.

The critical steps are, respectively, Step 2 and Step 3 in the flowchart:

Step 2: analyze the frequency scan and identify interference sources (signal detected; Figure 2).

A peak-detection algorithm is used to identify the presence of peaks, their prominence above the signal threshold, and their width, in the time-averaged frequency scan. Statistics of the signal power level within the ITU band are used to determine whether the peak is persistent, and as such it should be considered as unwanted signal, or transient, and as such should be neglected.

Peak detection is performed on the temporal average of signal power over frequency within the allocated frequency band. In this way, mean power over frequency within the allocated frequency band can be computed along with its standard deviation during the frequency scan. Any peak with a power level exceeding the noise floor level by a factor of one standard deviation is considered either as a potential service not be causing interference to, or an interference source to avoid Equation (2):

$$\text{Peak Power} \geq \text{NF}(f)_{band} + \sigma_{PL}, \tag{2}$$

where $\text{NF}(f)_{band}$ is the time-frequency averaged power within the band, with σ_{PL} its standard deviation during the frequency scan.

Step 3: adapt the acquisition parameters and resume operations (Adapt SSC; Figure 2).

Once a peak has been identified, additional parameters are required in order to determine the optimal operational settings to use for the following acquisition cycle. The peak frequency (f_P) and the peak width at half the power, (Δf_P, measured in Hz), are used to define the maximum usable bandwidth, BW_{max}, as follows:

$$BW_{max} = \max(f_P \pm \Delta f_P/2 - f_L, f_U - f_P \pm \Delta f_P/2), \quad (3)$$

where ($f_L, f_U, f_P \pm \Delta f_P/2$) are, respectively, the lower and upper limits of the allocated frequency band, and the band occupied by the interfering signal around the peak frequency f_P. In case BW_{max} equals the actual transmit bandwidth, no reduction is required and a new optimized center frequency for the transmitter can be determined as follows:

italic> $$TX_{opt} = (f_L + BW_{max}/2), \text{ if } f_P > f_L \ \& \ f_P \leq f_P + (f_U - f_L)/2 \quad (4)$$

$$TX_{opt} = (f_U - BW_{max}/2), \text{ if } f_P < f_U \ \& \ f_P \geq f_P - (f_U - f_L)/2 \quad (5)$$

Here it is assumed that the SeaSonde systems are not using the full available bandwidth but only a fraction of it is used during the acquisition cycle. This is the case for the SeaSonde systems deployed across Australia that use 50% of the available bandwidth.

There is the possibility that the transmit bandwidth also requires adjustments. If this is the case, the following alternative options can be discussed: (1) no adjustment is made, and acquisition is resumed without adaptation; (2) acquisition is stopped; (3) acquisition parameters are adjusted and acquisition resumed with reduced bandwidth. If this reduction is deemed acceptable, setting the new centre frequency in SeaSondeController and resuming acquisition in SeaSondeAcquisition is relatively straightforward.

3. Results and Discussion

In this section, some results of the algorithm described previously are summarized. For the sake of qualitative comparison, the same color scale is used across the frequency scans shown here.

The first spectrum scan was collected at Seal Rocks HFR station using the default parameters: the omnidirectional receive element, 256-point *fft*, 120 sweeps at 1 Hz sampling rate, resulting in a 2 min scan between 20190716T233810Z and 20190716T234009Z (Figure 3). The scan was processed and converted from voltage time series into signal power as described in Equation (1); the time- frequency-averaged signal level $NF(f)_{band}$, the corresponding standard deviation σ_{PL} and the resulting threshold value are, respectively: −114.5 dB, 2.9 dB, −111.6 dB.

The peak-finding algorithm tracks two peaks: a first one, located at 5.2599 MHz, with signal level −108.76 dB, and a second one at 5.2717 MHz, with signal level −115.46 dB. In Figure 3, they are identified by the red diamond markers.

The second peak (5.2717 MHz) is below the band-averaged signal level and the detection threshold, and is associated with continuous narrowband (~3 KHz) downwards chirping signals at the initial and final 10–15 s in the scan along with similar short-duration (~5 s) chirps repeating every 10–15 s. The origin of this signal is unknown.

The first peak (5.2599 MHz) is above the detection threshold and as such it is considered as a possible interference source. This signal is associated with the SeaSonde station at Red Head (site code: RHED), New South Wales, located approximately 100 km southwest of Seal Rocks. The half-power peak width Δf_P associated with this signal is 2.55 KHz, which is less than 25% the real bandwidth of the SeaSonde chirp in use at the RHED station.

Following the presence of the peak at 5.2599 MHz, the algorithm suggests an optimal transmit frequency of 5.2694 MHz (center value) with a bandwidth of 11.3 KHz for which no reduction is required.

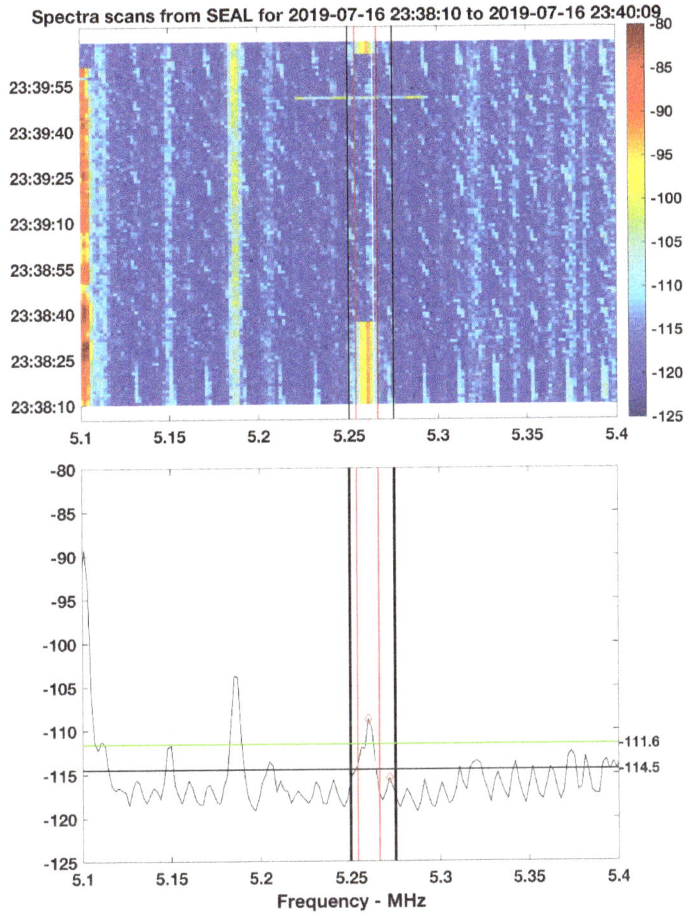

Figure 3. Frequency scan from SEAL HFR station collected between 20190716T233810Z–20190716T234009Z (upper panel) and time-averaged signal power (bottom panel). Red diamonds markers at 5.2599 MHz and 5.2717 MHz show the location of two potential interference sources. Vertical black lines indicate the 5.250–5.275 MHz ITU band in use to the HFR systems; red vertical lines show the band occupied by the RHED HFR station. The black horizontal line corresponds to the band-averaged signal level identify two different noise threshold. Units for signal level and frequency are dB and MHz, respectively.

As a second example, we consider the spectrum scan collected between 20190705T161835Z and 20190705T162034Z (Figure 4). This scan, collected again using the default values listed before (120 sweeps, 1 Hz sampling rate, 256 *fft* points) presents features similar to the case discussed previously, however it has also distinct features that were found to occur frequently in the region.

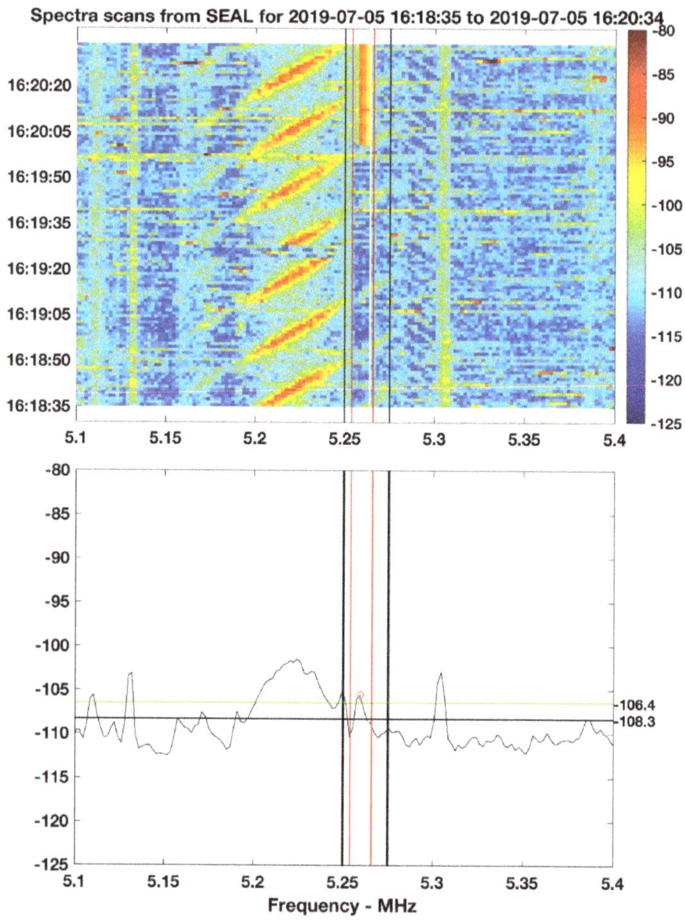

Figure 4. Frequency scan from SEAL HFR station collected between 20190705T161835Z–20190716T162034Z (upper panel) and corresponding time-averaged power (bottom panel). Vertical black lines indicate the 5.250–5.275 MHz ITU band in use to the HFR systems; red vertical lines show the band occupied by the RHED HFR station; green dotted lines show the allocated bandwidth and the center frequency for the new acquisition cycle.

Note that in general the average noise level in the entire spectrum is significantly higher than the previous example reported in Figure 4: time-frequency-averaged signal level $NF(f)_{band}$ and the detection threshold value have increased to, respectively: −108.3 dB and −106.4 dB. Interestingly enough, standard deviation σ_{PL} has on the opposite decreased to 1.9 dB, consistently with a persistently increased noise level within the band. The peak-finding algorithm tracks the peak associated with the RHED transmit signal, centered at 5.2596 MHz, starting at 20190705T162000Z (red marker in Figure 4), with a signal level of −105.5 dB, half-power peak width $\Delta f_P = 3.2$ kHz. Note also the intense upwards chirp centered at 5.225 MHz (origin unknown), which is entering the ITU frequency band.

Following the presence of the peak at 5.2596 MHz, the algorithm suggests an optimal transmit frequency of 5.2694 MHz (center value) with a reduced bandwidth of 9.1 KHz resulting in a transmit chirp in the frequency band 5.2648–5.2739 MHz. The optimized transmit bandwidth is 80.5% the bandwidth originally available at the site which results in a radial range resolution of 16.48 km.

A more complex case is presented in Figure 5, collected between 20190705T143835Z and 20190705T144034Z in which the chirp from the RHED station can again be noticed, along with two significant 50 kHz down-sweeping chirps centered at approximately 5.175 MHz and 5.225 MHz. The algorithm correctly identifies the peak associated with RHED station (signal level −102.8 dB) at 5.2576 MHz. This signal exceeds the detection threshold (−104 dB) and consequently the algorithm suggests the following parameters for the next data acquisition cycle: center transmit frequency 5.2694 MHz, optimal bandwidth 11.3 KHz, with upper-lower limits for the sweep between 5.2637–5.2750 MHz.

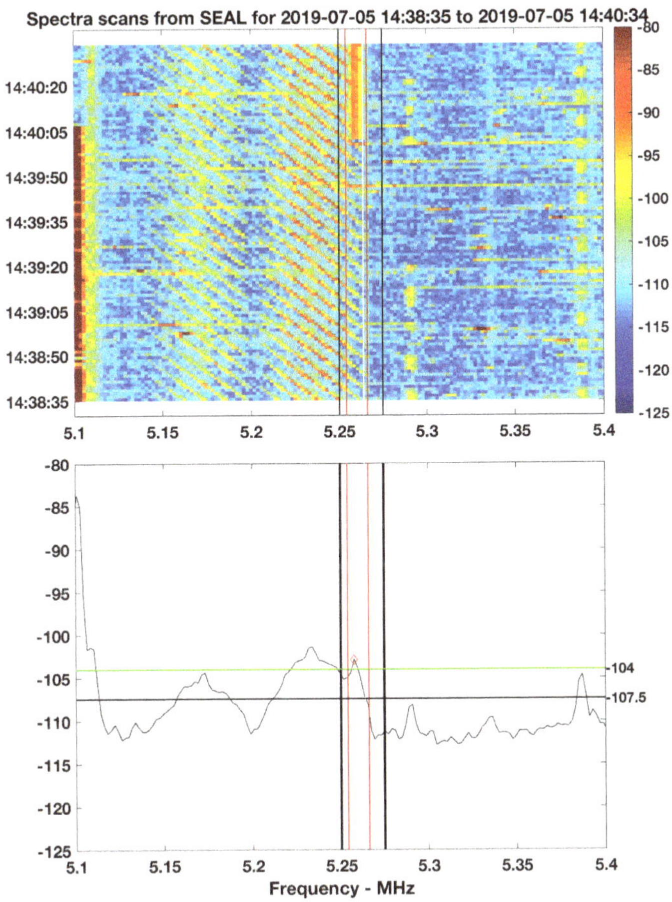

Figure 5. Frequency scan from SEAL HFR station collected between 20190705T143835Z–20190705T144034Z (upper panel) and corresponding time-averaged power (bottom panel). Vertical black lines indicate the 5.250–5.275 MHz ITU band in use to the HFR systems; red vertical lines show the band occupied by the RHED HFR station; green dotted lines show the allocated bandwidth and the center frequency for the new acquisition cycle.

Adjustments in the data acquisition parameters (centre frequency, bandwidth) are likely to impact on the data quality. Adjustments (i.e., reduction) in the transmit bandwidth (BW) after a peak is detected in the frequency scan would result in variations in the the radial range resolution (ΔR, km) as described in Equation (5):

$$\Delta R = c/(2^{*}BW), \quad (6)$$

where $c = 2.99 \times 10^8$ m/s the speed of light in a vacuum, BW the sweep width (Hz) of the transmit HFR chirp, and the factor 2 accounting for the two-way path between HFR and target. As an example, Figure 6 shows the decreased range resolution between two consecutive data acquisition cycles (20191026T000000Z–20191026T002000Z) after BW was reduced from 25 KHz to 12.5 KHz. Data were collected at the Seasonde site at Lancelin, WA, approximately 100 km North of Perth.

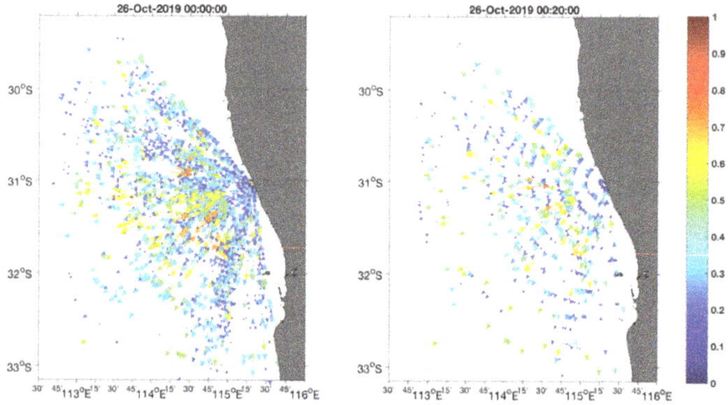

Figure 6. Example of the effects of bandwidth reduction between two consecutive radial maps at Lancelin SeaSonde station (Western Australia). To the left: radial map collected on 26 October 2019, 00:00 (UTC) with 25 kHz bandwidth, resulting in a radial range resolution of 6 km; to the right, radial velocity map collected on 26 October 2019, 00:20 (UTC) with 12.5 kHz bandwidth, resulting in a radial range resolution of 12 km. Operating frequency is 4.463 MHz. Unit for radial velocity is m/s.

The effects of the LBT on the operations of other system, for instance WERA HFRs deployed in Florida, have been investigated with great details in [14]. Authors suggest that changes in the local external noise levels will likely result in different bandwidths both spatially and temporally between paired sites in a HFR network. However, they also conclude that the application of the WERA LBT adaptive algorithm, along with a wide enough bandwidth to operate within, increased data coverage and improved the overall signal-to-noise ratio (SNR) conditions. In the example illustrated in Figure 6 relative to the SeaSonde system in Lancelin (WA), the 50% BW reduction corresponds to variations in the radial resolution from 6 km to 12 km within 20 min. SeaSonde systems typically require continuous transmission and averaging at a spectral level to increase the statistical significance of the Doppler spectra, decrease noise level, and improve the spectral inversion processes, and at a temporal and spatial level, to reduce errors in the final radial output. As such, it is likely that impacts of adjustments to the radar processing parameters, specifically bandwidth, are detrimental if repeated frequently. Depending on the operational settings and transmit band, a full measurement cycle is typically performed within 1 h to 3 h for the high-frequency (13 MHz, 25 MHz) and low-frequency (4 MHz, 5 MHz) systems, respectively. Additionally, low-frequency systems would be more affected than high-frequency counterparts by the limited bandwidth allocated by [5]: 150 KHz for the 25 MHz band, 100 KHz for the 13.5 MHz and 16.1 MHz band, 50 KHz for the 9.33 MHz band, and 50 KHz (25 KHz) for the 4 MHz (5 MHz) bands, respectively.

Other aspects that need to be accounted for are the losses in efficiency in both the transmitter and receiver antennas following changes in the transmit-receive bands. SeaSonde systems are particularly sensitive to the local environment and require calibrations of the so called antenna beam pattern in order to account for coupling with metal structures within the receive antenna near field. In principle, calibration should be repeated upon changes to the transmit frequency; however, this is unpractical, and it is commonly accepted within the HFR community that a calibration for a center frequency f_C is applicable within $\Delta f = \pm 100$–200 KHz (Brian Emery, personal communication). To some extent, this is an aspect in common with phased-array HFRs such as the WERAs in which the optimal spacing for the antenna elements in the receive array is strictly dependent on the wavelength of the transmitted signal. Changes in the center frequency f_C are likely to affect beam-forming capabilities. In both cases, more detailed investigations are needed in order to quantify effects on HFR accuracy. When it comes to the transmit antennas, changes in transmit frequency may possibly increase losses along the transmission line and as such decrease antenna radiation efficiency, cause distortions in the transmit signal, increase the fraction of power reflected at the radiator and trigger the power amplifier protection circuitry if available, or damage the amplifier. It is clear that a similar issue is platform-independent and would affect in similar ways both SeaSonde and WERA HFR systems.

Figure 7 shows a typical resonance curve for a standard low-frequency SeaSonde transmit antenna in terms of the voltage standing wave ratio (VSWR) spectrum. Data were collected in April 2019 during the most recent SeaSonde installation in Dongara-Port Denison, located approximately 400 km north of Perth. VSWR is commonly accepted as a measure of impedance matching of a load to the characteristic impedance of a transmission line and basically is a measure of antenna radiating efficiency with VSWR = 1 when 100% of the transmit power is radiated [16]. For this specific example, which refers to the 4.438–4.488 MHz ITU band, VWSR is in the range 2.4–1.9 resulting in 17%–9.6% of the transmit power being reflected along the transmission line. Typical transmit power barely exceed 10 W for the SeaSonde systems deployed in WA, and is below 3 W or less at the NSW deployments. Assuming the same VSWR values apply to both deployments, changes in transmit power are found in the range 8.3–9.04 W and 3.3–3.6 W respectively, and do not significantly affect overall system performances.

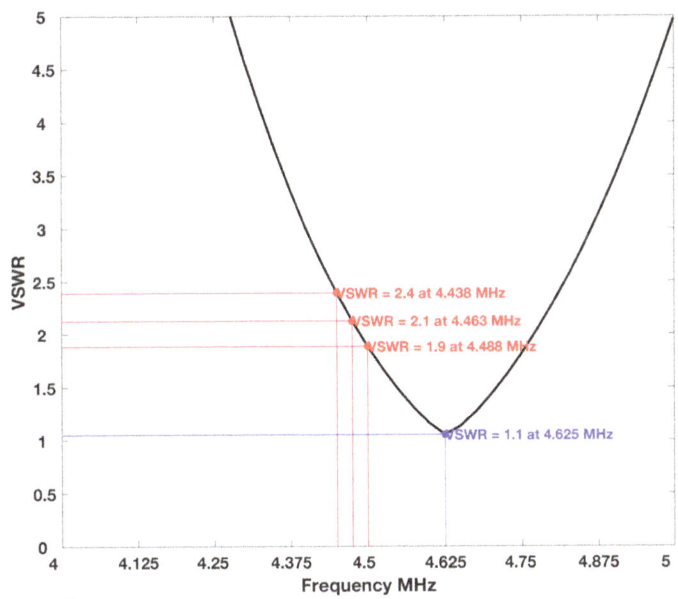

Figure 7. Voltage standing wave ratio (VSWR) data for the SeaSonde transmit antenna at Dongara—Port Denison HFR site, located approximately 400 km·N of Perth (WA). VSWR values for the 4.438 MHz, 4.463 MHz and 4.488 MHz frequency are reported in red on the VSWR curve. These frequencies correspond to the upper, center and lower frequencies for the ITU frequency band in use at the site. The minimum VSWR value is also reported in the plot, along with the corresponding frequency. Unit for frequency is MHz; VWSR is dimensionless.

One last aspect that needs to be accounted for, is the potential mutual cross-interference from a second SeaSonde system. This is the case illustrated across Figures 3–5 showing the chirp from the second station either at the beginning or the end of the frequency scan. The proprietary software is capable of handling in a very effective way several SeaSonde systems in the same frequency band through the proprietary GPS synchronization and band sharing capabilities [8]. The present peak identification algorithm appears on the opposite to interpret the chirp as a potential interference source. As such, an accurate timing is required across a HFR network, so to minimise or avoid false-alarm detection. This is a critical aspect since SeaSonde systems rely on commercial Macintosh systems in which the clocks can drift significantly.

4. Discussion

The operation of HFR systems in Australia is dependent upon the fulfillment of the ITU regulation, particularly for what concerns the band sharing capabilities, the interference mitigation measures and the reduction of the offband leakage. ITU Report M.2234 and Recommendation M.1874-1 [4,5] describe some of the possible mitigation measures, such as directional transmit antennas to reduce signal transmission over land or at unwanted directions, the use of pulse-shaping and the band sharing capabilities, and discuss of the possible implementation of the listen-before-talk and bandwidth adaptation mode and report possible significant impacts on data quality which is deemed unacceptable by the HFR operators. SeaSonde systems already implement the GPS syncronization capability, that allows for multiple sites to operate in the same band without mutual interference and allow for a more efficient use of the radio frequency spectrum. WERA systems typically operate with directional transmit arrays and have the capability to dynamically adapt their operating frequency and bandwidth in response to strong interference sources. Pulse shaping with longer ramping-deramping times, longer pulse width, and modified weighting windows have been satisfactorily deployed in the SeaSonde systems across Australia in order to reduce off-band leakage. Experimental directive antennas are used to limit the signal emission over land, and an interrupted transmit operation mode is also used to allow for radio silence periods between intermediate acquisition cycles. To further avoid interferences to primary users operating a 3 KHz channel at 5.264 MHz, center frequency and transmit width have been modified accordingly to 5.257 MHz and 11 KHz respectively, so to remain within the allocated ITU frequency band. The approximate 90 sec radio silence period between each acquisition cycles can be used to monitor the frequency band for either interferences or in case HFR transmission needs to be modified to account for primary users within the band.

Two main results are reported here. It is shown here that radio frequency scans can be acquired using the existing SeaSonde hardware, using the spectrum analyzer capabilities embedded in the proprietary software. No additional external hardware components are required. The resulting spectrum scans can be analyzed and the real-time acquisition parameters can be adapted in case of need, as shown in this work. However, an adaptation (i.e., reduction) of the transmit bandwidth would result in variable radial range resolution if performed too frequently, with consequent degradation of the quality of radial velocity data. It is as such suggested to perform an adaptation at the end of a full acquisition cycle. Spectral scans can still be collected at regular intervals for diagnostic purposes. It is also important that users and operators of the HFR systems are actively involved across all stages of HFR operations and that an informed decision is taken also in light of a fair-use policy for band sharing.

It is confirmed that operating HFR systems within the ITU frequency bands with a secondary-type license, that is on the basis of no interference to, and with no rights of protection from primary users, is complicated and may require extensive modifications to the hardware and acquisition parameters in order to comply with the national and international regulations. It is also confirmed that operating HFR systems within the ITU bands is prone to extensive interferences of unknown origin, especially in the low frequency portion of the radio frequency spectrum, as already reported previously [17].

Author Contributions: Conceptualization, S.C.; Methodology, S.C.; Software, S.C.; Formal Analysis, S.C.; Resources, S.C.; Data Curation, S.C.; Writing—Original Draft Preparation, S.C.; Writing—Review & Editing, S.C.; Project Administration, S.C.; Funding Acquisition, S.C. All authors have read and agreed to the published version of the manuscript.

Funding: This research was funded by Integrated Marine Observing System IMOS—Ocean Radar—2017–2019 grant number 53000300.

Acknowledgments: Data was sourced from the Integrated Marine Observing System (IMOS)—IMOS is supported by the Australian Government through the National Collaborative Research Infrastructure Strategy and the Super Science Initiative. The SeaSonde data used here were collected by the Ocean Radar Facility at the University of Western Australia at Seal Rocks and Red Head sites, New South Wales, Lancelin and Dongara - Port Denison, Western Australia. A special acknowledgement to Codar Ocean Sensors for useful indications on file formats and the support while implementing the procedure.

Conflicts of Interest: The authors declare no conflict of interest.

References

1. Roarty, H.; Cook, T.; Hazard, L.; Harlan, J.; Cosoli, S.; Wyatt, L.; Matta, K.S. The global high frequency radar network. *Front. Mar. Sci.* **2019**, *6*, 164. [CrossRef]
2. Cosoli, S.; Grcic, B.; de Vos, S.; Hetzel, Y. Improving data quality for the Australian high frequency ocean radar network through real-time and delayed-mode quality-control procedures. *Remote Sens.* **2018**, *10*, 1476. [CrossRef]
3. Cosoli, S.; de Vos, S. Interoperability of Direction-Finding and Beam-Forming High-Frequency Radar Systems: An Example from the Australian High-Frequency Ocean Radar Network. *Remote Sens.* **2019**, *11*, 291. [CrossRef]
4. Huang, W.; Gill, E.W. HF surface wave radar. In *Wiley Encyclopedia of Electrical and Electronics Engineering*; Webster, J.G., Ed.; John Wiley and Sons: Hoboken, NJ, USA, 2019.
5. International Telecommunication Union. Final Acts WRC-12, Resolution 612, Use of the radiolocation service between 3 and 50 MHz to support oceanographic radar operations. In Proceedings of the World Radiocommunication Conference, Geneva, Switzerland, 23 January–17 February 2012; Available online: https://www.itu.int/en/ITU-R/terrestrial/fmd/Documents/Res%20612.pdf (accessed on 20 April 2019).
6. International Telecommunication Union. *The Feasibility of Sharing Sub-Bands between Oceanographic Radars Operating in the Radiolocation Service and Fixed and Mobile Services within the Frequency Band 3–50 MHz*; ITU-R Radiocommunication Sector of ITU, Report ITU-R M.2234 (11/2011); Electronic Publication: Geneva, Switzerland, 2012; Available online: https://www.itu.int/en/ITU-R/terrestrial/fmd/Documents/R-REP-M.2234-2011-PDF-E.pdf (accessed on 20 April 2019).
7. International Telecommunication Union. *Technical and Operational Characteristics of Oceanographic Radars Operating in Sub-Bands within the Frequency Range 3–50 MHz*; ITU-R Radiocommunication Sector of ITU, Recommendation ITU-R M.1874-1, M Series, Mobile, Radiodetermination, Amateur and Related Satellite Services, 02/2013; Electronic Publication: Geneva, Switzerland, 2012; Available online: https://www.itu.int/en/ITU-R/terrestrial/fmd/Documents/R-REC-M.1874-1-201302-I!!PDF-E.pdf (accessed on 20 April 2019).
8. Barrick, D.E.; Lilleboe, P.M.; Teague, C.C. Multi-Station HF FMCW Radar Frequency Sharing with GPS Time Modulation Multiplexing. U.S. Patent US6856276B2, 15 February 2015. Available online: http://www.codar.com/images/about/patents/Patent_2005_GPS_granted.pdf (accessed on 20 April 2019).
9. SeaSonde Radial Site Application Guide for Sea Sonde Controller. Available online: http://www.codar.com/Manuals/GuidesToTheApplications/Guide_SeaSondeController.pdf (accessed on 20 April 2019).
10. Introduction to AppleScript Language Guide. Available online: https://developer.apple.com/library/archive/documentation/AppleScript/Conceptual/AppleScriptLangGuide/introduction/ASLR_intro.html (accessed on 20 April 2019).
11. De Vos, S.; Cosoli, S.; Munroe, J. Traveling Wave Antennae for directional low band HF RADAR Transmission. In Proceedings of the Radiowave Oceanography Workshop, Victoria, BC, Canada, 28–30 August 2019.
12. Rashid, I.; Simpson, S. The struggle for co-existence: Communication policy by private technical standards making and its limits in unlicensed spectrum. *Inf. Commun. Soc.* **2019**. [CrossRef]
13. Maglogiannis, V.; Naudts, D.; Shahid, A.; Moerman, I. An adaptive LTE listen-before-talk scheme towards a fair coexistence with Wi-Fi in unlicensed spectrum. *Telecommun. Syst.* **2018**, *68*, 701–721. [CrossRef]
14. Merz, C.R.; Liu, Y.; Gurgel, K.-W.; Petersen, L.; Weisberg, R.H. Effect of Radio Frequency Interference (RFI) Noise Energy on WERA Performance Using the "Listen Before Talk" Adaptive Noise Procedure on the West Florida Shelf. In *Coastal Ocean Observing Systems*; Academic Press: Cambridge, MA, USA, 2015.

15. Codar Ocean Sensors. *SeaSonde Radial Site Release 6*; Spectrum Analyzer File Format; Codar Ocean Sensors: Mountain View, CA, USA, 2011.
16. Gridley, J.H. *Principles of Electrical Transmission Lines in Power and Communication*; Elsevier: Amsterdam, The Netherlands, 2014; p. 265. ISBN 1483186032.
17. Middleditch, A.; Cosoli, S. Operational data management procedures for the Australian Coastal Ocean Radar Network. In Proceedings of the OCEANS 2016 MTS/IEEE Monterey, Monterey, CA, USA, 19–23 September 2016.

Journal of
Marine Science and Engineering

MDPI

Article

High-Frequency Radar Observations of Surface Circulation Features along the South-Western Australian Coast

Simone Cosoli *, Charitha Pattiaratchi and Yasha Hetzel

Ocean Graduate School and the UWA Oceans Institute, The University of Western Australia, 35 Stirling Highway, Perth, WA 6009, Australia; chari.pattiaratchi@uwa.edu.au (C.P.); yasha.hetzel@uwa.edu.au (Y.H.)
* Correspondence: simone.cosoli@uwa.edu.au; Tel.: +61-8-6488-7314

Received: 10 January 2020; Accepted: 4 February 2020; Published: 5 February 2020

Abstract: A new merged high-frequency radar (HFR) data set collected using SeaSonde and WERA (WEllen RAdar) systems was used to examine the ocean surface circulation at diurnal, seasonal and inter-annual time scales along the south-west coast of Australia (SWWA), between 29°–32° S. Merging was performed after resampling WERA data on the coarser SeaSonde HFR grid and averaging data from the two HFR systems in the area of common overlap. Direct comparisons between WERA and SeaSonde vectors in their overlapping areas provided scalar and vector correlation values in the range Ru = [0.24, 0.76]; Rv = [0.39, 0.83]; ρ = [0.44, 0.75], with mean bias between velocity components in the range [−0.02, 0.28] ms^{-1}, [−0.16, 0.16] ms^{-1} for the U, V components, respectively. The lower agreement between vectors was obtained in general at the boundaries of the HFR domains, where the combined effects of the bearing errors, geometrical constraints, and the limited angular field of view were predominant. The combined data set allowed for a novel characterization of the dominant features in the region, such as the warmer poleward-flowing Leeuwin Current (LC), the colder Capes Current (CC) and its northward extensions, the presence of sub-mesoscale to mesoscale eddies and their generation and aggregation areas, along with the extent offshore of the inertial-diurnal signal. The contribution of tides was weak within the entire HFR domain (<10% total variance), whilst signatures of significant inertial- and diurnal-period currents were present due to diurnal–inertial resonance. A clear discontinuity in energy and variance distribution occurred at the shelf break, which separates the continental shelf and deeper offshore regions, and defined the core of the LC. Confined between the LC and the coastline, the narrower and colder CC current was a feature during the summer months. Persistent (lifespan greater than 1 day) sub-mesoscale eddies (Rossby number O (1)) were observed at two main regions, north and south of 31.5° S, offshore of the 200 m depth contour. The majority of these eddies had diameters in the range 10–20 km with 50% more counter clockwise rotating (CCW) eddies compared to clockwise (CW) rotating eddies. The northern region was dominated by CCW eddies that were present throughout the year whilst CW eddies were prevalent in the south with lower numbers during the summer months.

Keywords: high frequency radar; sea surface temperature; surface currents; south-west Australia

---

1. Introduction

Surface circulation along the eastern margin the southern Indian Ocean is very different compared to other ocean basins, with the poleward Leeuwin Current (LC) transporting warmer water (Figure 1) and promoting a downwelling system [1]. The eastern margins of all other ocean basins comprise equatorward cold water currents associated with upwelling. The anomalous ocean currents together with diurnal tides with low range [2], strong sea breezes [3], resonance at the critical latitude [4], high air–sea heat and freshwater fluxes [5] provide a unique environment. In addition, the LC is associated with elevated levels of mesoscale eddy activity compared with other eastern boundary current regions [6,7]. The region is dominated by two major currents (Figure 1): the warmer Leeuwin Current (LC), flowing southwards along the shelf break, and the colder Capes Current (CC), flowing northwards along the continental shelf inshore the LC [1,8].

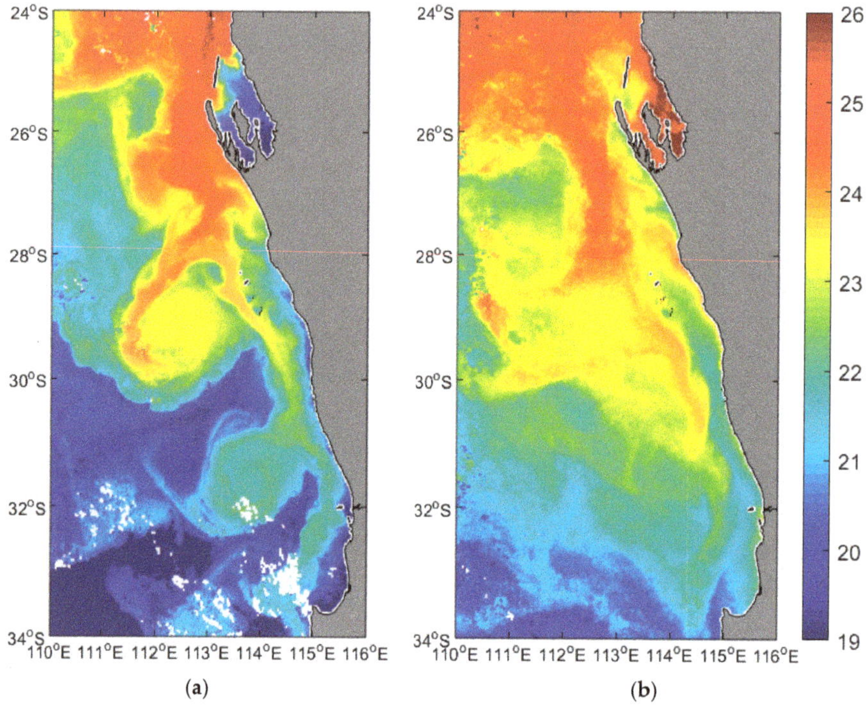

Figure 1. Satellite derived sea surface temperature distribution (°C) during: (**a**) winter (28 May 2018); and, (**b**) summer (23 February 2016) indicating the eddy systems of the Leeuwin Current (LC) and colder Capes Current (CC). Sea surface temperature (SST) data originated from the multi-sensor Australian Integrated Marine Observing System (IMOS) L3S foundation SST product.

The LC is a poleward eastern boundary current driven by an alongshore pressure gradient, generated through the transport of warmer, lower salinity water into the tropical Indian Ocean by the Indonesian Throughflow [9–11]. It is augmented with contributions from the central South Indian Counter Current (SICC) and the Holloway Current in the shelf region around 20°–23° S and from southern SICC at around 32° S. LC flows southwards as a relatively narrow (~100 km wide), shallow (upper 250 m) current along the continental shelf break (~200 m) to the southern tip of Tasmania [11,12]. The LC has a seasonal pattern with intensification during Austral autumn/winter (April–September) and a weakening during the Austral summer months (November–March). This seasonal variability is controlled by seasonal changes in the wind stress: (1) during the Australasian monsoon season (January to March) to the north of Australia [13]; and, (2) equatorward wind stress within the study region [3,9]. The LC is also influenced by El Niño and La Niña cycles with the current being stronger (weaker) during La Niña (El Niño) events due to fluctuations in the ITF and equatorial monsoonal winds [10,11,14]. Typical features of the LC include meanders, eddies, and filaments [15] and are highlighted in Figure 1. One of the major features of the LC is the presence of a highly energetic meso-scale eddy field with diameters >200 km and lifetimes of one month or more [15–18]. Anti-clockwise (warm-core) and clockwise (cold-core) rings are generated by the LC at preferential locations (Figure 1) that are associated with changes in the coastline orientation, and they migrate westward into the interior of the southern Indian Ocean [6]. The formation of these eddies has been postulated to be from different mechanisms such as instabilities through meanders, shear between the LC and Leeuwin Undercurrent/Capes Current and bathymetric features [16–19] and are located at ~28° S and ~30.5° S (Figure 1). These eddies are important circulation features, as they influence the heat and momentum balance [7]; regional upwelling and therefore contribute to local biological activity [16,19,20] and, influence the coastal wave climate [21].

The CC is a wind driven current that transport upwelled cooler, salty waters northwards inshore of the LC (Figure 1b) originating at ~34° S and extending as far north as 29° S [8]. It is mainly present during the summer months (November-March) when the upwelling favorable equatorward winds are the strongest. Here, due to Ekman dynamics, the upper layers move offshore in response to the northward wind stress pushing the LC offshore [7,8]. The CC is usually visible as a band of colder water on the continental shelf (Figure 1b).

Wind action is the primary driving mechanism for surface current variability on the continental shelf. The study region is micro-tidal, dominated by diurnal tides and with a mean spring tidal range of ±0.6 m [2]. The inertial period is around 24 h. The strong land–sea breeze (LSB) regime that characterizes this region [3,4,22] results in strong counterclockwise diurnal currents in the upper-ocean dynamics through the diurnal–inertial resonance at critical latitudes, that can penetrate below the mixed-layer depth [4]. Extra-tropical storms or tropical cyclones may also contribute to the intensification of upper layer current variability in the region [23].

The study region consists of several distinct bathymetric features (Figure 2): (1) an upper continental shelf terrace to a depth of 100 m; (2) rapid increase in water depth between the 100 and 200 m depth contours; (3) the main shelf break at the 200 m isobath; and, (4) presence of the Perth canyon at 32° S with maximum depths to 4000 m.

High-frequency radar (HFR) measurement of surface currents in the coastal ocean has become a standard and cost-effective component for ocean observing systems globally [24]. They were developed more than three decades ago [25,26] and are now widely used with a high level of accuracy [27,28] for a variety of purposes, including ocean current forecasting [29] and long-term trends in surface circulation in coastal areas [30]. HFR systems allow for the synoptic observation of large-scale coastal circulation features with temporal and spatial resolution not readily available with conventional instrumentation a high level of accuracy [27]. As part of the Australian Integrated Marine Observing System (IMOS), HFR systems have been deployed along the south-west Australian coast since 2009.

Previous studies in the region have mainly used satellite imagery [6,8,15], shipborne measurements [1,7,31] and, numerical modelling [17,19]. Mihanovic et al. [4] used oceanographic moorings, satellite sea surface temperature (SST) and HFR data in the southern section of the study region to examine the diurnal–inertial resonance at the critical latitude during the summer months. Studies that have examined the diurnal, seasonal and eddy dynamics in the study region have been limited either by low spatial and temporal resolution (e.g., satellite data), or restricted by small spatial domains (e.g., moorings or single HFR systems).

We build on previous work by merging data from two HFR systems that doubles the spatial coverage whilst retaining the high temporal and spatial information. The aim of this paper is to describe the surface oceanographic features along the coast of south-west Western Australia (SWWA) using the new merged near-surface HFR derived currents. The emphasis is on characterising the: (1) surface currents at diurnal (including inertial) lower frequency (>4 days) and seasonal time scales; and, (2) dynamics of the eddy field. The merged surface current data were collected over a 33-month period (January 2015 to September 2017) and covered a region 180 km offshore and 250 km alongshore at a resolution of 6 km.

This paper is arranged as follows: Section 2 describes the data collection and analysis techniques. Results and discussion are presented in Sections 3 and 4. The conclusions are presented in Section 5.

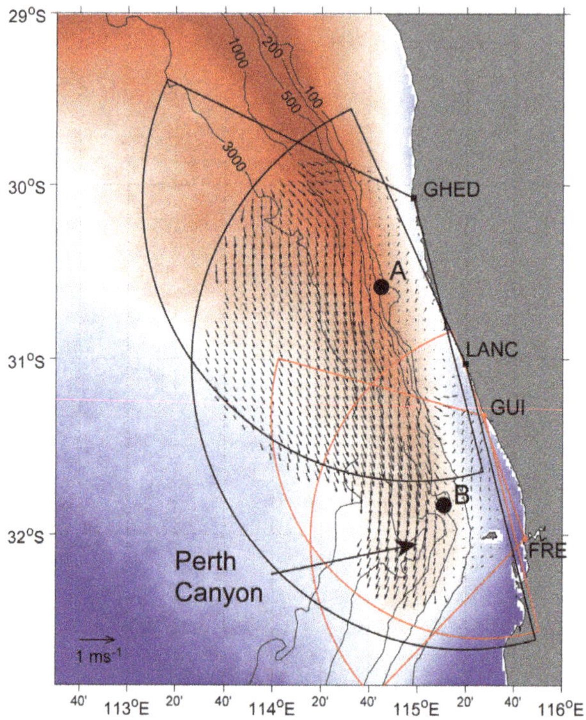

Figure 2. The study region showing the high-frequency radar (HFR) coverage region. Locations marked A and B correspond to the points with maximum data return for the Seasonde (**A**) and WERA (**B**) HFR nodes, respectively. The time averaged HFR currents are overlain on SST data for the period January 2015–September 2017.

2. Materials and Methods

Surface current data (top 2 m of the water column) collected by long-range HFR systems (Figure 2) over the 33-month period January 2015 to September 2017 were used in this study. HFR current data were sourced through commercial direction-finding (SeaSonde) and phased-array (WERA) HFR systems, provided by Codar Ocean Sensors (COS) and Helzel MessTechnik, respectively.

Two SeaSonde systems, located at Lancelin (LANC; 31°01.588′ S, 115°19.708′ E) and Green Head (GHED; 30°04.393′ S, 114°58.003′ E), were used to collect ocean currents at hourly intervals with ranges up to 200 km offshore with a spatial resolution of 6 km. Both stations operated at a central frequency of 4.463 MHz (measurement depth is approximately 2.5 m) with a 25 kHz bandwidth, 512-point *fft*, 1 Hz sweep rate, 20 min Doppler spectra averaging with 15-min output rate. Calibrated antenna patterns are used at both stations. Two WERA systems, located at Guilderton (GUI; 31°20.4960′ S, 115°29.3640′ E) and Port Beach, Fremantle (FRE; 32°.0198′ S, 115°44.7480′ E) provided hourly ocean surface currents at ranges up to 180 km offshore at 4 km spatial resolution. Both sites used 4-element transmit and receive arrays of 16 equally spaced elements. Systems operated at 9.335 MHz (sampling depth approximately 1.3 m) with 33.3 kHz bandwidth, 1024-point *fft* and 0.26 s sweep rate, and an integration time of 5 min. Radial velocity data were collected on a Cartesian grid with 4 km spacing every 10 min in alternate mode. The same Cartesian grid was used to compute hourly surface current vectors from the hourly-averaged radial data.

Sea-surface temperature (SST) data originate from quality-controlled, merged, day and night, multi-sensor L3S foundation SST product from IMOS level L3S gridded single-sensor composites

(IMOS-SRS SATELLITE-SST L3S-01 day composite). SST data for the WA coast were downloaded from the AODN (Australian Ocean Data Network) data portal (https://portal.aodn.org.au). HRPT AVHRR (High Resolution Picture Transmission/Advanced Very High Resolution Radiometer) SST (sea surface temperature) retrievals are produced by the Australian Bureau of Meteorology as a contribution to the Integrated Marine Observing System [32].

The analysis approach used hourly, daily, monthly and seasonally-averaged delayed-mode quality controlled (DMqc) data [33], and the most appropriate statistical tools were used to characterize the dominant circulation features in the study region.

Smoothed estimates of the power density distribution (PSD) of surface HFR currents were derived from hourly HFR data in order to partition current variance over resolvable frequencies at each HFR grid cell. Spectra of hourly HFR observations were derived following the Welch-modified periodogram method [34] using 1024-point segments (approximately 42-days long segments), a Hanning window and 50% overlap between the data segments. Band-averaged energy levels were then extracted by integrating the resulting PSD curves in the selected frequency bands: diurnal-inertial frequency [0.038, 0.046] cph (cycles per hour), corresponding to periods in the range [21 h 44 m 20.8 s, 26 h 18 m 56.8 s]; and, low-frequency band, with frequency <0.0088 cph, or periods longer than *4d*. Rotary coefficients (C_R; Equation (1); [35]), were then derived from the smooth PSD estimates of the clockwise (CW)—counterclockwise (CCW) spectra:

$$C_R = \frac{PSD_+(\omega) - PSD_-(\omega)}{PSD_+(\omega) + PSD_-(\omega)}, \tag{1}$$

where $PSD_\pm(\omega)$ are, the CCW and CW spectral estimates at each HFR grid cell at the frequency ω respectively. C_R quantifies the degree of polarization of current observations at each frequency band; for purely rectilinear flow, C_R equals to zero, and $C_R = \pm 1$ for purely circular (counterclockwise-clockwise) motion.

The MATLAB *t tide* package [36] was used to quantify the contribution of tidal forcing to the current variance. Synthetic time series at each grid cell included in the analysis are derived using only constituents with signal-to-noise ratios greater than 2.

For spectral analysis, gaps in the time series were linearly interpolated before the *fft* calculations, whilst no interpolation was performed before the tidal analyses. Grid points with at least 70% data return were used for both analyses, in order to increase the frequency resolution and reduce biases from missing data.

The merged SeaSonde—WERA data set was obtained as follows. DMqc procedures described in [33,37] were applied separately to SeaSonde and WERA current vector data at each HFR grid point. DMqc flags were applied to the hourly data sets to exclude anomalous currents from the records. We then computed daily-averaged currents fields separately for the SeaSonde and WERA HFR data sets using grid points with at least 50% data return (after removing spikes). Daily SeaSonde and WERA currents were remapped on a common longitude–latitude grid by searching velocity data within 2 km from each grid cell, and averaging separately the U, V components of the velocity vectors. In case of overlap between SeaSonde and WERA currents, U, V from both HFR genres were averaged.

Validation of SeaSonde and WERA HFR data in the region against current meter data for the same data period were performed previously [33,38], showing that both HFR types captured the variability sampled by the current meters consistent with accuracy levels reported in the literature. The interoperability of the two HFR genres in the region was also demonstrated for a shorter time period (1 month), showing that radial data from SeaSonde can be merged successfully with radial data from WERA systems with no significant performance losses [38]. Extending the comparisons to the overlapping areas between the two grids, scalar (vector [39]) correlation (Equation (2); Equation (4)) values for the U, V components were in the range Ru = [0.24, 0.76]; Rv = [0.39, 0.83]; ρ = [0.44, 0.75]:

$$R = \frac{\sum_{i=1}^{n}(x_i - \overline{x})(y_i - \overline{y})}{(n-1)\sigma_x \sigma_y} \tag{2}$$

$$\sigma_x = \sqrt{\frac{\sum_{i=1}^{n}(x_i - \overline{x})^2}{n-1}} \tag{3}$$

$$\rho = \frac{u_1 u_2 + v_1 v_2 + i u_1 v_2 - v_1 u_2}{u_1^2 + v_1^{21/2} u_2^2 + v_2^{21/2}} \tag{4}$$

$$bias = \frac{\sum_{i=1}^{n}(x_i - y_i)}{n} \tag{5}$$

In Equations (4), ($u_{1,2}$, $v_{1,2}$) stand for the zonal and meridional components of the SeaSonde (WERA) HFR velocity components respectively, i identifies the imaginary unit, and < > represents the ensemble average operator.

Values for U, V biases (Equation (4)) were in the range $bias_U = [-0.02, 0.28]$ ms^{-1}, $bias_V = [-0.16, 0.16]$ ms^{-1}. Statistics reported here are relative to a sample of 67 grid points, with a lowest population (number of common data) in excess of 500. It must be pointed out that lower agreement between vectors was obtained in general at the boundaries of the SeaSonde domain, where the combined effects of the bearing errors [33,38], the geometrical constraints common to both systems, and the limited angular field of view of the WERA systems were predominant.

For the following analyses, the HFR and SST data were seasonally averaged for 2015, 2016 and 2017. Here, the seasons were defined as: austral summer: December, January, February (DJF); autumn: March, April, May (MAM); winter: June, July, August (JJA); spring: September, October, November (SON). The daily-, monthly-, and seasonally-, averaged, merged SeaSonde-WERA HFR currents were then used for detailed analysis.

Ocean motions can be divided into four classes based on horizontal length scales [40]: (1) large scale motions that are > 300 km; (2) meso-scale motions are related to eddies with diameters 50–300 km; (3) sub-mesoscale motions have horizontal scales 1–50 km; and, (4) small-scale motions have scales <1 km. With HFR coverage, extending 100–170 km offshore with a resolution of 6 km, only sub-mesoscale motions could be resolved. Statistics of eddies (lifespan, dimensions, locations, rotation sense, vorticity, eddy Rossby number) were extracted from the daily-averaged, merged SeaSonde—WERA data set. The vector geometry approach described in [41] was used to identify eddies and their properties in the merged data set. Sub-mesoscale eddies that have time scales of O(days) are commonly observed in the region, particularly in the shear region between the LC and the CC currents, and only persistent eddies (minimum lifespan 1 day) could be detected in the data set used here.

Relative vorticity (ζ_{HF}; Equation (6)), divergence (ω_{HF}; Equation (7)) and Rossby number (R_o; Equation (8)) were computed from the merged, daily-averaged HFR currents at each grid point by a locally least-squares fitting a velocity plane to each grid point using current measurements from nearby locations [42]. Planetary vorticity (f) is defined in Equation (9), with $\Omega = 7.2921 \times 10^{-5}$ radians/s the rotation rate of Earth and θ the latitude.

$$\zeta_{HF} = \frac{\partial v}{\partial x} - \frac{\partial u}{\partial y}, \tag{6}$$

$$\omega_{HF} = \frac{\partial u}{\partial x} + \frac{\partial v}{\partial y}, \tag{7}$$

$$Ro = \frac{\zeta_{HF}}{f}, \tag{8}$$

$$f = 2\Omega \, sin\theta, \tag{9}$$

3. Results

3.1. The Mean State

The mean HFR currents over the study period (January 2015 to September 2017) indicated southward currents, offshore the 200 m contour over the whole HFR coverage region associated with warmer SSTs (Figure 2). This is the Leeuwin Current. In the northern region of HFR coverage, the range of the SeaSonde system extended further offshore compared to the southern region where the range of WERA was lower. Inshore of the 100 m depth contour, the currents were variable but the majority contained a northward component and were associated with colder water, particularly in the southern region of HFR coverage (Figure 2). This is the signature of the wind-driven Capes Current, most prominent during the summer months. There was also a northward component of wind stress, during the winter months that resulted in northward currents.

The mean cross-shelf profile of HFR currents and SST both indicated a Gaussian shape with weaker currents inshore, increasing to a maximum (the LC 'core') and then decreasing further offshore (Figure 3). The SST profiles indicated colder water inshore, increasing offshore to a maximum and then decreasing further offshore (Figure 3b). The peak in the southward velocity maxima was located further offshore compared to the SST maxima. For example, at 31° S, the peak in velocity (-0.17 ms^{-1}) was located at 114.5° E whilst for SST it was at 114.8° E (Figure 3). The LC flows parallel the depth contours (Figure 2). However, the depth contours (and the shoreline) is oriented 337° with respect to north. Therefore, the 200 m depth contour is shifted to the east as we move south. This is reflected in the cross-shelf profiles of both HFR currents and SST with the maxima migrating eastward (Figure 3). Profiles of HFR currents also indicated northward mean currents at the two southern latitudes (0.07 ms^{-1} at 32° S and 0.025 ms^{-1} at 31.7° S) whist at 31° S, although the currents so not indicate a positive component mean currents were close to zero (-0.02 ms^{-1}). The maximum velocity of the LC 'core' increased southward, from -0.17 ms^{-1} at 31° S and 31.7° S to -0.22 ms^{-1} at 32° S (Figure 3a). In contrast, the SST maximum decreased from 21.35 °C to 21.06 °C at 31° S (Figure 3b).

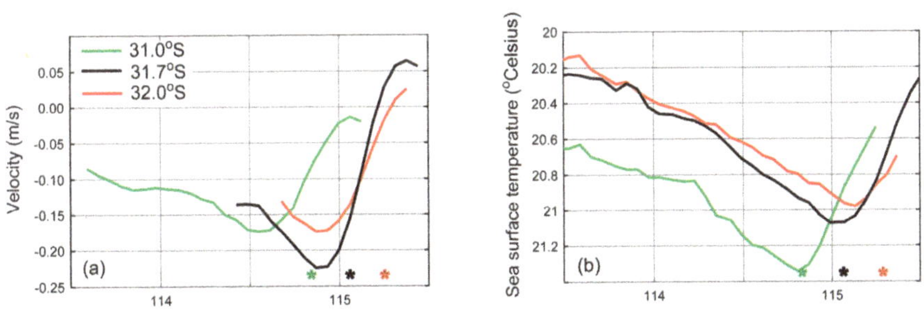

Figure 3. Mean cross-shelf profiles of (**a**) velocity; and, (**b**) sea surface temperature at different latitudes. The asterisk denotes the location of the 200 m depth contour. Units for the x-axis are °E.

3.2. Seasonal and Inter-Annual Variability

Seasonally averaged HFR and SST data indicated the typical mean circulation structures in the region with the relatively warmer LC as a southward current and the CC as a weaker, colder coastal current (Figures 4–6). The presence of colder water adjacent to the coast during the autumn and winter months (Figures 4c, 5c and 6c) is due to heat loss [5]. In general, the LC was stronger (mean speeds up to 0.50 ms^{-1}), during autumn/winter and weaker during spring/summer.

During the summer months, a band of colder water associated with northward currents is the CC (Figures 4a, 5a and 6a). Note that the water adjacent the shoreline is warmer due to summer solar heating and the band of colder water, associated with the upwelling CC, is located further offshore but inshore of the warmer LC. The boundary between the colder and warmer water corresponded with a change in current direction from northward to southward with strong horizontal velocity gradients. This was also highlighted by the cross-shelf velocity profiles at different latitudes where northward (positive) current component were recorded inshore (Figure 7a). At 31.0° S, the LC was present as a broader current with the strongest currents west of 114.5° E whilst at 31.7° S and 32.0° S the cross-shelf profiles indicated positive and negative maxima associated with the LC and CC (Figure 7a).

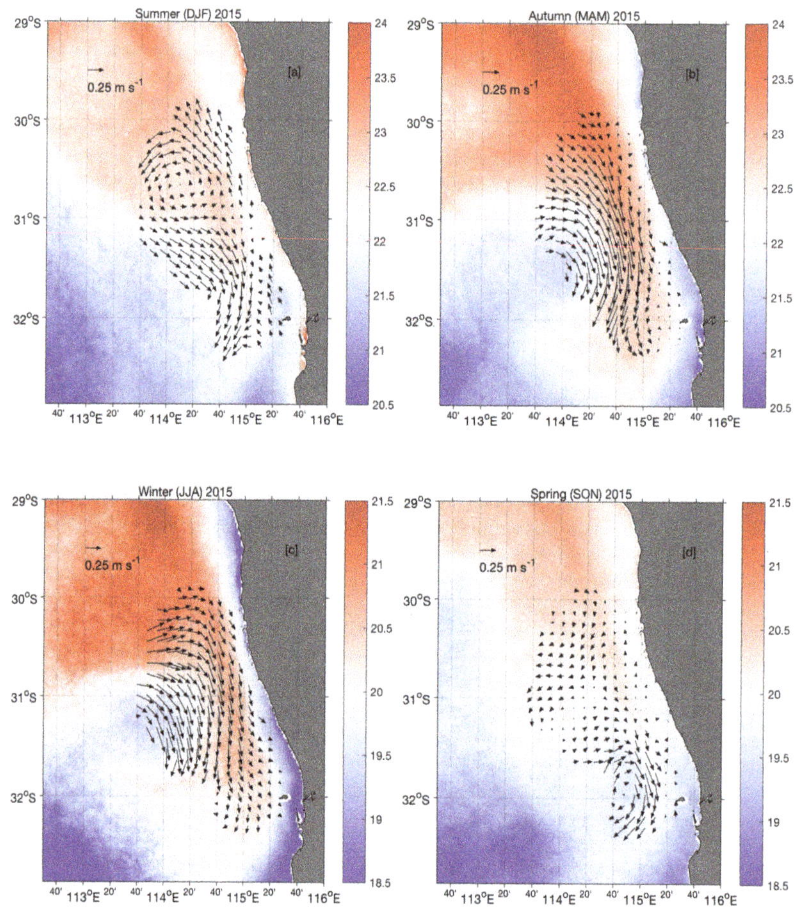

Figure 4. Seasonally-averaged HFR and SST data for 2015: (**a**) Summer (DJF); (**b**) Autumn (MAM); (**c**) Winter (JJA); (**d**) Spring (SON). Note changes in SST scale for summer/autumn (**a**,**b**) and winter/spring (**c**,**d**). Vector data have been subsampled every two grid point to increase readability. SST data originated from multi-sensor L3S foundation SST product.

During the winter months, the LC was the most dominant feature mostly associated with the warmer water. Although the HFR coverage captured the 'core' of the current, the larger meso-scale eddies were not captured fully. For example, in 2015, there was a large mesoscale clockwise eddy centered around 31.5° S, 114.0° E during autumn and winter that was only captured partially (Figure 4b,c) although altimetry data indicated the presence (and the persistence) of this eddy through autumn to winter (http://oceancurrent.imos.org.au/). Similar partial capture of a clockwise mesoscale eddy was also recorded during winter months of 2016 and 2017 (Figures 5c and 6c). Cross-shelf velocity profiles at different latitudes indicated strengthening of the southward (negative) current component, compared to those during the summer at all three latitudes (Figure 7). In particular, at 31.0° S, there was a pronounced 'core' in the LC with a maximum at 114.6° E, a significant onshore migration during winter compared to the summer (see Figure 7a,b).

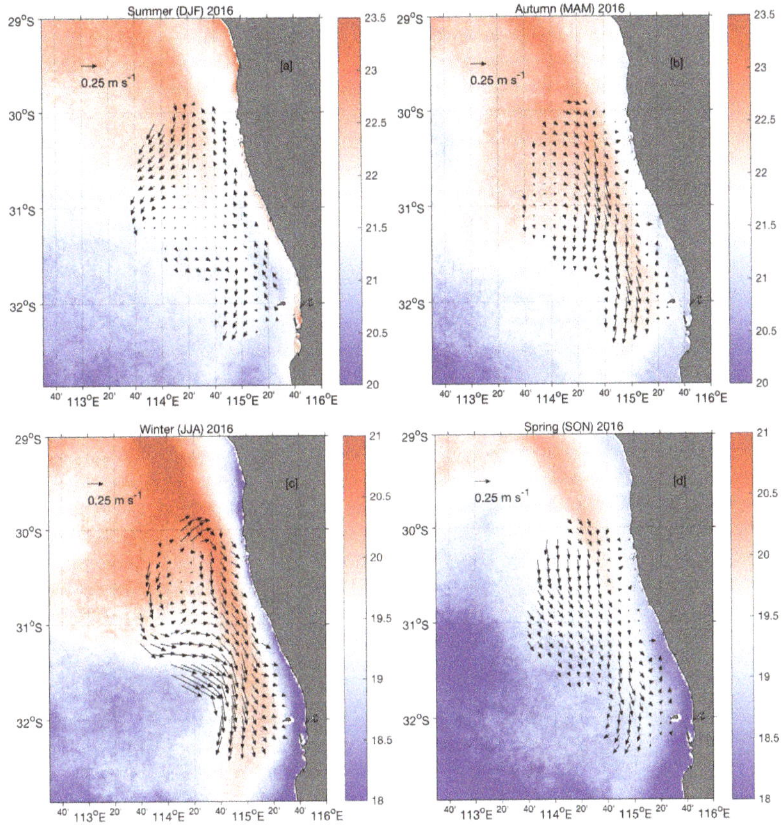

Figure 5. Seasonally-averaged HFR and SST data for 2016: (**a**) Summer (DJF); (**b**) Autumn (MAM); (**c**) Winter (JJA); (**d**) Spring (SON). Note changes in SST scale for summer/autumn (**a**,**b**) and winter/spring (**c**,**d**). Vector data have been subsampled every two grid point to increase readability. SST data originated from multi-sensor L3S foundation SST product.

Autumn months indicated the strengthening of the LC between summer and winter (Figures 4b, 5b and 6b). The currents during 2015 autumn were stronger when compared to 2016 and 2017. The spring months represented weakening of the LC from winter to summer (Figures 4d, 5d and 6d) and the currents during 2017 spring were stronger when compared to 2015 and 2016. The autumn and spring patterns indicated the presence of anti-clockwise eddies over the Perth canyon (approximately 32° S) e.g., spring 2015 and autumn 2017. A large meander with the LC flowing in an 'S' shaped pattern was evident during spring 2017 (Figure 6d).

It is well documented that LC is weaker during El Niño events and stronger during La Niña events [11,14]. HFR data, presented here, were collected during the strong El Niño event of 2015-2016 and thus represented weaker LC conditions. Interestingly, there were differences in the cross-shelf current profiles where at latitude 31° S the annual mean currents recorded in 2017 were almost double the magnitude of 2015 with maximum intensity further offshore (Figure 8a). In contrast, at latitudes 31.7° S and 32.0° S there was not a large difference between the years with the maximum currents recorded in 2016 (Figure 8b,c). This could be due to the inflow from the southern South Indian Counter

Current into the LC that occurs around 32° S, as well as other factors such as eddy activity or other larger scale features outside the HFR domain.

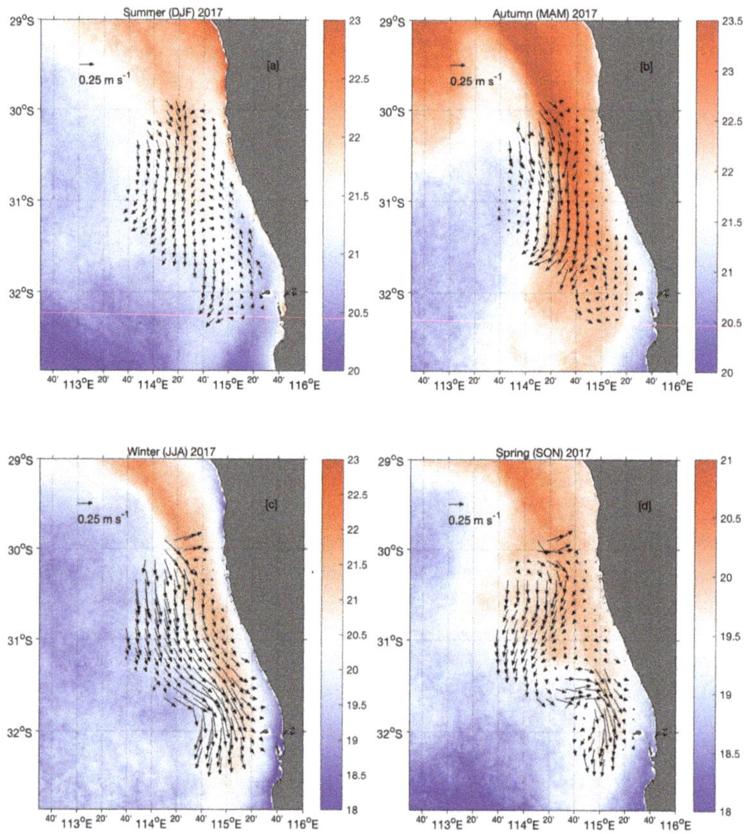

Figure 6. Seasonally-averaged HFR and SST data for 2017: (**a**) Summer (DJF); (**b**) Autumn (MAM); (**c**) Winter (JJA); (**d**) Spring (SON). Note changes in SST scale for summer/autumn (**a**,**b**) and winter/spring (**c**,**d**). Vector data have been subsampled every two grid point to increase readability. SST data originated from multi-sensor L3S foundation SST product.

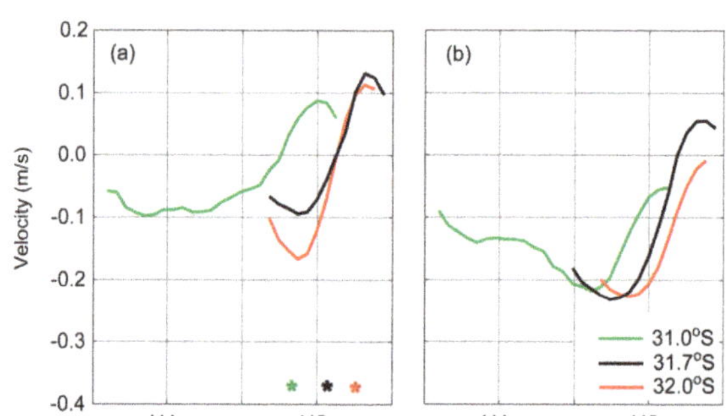

Figure 7. Mean cross-shelf profiles of HFR velocity at three different latitudes: at 31° S (green); 31.7° S (black); and, 32° S (red). Panel (**a**) refers to summer. Panel (**b**) refers to winter. The asterisks denote the locations of the 200 m depth contour. Units for the x-axis are °E.

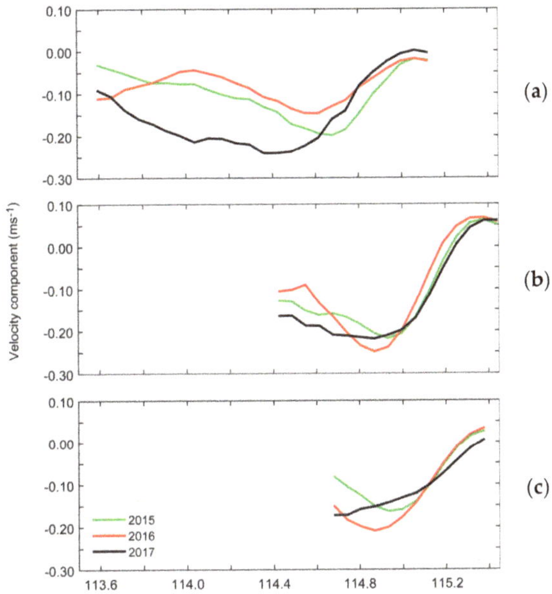

Figure 8. Mean cross-shelf profiles of HFR velocity for 2015, 2016 and 2017 at (**a**) 31° S; (**b**) 31.7° S; and, (**c**) 32° S. Units for the *x*-axis are °E.

3.3. Frequency Domain Analyses

Within the HFR domain, rotary spectra of hourly current vectors indicated common features that included energetic low-frequency components, well-defined peaks in the inertial band, (the inertial period ranged between 21 h 45 m and 24 h 17 m within the HFR domain), diurnal frequency band (K_1 and S_1 tidal constituents or wind), and limited energy within the semidiurnal tidal frequency band (M_2, S_2). These characteristics are highlighted in the spectra (Figure 9) at the shelf break in the Turquoise coast (location 'A' in Figure 2) and the Rottnest shelf (location 'B' in Figure 2). The points A and B were selected based on their data return (in excess of 90% over the analysis period at both locations), to minimize the effects of gaps and data interpolation.

Both systems yielded similar spectral energy level at lower frequencies. However, there were differences in both the high and low frequency tails of the spectra (Figure 9): (1) at higher frequencies, SeaSonde currents contained significantly higher noise levels than that for WERA data (Figure 9a); (2) at the low-frequency bands, both SeaSonde and WERA spectra showed a predominant counterclockwise (CCW) component over the clockwise (CW) spectrum (Figure 9b) although the difference between the CCW and CW peaks were larger for SeaSonde data; and, (3) at the diurnal frequency band, the CCW spectral component was more energetic in the WERA currents and exhibited a stronger polarization (CR = 0.98 at a 24 h period) than SeaSonde data (CR = 0.86 at a 24 h period) over a broader frequency range (Figure 9c). Different variance levels within the CW and CCW spectra and the corresponding rotary coefficients (CR) at point locations (A and B, Figure 2) implied dominance of quasi-circular motions within the diurnal-inertial frequency band (Figure 9b).

Similar features were observed along two zonal (East-West), cross-shore transects centered at 30°48′ S (Figure 10a) and 32° S (Figure 10b), respectively. A different behavior between the two transects can be identified in the semidiurnal tidal frequency band, with CW variance increasing (decreasing) in the offshore direction at 30°48′ S (32° S; Figure 10a,b). Polarization at this frequency band was negative (CW) for both transects but changed to positive (CCW) at higher range (>60 km offshore).

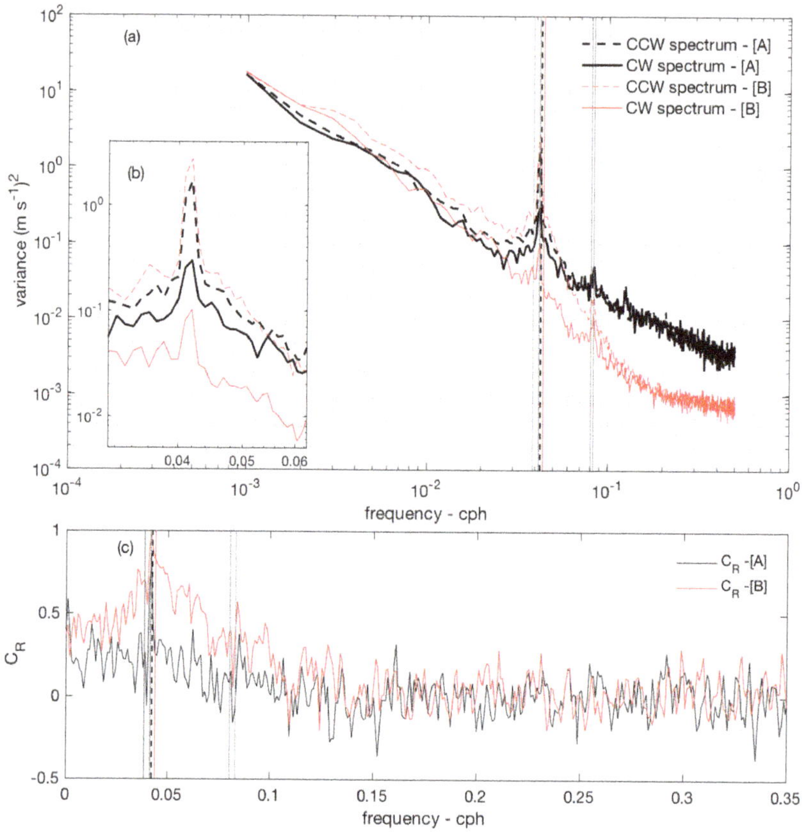

Figure 9. (**a**) Rotary power spectra of hourly surface currents at locations A (Seasonde) and B (WERA) (Figure 2) within the HFR domain showing the dominant spectral features in the region. Vertical lines identify the semidiurnal and diurnal frequency bands (thin grey lines), respectively; and, the inertial frequency at the two locations. (**b**) Same as (**a**), but for the diurnal frequency band. (**c**) Rotary coefficients (C_R) spectra derived from for (**a**). Units are $m^2s^{-2}cph^{-1}$ and cycles per hour (cph) for rotary spectra and frequency axis; C_R is dimensionless.

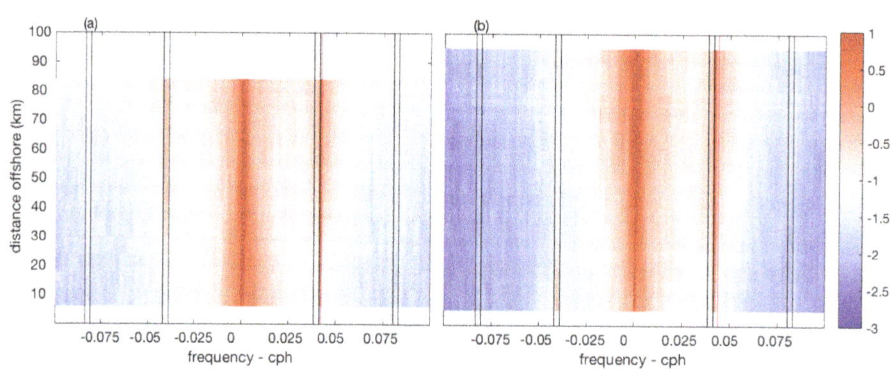

Figure 10. *Cont.*

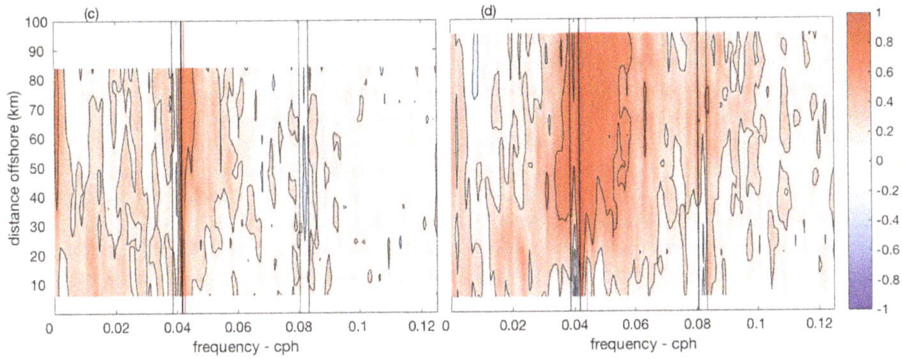

Figure 10. (**a**) Rotary power and rotary coefficient (C_R) spectra of hourly surface currents along two zonal (offshore) transects at 30°48′ S (**a**,**c**); and, 32° S (**b**,**d**). Vertical lines mark the semidiurnal, diurnal and inertial frequency bands at the two latitudes. Units are $m^2s^{-2}cph^{-1}$ and cycles per hour (cph) for rotary spectra and frequency axis; C_R is dimensionless.

Variance levels of the currents increased both southwards and offshore, particularly within the low- and diurnal- frequency bands, with a distinct discontinuity around 30-40 km from the coast, approximately at the location of the 200 m depth contour (e.g., shelf break, Figure 2), indicating spatial distribution of near-inertial and diurnal variance and the low-frequency band (Figure 11). Here, the near inertial to diurnal band spans [0.038, 0.046] cph, corresponded to periods in the range [21 h 44 m 20.8 s, 26 h 18 m 56.8 s]. The study region experiences strong diurnal–inertial resonance due to strong diurnal sea breezes and the critical latitude that generate near-inertial waves that extend to depths >200 m [4]. Higher variance in deeper water (Figure 11a), therefore, reflect that the diurnal–inertial resonance being more effective in deeper water (see also [4]). The low-frequency band was defined as frequencies below 0.0088 cph (periods longer than 4 days) and the distribution of variance indicated a band of higher values between the 200 m and 3000 m depth contours and is representative of the LC variability (Figure 11b).

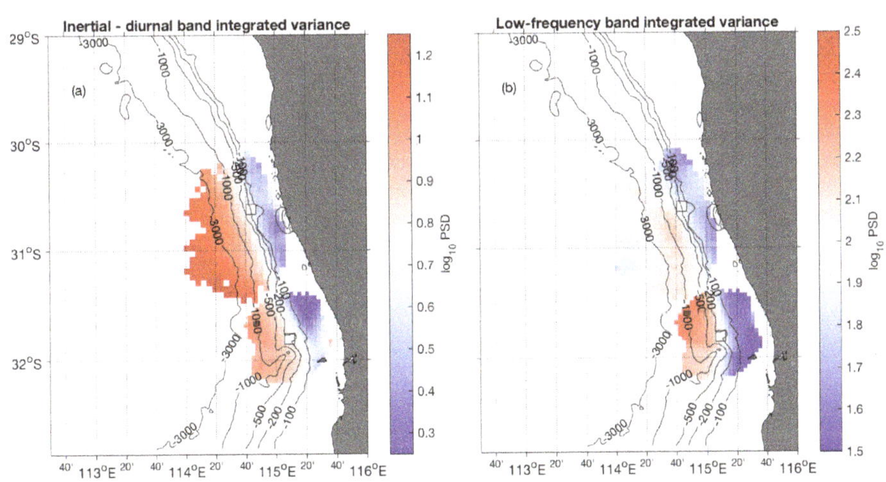

Figure 11. Spatial distribution of band-integrated current variance for (**a**) diurnal-inertial, corresponding to periods in the range [21 h 44 m 20.8 s, 26 h 18 m 56.8 s]; and, (**b**) low-frequency (periods longer than 4 days). Note the discontinuity along the 200 m isobath, separating the shelf break and the deep-water regions. Units for band-integrated variance is $\log_{10}$ ($m^2s^{-2}cph^{-1}$).

Tidal currents were generally negligible, consistent with microtidal conditions in this region. Semidiurnal tides and diurnal tides represented 4%–10% of the total variance, with typical amplitudes that barely exceeded 0.02 ms^{-1} for semidiurnal constituents within the HFR coverage. Amplitudes of diurnal constituents were generally larger but they most likely reflected contamination from the strong wind-driven currents in this frequency band rather than being 'true' tidal oscillations.

3.4. Eddy Statistics

The LC is associated with an energetic meso-scale eddy field with maximum eddy diameters > 200 km. However, the HFR coverage was limited to examining the sub-mesoscale (eddies with diameter < 50 km). The automated eddy identification method revealed ~300 eddies with at least 1-day persistence over the period 2015–2017. There was a prevalence of counterclockwise-rotating (CCW, 177 cases) over clockwise-rotating eddies (CW, 120 cases). Eddies appeared to persist at the same location for extended periods, with typical drift distance of ~10 km around their center; however, in some cases they were advected with the mean flow and travelled in a southwards direction with speeds up to 8 km/day. Their occurrence was limited to deeper water; offshore 200 m depth contour (Figure 12a). There were two main regions for the aggregation of eddies (Figure 12a): (1) between latitudes 30° S and 31.5° S where larger (50–60 km diameter; Figure 12b), CCW eddies were prevalent. Here, the eddies were spread over a wider region; and, (2) offshore the Rottnest island, associated with the Perth canyon (Figure 2), where both isolated or paired CW and CCW eddies were observed. The Perth canyon region was where the most persistent eddies were found concentrated in a smaller region. This is most likely due to the topographic effect of the canyon and the horizontal shear associated with the LC and the CC.

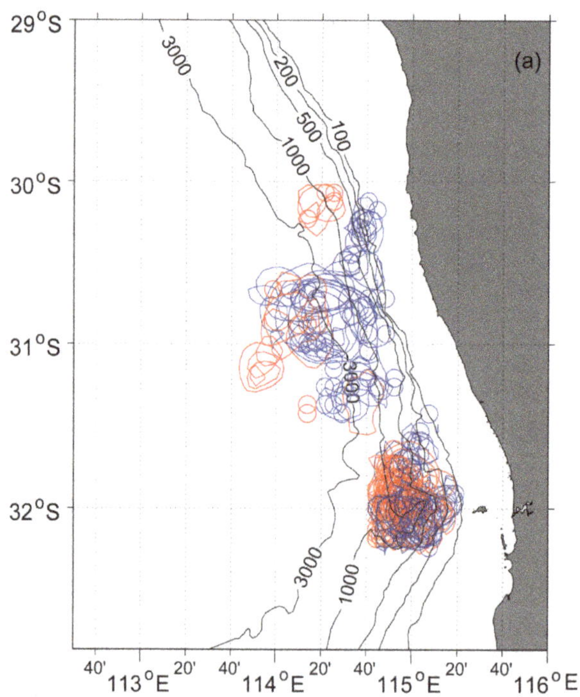

Figure 12. *Cont.*

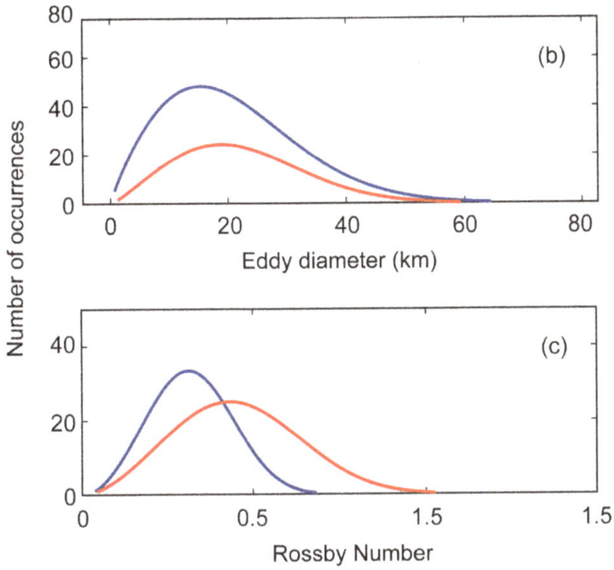

Figure 12. (**a**) Shapes and locations of persistent (lifespan longer than 1 day) eddies along the study region; (**b**) probability density distribution (pdf) curves for eddy diameters (km); (**c**) probability density distribution (pdf) curves for Rossby number (R_o; Equation (8)). Blue and red curves represent counterclockwise (CCW) and clockwise (CW) eddies.

Overall, there were 50% more CCW eddies compared to CW with eddies with diameter 15–25 km being the most prevalent (Figure 12b). Maximum eddy diameters were up to 60 km. The Rossby numbers (R_o) were up to 1.5 (mean value $R_o = 0.46$) for CW eddies (Figure 12c), and to $R_o = 0.64$ for CCW eddies (mean value: $R_o = 0.32$).

Time series of monthly number of days with eddies over the period January 2015 to September 2017 highlighted the seasonal and inter-annual variability (Figure 13). It should be noted that the *y*-axis in Figure 13 does not represent the number of unique eddies but rather number of days in each month ('eddy-days') where an eddy was present. For example, October 2015 indicates >20 days of CW eddy activity in the southern region. This was due to a single eddy that persisted for most of this month that also dominated the seasonal currents (see Figure 4d). However, even neglecting October 2015 it is obvious that 2015 had more eddies compared to 2016 and 2017. In the northern region (north of 31.5° S) there were more days with CCW eddies compared to CW eddies (Figure 13b). There was also only a small month-to-month variability in the northern region (except for the second half of 2016 where there only a few 'eddy-days'). In contrast in the southern region, there was a quasi-seasonal signal with fewer 'eddy-days' during the summer (Figure 13c).

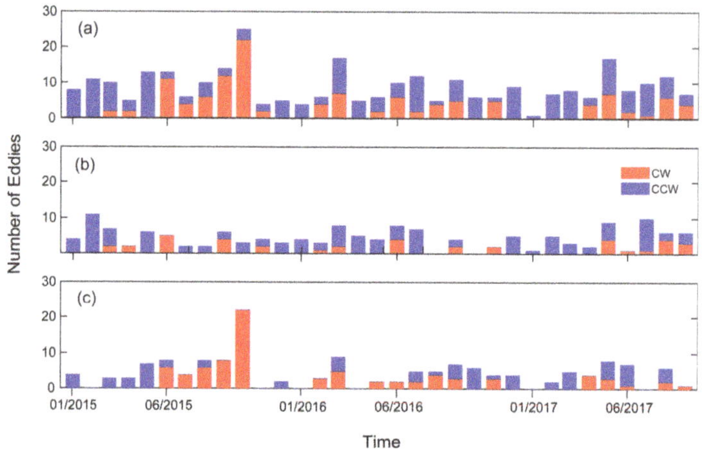

Figure 13. Time series of monthly number of days with eddies over the period January 2015 to September 2017: (**a**) total for the study region; (**b**) for the region north of north of 31.5° S; and (**c**) for the region south of 31.5° S. Blue and red bars represent counterclockwise (CCW) and clockwise (CW) eddies respectively.

4. Discussions

Ocean current measurement using HFR is well established globally over more than three decades and is one of the few techniques that provide near-real time synoptic surface currents in the coastal ocean [24,27]. Currently, two HFR systems are available commercially: SeaSonde by Codar Ocean Sensors (USA) and the WERA by Helzel MessTechnik (Germany). Although they both use HFR signals and Bragg scattering as the fundamental basis for current measurements, they use contrasting methods to estimate the surface currents. The SeaSonde system uses a direction-finding technique whilst the WERA uses a phased-array system. Although these systems are deployed at many locations [24] globally, the West Australian coast is the only region in which the two different systems are co-located and provides data along a 270 km of the coastline (Figure 2). The systems operate at different frequencies and resolutions: the SeaSonde at 4.463 MHz and 6 km resolution with WERA at 9.335 MHz and 4 km. It was previously shown that data from the two major HFR genres do map accurately the observed variability of the ocean currents [33,38], and that data from the two platforms can be successfully merged into a combined product. Here, we show that the combined data sets, especially at the boundaries of the two domains, have the potential to greatly increase the capabilities of coastal observing networks, in general, and along the southwestern Western Australia (SWWA) coastal region, in particular. Previous research used WERA HFR data to focus on the smaller domain in the Rottnest Shelf [4,5], and neglected the area to the North, which has proven to be important for the mesoscale eddy field. The merged SeaSonde–WERA surface current data, mapped onto a 6 km grid, provided robust and consistent current measurements with negligible differences at the diurnal, seasonal and inter-annual time scales. Here we used a three-year record to document unreported features in the SWWA region, and plan to extend the data set to the time period 2010–2019. This data set can be used for a variety of purposes, such as calibration and validation of ocean circulation models, characterization of long-term trends, and in support of fishery management.

There was seamless transition between the two systems as evidenced by the mean (Figure 2), seasonal currents (Figures 4–6) and the variance (Figure 11). However, there were differences between the two systems due to their local configuration: the maximum horizontal ranges were 170 km and 105 km for the SeaSonde and WERA systems, respectively. This was due to the operating frequencies and grid resolution (see above). There were also differences in both the high- and low-frequency tails of the spectra from each system with SeaSonde currents having higher noise levels at higher frequencies whilst at the diurnal frequency band, the CCW spectral component was more energetic in the WERA currents (Figure 9).

The merged SeaSonde–WERA surface current data over the study period (January 2015 to September 2017) reflected the two major current systems in the region: Leeuwin (LC) and Capes (CC) Currents with the latter present mainly during the summer months. The southward flowing LC was associated with warmer water and was located offshore of the 200 m contour (Figure 2). Inshore of the 100 m depth contour, the currents were more variable. The northward flowing CC was present in the summer months, inshore of the LC, and was associated with colder water. The LC and CC systems are well documented along this region of the coast and the HFR data accurately reflected their presence and variability [1,7,8,11,12]. The 'core' of the LC was generally located offshore of the 200 m contour with the maximum velocity increasing from 31° S to 32° S most likely due to inflow from the South Indian Counter Current (SICC). This is consistent with the long-term numerical simulations [3] that indicated inflow from the southern arm of the SICC into the LC at around 32° S (see also [43,44]). The seasonal variability of the LC was well captured by the HFR data with higher currents (mean speeds up to 0.50 ms^{-1}), during the winter months whilst weaker currents were present during the summer consistent with previous studies. In general, the LC was stronger during autumn/winter and weaker during spring/summer. The CC was mainly a summer feature with strong horizontal gradients at the interface between the southward flowing LC and the northwards flowing CC (see also [8]).

The study period coincided with the 2015–2016 strong El Niño event and represented conditions when the LC was weaker [11,14]. Nevertheless, there were inter-annual differences with the maximum currents recorded in 2017 almost double that recorded in 2015 (Figure 8a) as the system was adjusting to neutral ENSO conditions in 2017.

The study region has many processes that contribute to diurnal variability that include tides, inertial oscillations and sea breezes. Although the K_1 and O_1 diurnal tidal constituents dominate the tidal signal [45], analysis of the HFR data indicated that the tidal currents were generally negligible representing <10% of the total variance. The study region experiences one of the strongest and persistent diurnal coastal land sea breeze (LSB) systems globally [3]. The inertial period over the HFR coverage region varies between 21 h 45 m and 24 h 17 m. When the period of local wind forcing is close to the local inertial period as is the case in the study region, a resonance condition occurs. These resonant wind-current responses have been addressed in a variety of field and theoretical studies and have been shown to enhance the upper ocean velocity field and vertical mixing [4,46]. Using HFR and oceanographic mooring data, [4] documented the presence of near-inertial waves in the study region that were generated through diurnal–inertial resonance. During the austral summer, when southerly winds and the LSB system dominated the wind regime, strong counterclockwise diurnal motions with surface amplitudes >0.3 ms^{-1}, penetrated to 300 m depth with diurnal vertical isotherm fluctuations up to 60 m with the strongest response in water depths >200 m [4]. The HFR data indicated strong CCW oscillations throughout the study region (Figure 10) with higher variance at the diurnal band in the deeper water (>200 m) consistent with the results presented by [4].

Eastern margins of ocean basins consist of large scale upwelling and equatorward flow except off the West Australian coast where the Leeuwin Current (LC) transports warmer water poleward. The LC is present due to the Indonesian Throughflow that transports water between the Pacific and Indian oceans through the Indonesian Archipelago. The LC is associated with high eddy kinetic energy that is much higher than that of other eastern boundary currents [7]. The highest eddy kinetic energy (>600 cm^2s^{-2}) occurs offshore of the main current flow (~200 m depth contour) between 28.75 and 32° S (Figure 12a). The LC supports a highly energetic eddy field from the sub-mesoscale to mesoscale with eddy diameters ranging from <20 km to >200 m [18]. Fang and Morrow [6], using altimeter data, found that majority of the mesoscale eddies were generated between 28° and 31° S and propagated into the interior of the Indian Ocean. Over a 6 year period, [6] found that 3–9 large eddies were spawned annually. These mesoscale eddies have diameters >100 km and were not wholly identified in the HFR data. However, there was evidence for the presence of these eddies, during the winter months, where part of the eddy was present in the HFR data (Figures 4c, 5c and 6c). HFR data, together with the automated eddy detection algorithm, were able to identify CW and CCW sub-mesoscale eddies with diameters up to 60 km with the majority in the range 10–20 km (Figure 12b). There were 50% more CCW eddies compared to CW eddies. The 177 eddies identified by the automated eddy detection algorithm had Rossby numbers (R_o) of order 1 confirming that these eddy features were part of the sub-mesoscale motions. There were two main regions of eddy presence, separated to the north and south of 31.5° S. The northern region was dominated by CCW eddies and occurred throughout the year. In contrast, the southern region was dominated by CW eddies and was subject to seasonal variability with lower numbers during the summer months (Figure 13). The differences in the two regions may be explained by the flow patterns associated with the LC and interaction with topography. The LC follows that 200 m depth contour which is similar to the changes in the shoreline orientation. At locations where the shoreline orientation changes, the LC becomes unstable and mesoscale eddies are formed [15]. This is highlighted in Figure 1a where two large CCW eddies are seen to be spinning off the LC at 27.5° S (eddy center at 27.5° S) and 30.5° S (eddy center at 31.5° S) respectively. The circulation around the southern eddy centered at 30.5° S 31.5° S follows an 'S' shaped pattern and are reflected in the mean seasonal HFR currents (Figures 5c and 6c,d) and shown schematically in Figure 14. Here, the LC pathway separates the CCW and CW eddies to the north and south, respectively. This is the reason for the prevalence of CCW (CW) in the north (south). In the south, the concave shape of the coastline associated with the Perth canyon (Figure 2) promotes a CW eddy as reflected in HFR currents (e.g., Figures 4d, 5c and 6c,d). However, the LC pathway changes often with the whole system shown in Figure 14 shifting to the north and thus changing the orientation of eddies at different locations.

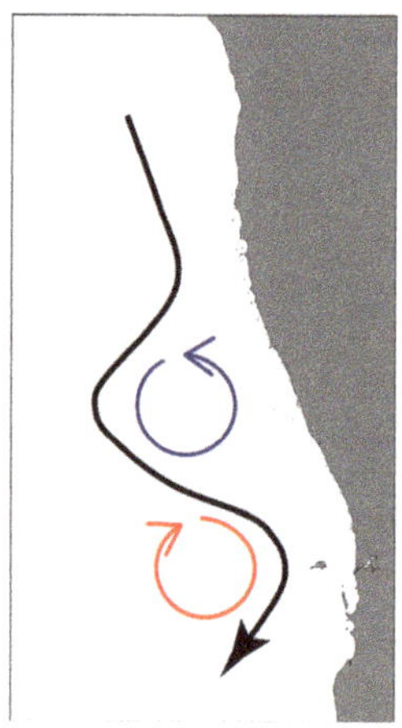

Figure 14. Schematic of the typical Leeuwin Current flow path (black line) and the generation of CCW (blue) and CW (red) vorticity and thus sub-mesoscale eddies.

5. Conclusions

A multi-year merged HFR data collected using SeaSonde and WERA systems, along the south-west Australia (SWWA) between 29°–32° S were used to examine the ocean surface circulation at diurnal, seasonal and inter-annual time scales. The results indicated that:

- SeaSonde and WERA HFR data can be merged successfully with the potential to increase the extent of the coastal ocean under monitoring.
- The warmer poleward-flowing Leeuwin Current (LC) was the dominant feature of the circulation with a strong seasonal signal: stronger during winter and weaker during summer. The colder Capes Current (CC), located inshore of the LC, was mainly present during the summer months.

- We observed a zonal migration of the core of the Leeuwin Current and differences in magnitude between years, however more investigation is needed to determine the driving mechanisms and long term variability.
- The contribution of tides to the circulation was small (<10% total variance). The energy contained in the diurnal period currents was dominant due to diurnal–inertial resonance generated strong sea breezes and inertial period at the critical latitude.
- A clear discontinuity in energy and variance distribution occurred at the shelf break that separated the continental shelf and deeper offshore regions for both diurnal and low-frequency bands reflecting reflected the diurnal–inertial resonance and the LC, respectively.
- Persistent (lifespan greater than 1 day) sub-mesoscale eddies (Rossby number O(1)) were observed at two main regions, north and south of 31.5° S, offshore of the 200m depth contour. Majority of these eddies had diameters in the range 10–20 km with 50% more counter clockwise rotating (CCW) eddies compared to clockwise (CW) rotating eddies. The northern region was dominated by CCW eddies that were present throughout the year whilst CW eddies were prevalent in the south with lower numbers during the summer months. It is planned to use the merged data set to study the generation mechanisms and the propagation of the eddies within the study area.

Author Contributions: Conceptualization, S.C., C.P., and Y.H.; methodology, S.C.; software, S.C.; formal analysis, S.C.; investigation, S.C., C.P., and Y.H.; resources, S.C. and C.P.; data curation, S.C.; writing—original draft preparation, S.C., C.P., and Y.H.; writing—review and editing, S.C., C.P., and Y.H.; visualization, S.C.; project administration, S.C.; funding acquisition, S.C. and C.P. All authors have read and agreed to the published version of the manuscript.

Funding: This research was funded by Integrated Marine Observing System IMOS—Ocean Radar—2017–2019 grant number 53000300.

Acknowledgments: Data were sourced from the Integrated Marine Observing System (IMOS)—IMOS is supported by the Australian Government through the National Collaborative Research Infrastructure Strategy and the Super Science Initiative. Subsurface current data were collected by the IMOS-Australian National Mooring Network (ANMN) Facility. The HFR data used here were collected by the Ocean Radar Facility at the University of Western Australia. HRPT AVHRR SSTfnd retrievals are produced by the Australian Bureau of Meteorology as a contribution to the Integrated Marine Observing System—an initiative of the Australian Government being conducted as part of the National Collaborative Research Infrastructure Strategy and the Super Science Initiative. The imagery data were acquired from NOAA spacecraft by the Bureau, Australian Institute of Marine Science, Australian Commonwealth Scientific and Industrial Research Organization, Geoscience Australia, and Western Australian Satellite Technology and Applications Consortium.

Conflicts of Interest: The authors declare no conflict of interest.

References

1. Pattiaratchi, C.B.; Woo, M. The mean state of the Leeuwin Current system between North West Cape and Cape Leeuwin. *J. R. Soc. West. Aust.* **2009**, *92*, 221–241.
2. Pattiaratchi, C.B.; Hollings, B.; Woo, M.; Welhena, T. Dense shelf water formation along the south-west Australian inner shelf. *Geophys. Res. Lett.* **2011**, *38*, L10609. [CrossRef]
3. Pattiaratchi, C.B.; Hegge, B.; Gould, J.; Eliot, I. Impact of sea-breeze activity on nearshore and foreshore processes in southwestern Australia. *Cont. Shelf Res.* **1997**, *17*, 1539–1560. [CrossRef]
4. Mihanovic, H.; Pattiaratchi, C.B.; Verspecht, F. Diurnal sea breezes force near-inertial waves along Rottnest continental shelf, Southwestern Australia. *J. Phys. Oceanogr.* **2016**, *46*, 3487–3508. [CrossRef]
5. Mahjabin, T.; Pattiaratchi, C.B.; Hetzel, Y. Wind effects on dense shelf water cascades in south-west Australia. *Cont. Shel. Res.* **2019**, *189*. [CrossRef]
6. Fang, F.; Morrow, R. Evolution, movement and decay of warm-core Leeuwin Current eddies. *Deep Sea Res. Part II Top. Stud. Oceanogr.* **2003**, *50*, 2245–2261. [CrossRef]
7. Gersbach, G.H.; Pattiaratchi, C.B.; Ivey, G.N.; Cresswell, G.R. Upwelling on the south-west coast of Australia—Source of the Capes Current? *Cont. Shelf Res.* **1999**. [CrossRef]
8. Pearce, A.; Pattiaratchi, C.B. The Capes Current: A summer countercurrent flowing past Cape Leeuwin and Cape Naturaliste, Western Australia. *Cont. Shelf Res.* **1998**, *18*, 401–421. [CrossRef]

9. Smith, R.L.; Huyer, A.; Godfrey, J.S.; Church, J.A. The Leeuwin Current off Western Australia. *J. Phys. Oceanogr.* **1991**, *21*, 813–827. [CrossRef]
10. Feng, M.; Meyers, G.; Pearce, A.; Wijffels, S. Annual and inter-annual variations of the Leeuwin Current at 32oS. *J. Geophys. Res. Oceans* **2003**, *108*, 3355. [CrossRef]
11. Wijeratne, E.M.S.; Pattiaratchi, C.B.; Proctor, R. Estimates of surface and subsurface boundary currents transport around Australia. *J. Geophys. Res. Oceans* **2018**, 3444–3466. [CrossRef]
12. Ridgway, K.R.; Condie, S.A. The 5500-km-long boundary flow of western and southern Australia. *J. Geophys. Res. Oceans* **2009**, *109*, C04017. [CrossRef]
13. Ridgway, K.R.; Godfrey, J.S. The source of the Leeuwin Current seasonality. *J. Geophys. Res. Oceans* **2015**, *120*, 6843–6864. [CrossRef]
14. Pattiaratchi, C.B.; Buchan, S.J. Implications of long- term climate change for the Leeuwin Current. *J. R. Soc. West. Aust.* **1991**, *74*, 133–140.
15. Pearce, A.F.; Griffiths, R.W. The mesoscale structure of the Leeuwin Current: A comparison of laboratory models and satellite imagery. *J. Geophys. Res.* **1991**, *96*, 16739. [CrossRef]
16. Waite, A.M.; Thompson, P.A.; Pesant, S.; Feng, M.; Beckley, L.E.; Domingues, C.M.; Gaughan, D.; Hanson, C.E.; Holl, C.M.; Koslow, T. The Leeuwin Current and its eddies: An introductory overview. *Deep Sea Res. Part II Top. Stud. Oceanogr.* **2007**, *54*, 789–796. [CrossRef]
17. Meuleners, M.J.; Pattiaratchi, C.B.; Ivey, G.N. Numerical modelling of the mean flow characteristics of the Leeuwin current system. *Deep Sea Res. Part II Top. Stud. Oceanogr.* **2007**, *54*, 837–858. [CrossRef]
18. Rennie, S.J.; Pattiaratchi, C.; McCauley, R.D. Eddy formation through the interaction between the Leeuwin Current, Leeuwin Undercurrent and topography. *Deep Sea Rese Part II Top. Stud. Oceanogr.* **2007**, *54*, 818–836. [CrossRef]
19. Rennie, S.J.; Pattiaratchi, C.B.; McCauley, R.D. Numerical simulation of the circulation within the Perth submarine canyon, Western Australia. *Cont. Shelf Res.* **2009**, *29*, 2020–2036. [CrossRef]
20. Rennie, S.J.; Hanson, C.E.; McCauley, R.D.; Pattiaratchi, C.B.; Burton, C.; Bannister, J.; Jenner, C.; Jenner, M.-N. Physical properties and processes in the Perth Canyon, Western Australia: Links to water column production and seasonal pygmy blue whale abundance. *J. Mar. Syst.* **2009**, *77*, 21–44. [CrossRef]
21. Wandres, M.; Wijeratne, E.M.S.; Cosoli, S.; Pattiaratchi, C.B. The effect of the Leeuwin Current on offshore surface gravity waves in southwest western Australia. *J. Geophys. Res. Oceans* **2017**, *122*, 9047–9067. [CrossRef]
22. Gallop, S.L.; Verspecht, F.; Pattiaratchi, C.B. Sea breezes drive currents on the inner continental shelf off southwest Western Australia. *Ocean Dyn.* **2012**, *62*, 569–583. [CrossRef]
23. Verspecht, F.; Pattiaratchi, C.B. On the significance of wind event frequency for particulate resuspension and light attenuation in coastal waters. *Cont. Shelf Res.* **2010**, *30*, 1971–1982. [CrossRef]
24. Roarty, H.; Cook, T.; Hazard, L.; George, D. The Global High Frequency Radar Network Front. *Mar. Sci.* **2019**, *6*, 164. [CrossRef]
25. Barrick, D.E.; Evans, M.W.; Weber, B.L. Ocean surface currents mapped by radar. *Science* **1977**, *198*, 138–144. [CrossRef]
26. Hammond, T.M.; Pattiaratchi, C.B.; Eccles, D.; Osborne, M.J.; Nash, L.A.; Collins, M.B. Ocean Surface Current Radar (OSCR) vector measurements on the inner continental shelf. *Cont. Shelf Res.* **1987**, *7*, 411–431. [CrossRef]
27. Paduan, J.D.; Washburn, L. High-frequency radar observations of ocean surface currents. *Annu. Rev. Mar. Sci.* **2013**, *5*, 115–136. [CrossRef]
28. Lipa, B.; Barrick, D.; Whelan, C. A Quality Control Method for Broad-Beam HF Radar Current Velocity Measurements. *J. Mar. Sci. Eng.* **2019**, *7*, 112. [CrossRef]
29. Ren, L.; Miao, J.; Li, Y.; Luo, X.; Li, J.; Hartnett, M. Estimation of Coastal Currents Using a Soft Computing Method: A Case Study in Galway Bay, Ireland. *J. Mar. Sci. Eng.* **2019**, *7*, 157. [CrossRef]
30. Lorente, P.; Piedracoba, S.; Sottillo, M.G.; Alvarez-Fanjul, E. Long-Term Monitoring of the Atlantic Jet through the Strait of Gibraltar with HF Radar Observations. *J. Mar. Sci. Eng.* **2019**, *7*, 3. [CrossRef]
31. Twomey, L.J.; Waite, A.M.; Pez, V.; Pattiaratchi, C.B. Variability in nitrogen uptake and fixation in the oligotrophic waters off the south west coast of Australia. *Deep Sea Res. Part II Top. Stud. Oceanogr.* **2007**, *54*, 925–942. [CrossRef]

32. Griffin, C.; Beggs, H.; Majewski, L. *GHRSST Compliant AVHRR SST Products over the Australian Region*; Version 1, Technical Report; Bureau of Meteorology: Melbourne, Australia, 2017; p. 151.
33. Cosoli, S.; Grcic, B.; de Vos, S.; Hetzel, Y. Improving Data Quality for the Australian High Frequency Ocean Radar Network through Real-Time and Delayed-Mode Quality-Control Procedures. *Remote Sens.* **2018**, *10*, 1476. [CrossRef]
34. Welch, P.D. The use of fast-Fourier transform for the estimation of power spectra: A method based on time-averaging over short, modified periodograms. *IEEE Trans. Audio Electroacoust.* **1967**, *2*, 70–73. [CrossRef]
35. van Haren, H. On the polarization of oscillatory currents in the Bay of Biscay. *J. Geophys. Res.* **2003**, *108*, 3290. [CrossRef]
36. Pawlowicz, R.; Beardsley, B.; Lentz, S. Classical tidal harmonic analysis including error estimates in MATLAB using T_TIDE. *Comput. Geosci.* **2002**, *28*, 929–937. [CrossRef]
37. Cosoli, S.; Grcic, B. *Quality Control Procedures for IMOS Ocean Radar Manual*; Integrated Marine Observing System Version 2.0; UNESCO/IOC Project Office for IODE: Oostende, Belgium, 2019. [CrossRef]
38. Cosoli, S.; de Vos, S. Interoperability of Direction-Finding and Beam-Forming High-Frequency Radar Systems: An Example from the Australian High-Frequency Ocean Radar Network. *Remote Sens.* **2019**, *11*, 291. [CrossRef]
39. Kundu, P.K. Ekman veering observed near the ocean bottom. *J. Phys. Oceanogr.* **1976**, *6*, 238–242. [CrossRef]
40. McWilliams, J.C. A survey of submesoscale currents. *Geosci. Lett.* **2019**, *6*. [CrossRef]
41. Nencioli, F.; Dong, C.; Dickey, T.; Washburn, L.; McWilliams, J.C. A vector geometry based eddy detection algorithm and its application to a high-resolution numerical model product and high-frequency radar surface velocities in the Southern California Bight. *J. Atmos. Ocean. Technol.* **2010**, *27*, 564–579. [CrossRef]
42. Sanderson, B.G. Structure of an eddy measured with drifters. *J. Geophys. Res.* **1995**, *100*, 6761–6776. [CrossRef]
43. Menezes, V.V.; Phillips, H.E.; Vianna, M.L.; Bindoff, N.L. Interannual variability of the South Indian Countercurrent. *J. Geophys. Res. (Oceans)* **2016**, *121*, 3465–3487. [CrossRef]
44. Lambert, E.; Le Bars, D.; de Ruijter, W.P.M. The connection of the Indonesian Throughflow, South Indian Ocean Countercurrent and the Leeuwin Current. *Ocean Sci.* **2016**, *12*, 771–780. [CrossRef]
45. Pattiaratchi, C.B.; Eliot, M. Sea level variability in south-west Australia: From hours to decades. In *Coastal Engineering, Proceedings of the Thirty-First International Coastal Engineering Conference, Rio de Janeiro, Brazil, 31 August–5 September 2008*; Smith, J.M., Ed.; World Scientific Publishing: Singapore, 2008; pp. 1186–1198.
46. Simpson, J.H.; Hyder, P.; Rippeth, T.P. Forced oscillations near the critical latitude for diurnal-inertial resonance. *J. Phys. Oceanogr.* **2002**, *32*, 177–187. [CrossRef]

Journal of
Marine Science and Engineering

MDPI

Article

Surface Currents Derived from SAR Doppler Processing: An Analysis over the Naples Coastal Region in South Italy

Virginia Zamparelli [1,*], Francesca De Santi [2], Andrea Cucco [3], Stefano Zecchetto [4], Giacomo De Carolis [2] and Gianfranco Fornaro [1,5]

1 IREA-CNR, Institute for Electromagnetic Sensing of the Environment-National Research Council of Italy, 80128 Naples, Italy; fornaro.g@irea.cnr.it

2 IREA-CNR, Institute for Electromagnetic Sensing of the Environment-National Research Council of Italy, 20133 Milan, Italy; desanti.f@irea.cnr.it (F.D.S.); decarolis.g@irea.cnr.it (G.D.C.)

3 IAS-CNR, Institute for the Study of Anthropic Impacts and Sustainability in Marine Environment-National Research Council of Italy, 09170 Oristano, Italy; andrea.cucco@cnr.it

4 ISP-CNR, Institute of Polar Sciences-National Research Council of Italy, 35127 Padova, Italy; stefano.zecchetto@cnr.it

5 CNIT-National Inter-University Consortium for Telecommunications Research Unit IREA, 80128 Naples, Italy

* Correspondence: zamparelli.v@irea.cnr.it; Tel.: +39-081-7620629

Received: 4 February 2020; Accepted: 10 March 2020; Published: 15 March 2020

Abstract: Several studies have shown the capabilities of Synthetic Aperture Radar to map sea currents in ocean regions mainly characterized by large flows. We consider the well known method based on the analysis of the Doppler Centroid. The Doppler, as, in general, the scattering from the sea, is sensitive to several phenomena, occurring between the upper ocean and atmospheric boundary layers. To investigate such phenomena, we considered the combined use of both ENVISAT calibrated amplitude and Doppler data in conjunction with hindcast wind information provided by atmospheric models as well as Wind and Doppler Geophysical Model Functions (W/D-GMF) developed, in the literature, for C-Band systems. This integrated analysis for the interpretation of the Doppler surface currents measurements was carried out on a case study located in the Mediterranean Sea which is characterized by a general low circulation regime: specifically, the coastal region around the city of Naples. In this case study, we show that generally, wind plays a direct significant role in the observed Doppler surface current. The availability of an oceanographic numerical model for one of the analyzed cases also allowed us to attempt to interpret the effect of the typical thermohaline circulation pattern on the Doppler anomaly.

Keywords: synthetic aperture radar; Doppler anomaly; sea surface currents

1. Introduction

Driven by the development of geophysical applications, investments in the construction and launch of satellites equipped with high-resolution imaging radars, specifically Synthetic Aperture Radars (SAR), have been multiplied in recent years. In addition to operating at different frequencies, SAR sensors are characterized by the availability of different operating, modes thus giving the possibility to trade-off the resolution and coverage. On one hand, very high-resolution—up to a sub-metric level at X-Band (e.g., COSMO/Skymed and TerraSAR-X/Tandem-X)—data are available with a limited coverage of a few tens of kilometers. On the other hand, by exploiting scanning beams with electronic steering of the antenna, high-resolution modes with decametric resolution at C-Band allow to achieve coverage approaching (ERS and ENVISAT) and even far exceeding (Sentinel-1) hundreds of kilometers.

Above all, the increase in the number of available SAR sensors is significantly reducing the access area and the revisiting time, thus expanding the SAR data archives that frequently can go back up to 20 years for many regions of the Earth's globe [1–3].

Sea surface scattering shows large variation over repeated orbits, depending on the sea state conditions. When compared to the return from terrestrial regions, the signal scattered back from the sea shows a larger (relative) dynamic, thus providing a potentially valuable source of information for understanding several geophysical processes. However, the observed scenario is, in this case, rapidly changing and influenced, much more than on the land, by the presence of processes occurring in the atmosphere, thus making the interpretation of the signal more complex.

SAR image formation mainly depends on the small gravity waves, also referred to as ripples: the occurrence of Bragg conditions coupled with the wavelength of the illuminating (microwave) radiation offers favorable conditions for an increase of backscattering. Ripples are generated by the wind and are modulated by longer waves thus providing the possibility to analyze the images to derive information on both the wind and the sea surface waves. [4]. The large sensitivity of the sea scattering makes it possible for SAR images to distinguish oceanographic features related to wind waves and swell as well as ocean internal waves and sea currents [5–9]. These phenomena have different spatial scales and spatial layouts and often, are present together in a single SAR image covering swaths of several tens of kilometers.

Modulation of the radar backscatter due to the wind reflects the spatial properties of the wind itself, which depends on factors such as wind speed, air stability conditions, and interaction with orography in coastal regions. The extraction of the spatial properties of the wind from SAR images has been extensively studied [10,11] and well demonstrated.

Another "hot-point" in the area of extraction of information from sea SAR images, which is the main interest of this work, is related to sea currents [12,13]. The process of hydrodynamic modulation of the sea-surface roughness allows, for example, sea surface currents signatures to be observable in SAR images, as well established in the literature, see, for instance [4,12–14]. The extraction of quantitative parameters on currents from the amplitude image alone is, however, a rather complex issue. Many variables are involved for the backscattering determination, such as wind vector, bathymetry, etc. [15]. For this reason, similarly to other SAR applications, the phase information may provide valuable information [4,12].

The possibility to observe the surface ocean currents by SAR images dates back to the eighties, when scientists associated some of the large scale features imaged by SAR with mesoscale oceanic eddies [16] and ocean current fronts [17].

Since then, a quantitative measurement of the radial component of the sea surface currents can be achieved from a single SAR image using spectral analysis. Specifically, it can be obtained via the estimation of the residual Doppler shifts of the received radar echoes, the Doppler Centroid Anomaly (DCA) approach [4], or by means of interferometric acquisitions with two antennas displaced in the along-track direction, Along-Track Interferometry (ATI) approach [13,18]. Both approaches have been attempted to get the radial ocean current, especially over regions of well known strong currents, as in the Gulf Stream [12,19] offshore of the USA and in the Agulhas Greater Current offshore of South Africa [20]. In [4], other experiments have been presented in coastal areas of the Gulf of Normandy, which is characterized by intense tidal currents.

Based on many successful experiments, steps aimed at providing operational services for coastal ocean monitoring have been proposed in recent years [21] and in some cases, implemented. However, regarding currents, the existing examples of applications are mainly limited to the above-cited sites characterized by significant water circulation phenomena.

This work focuses on the DCA, which, similarly to the phase offset for ATI, is induced by the movement of the scattering centers, thus providing a direct measurement of the radial component of the sea surface currents [4].

In particular, we applied the technique of retrieving the radial surface sea current from the DCA in the Mediterranean Sea, a semi-enclosed sea where the sea currents are weak and the geophysical phenomena in coastal areas are strong, due to the interactions between the airflows and the orography. In this respect, this work acts as a feasibility study.

At present, the topic of deriving sea surface currents from SAR involves many issues, both from the engineering and geophysical points of view. The former because of the typically large non-geophysical velocities associated with to the satellite attitude and the SAR instrument imaging characteristics; the latter because of the difficulty to split the different phenomena contributing to the DCA, such as the wind, the waves and the currents [22]. For the investigation of the above open issues, we considered the DCA method and we focused on a case study involving the Naples bay and the adjacent coastal region. Among the Italian coastal zones, the Gulf of Naples and the surrounding coastal regions are particularly interesting as first of all they are areas influenced by numerous environmental, socio-economic, and interacting cultural factors (strong anthropogenic impact, intense maritime traffic, the presence of polluted rivers, tourist and economic activities and the presence of four protected areas) [23]. The area is also interesting from an oceanographic standpoint as both coastal wind-driven and open-sea currents influence the local hydrodynamics [24–26]. In particular, several factors make this an area of general oceanographic interest, among them: (i) the presence of the Gulf of Naples, characterized by dimensions comparable to the internal Rossby deformation radius; (ii) the presence of extended shallow water areas at the edge of the continental slope; (iii) a coastal circulation shaped by both the interaction between the larger scale flows with the coastal topography and by the local wind forcing [27].

The peculiar dynamics of water circulation in this area qualify this region as a natural challenging laboratory both for investigating oceanographic phenomenology [27] and for testing the skill of ocean and atmospheric observational platforms as well as numerical models [23]. Furthermore, the choice of the zone is driven by the availability of a large historical data archive regularly acquired over the years. As also better specified in the following sections, we focus in particular on the use of the ENVISAT SAR archive due to the available larger azimuth bandwidth compared to the most recent data gathered by the Sentinel-1 sensor.

Wind fields have been estimated from calibrated SAR data by using the empirical geophysical model function CMOD developed at Institut Francais de Recherche pour l'Exploitation de la Mer (IFREMER), which relates the SAR backscatter to the actual wind vector. SAR wind inversion used the hindcast information provided by the WRF-ARW atmospheric model as reference wind vector. Moreover, the SAR estimated velocity fields were compared with the Doppler model developed by [28].

The paper is organized as described in the following section. The detailed description of the technique for measuring sea radial velocity based on the DCA estimation is presented in Section 2. In Section 3, an overview of the Bayesian approach used to extract the wind information from the SAR data is provided. In Section 4, we describe the Doppler modeling of the wind component. Section 5 is devoted to the description of the overall data analysis method. Section 6 provides details on the analyzed area and motivates the dataset choice. Section 7 is devoted to the description of the experimental results over all the different considered dates. Conclusions are finally addressed in Section 8.

2. Doppler Anomaly Extraction

The DCA technique, described in [4], is based on the measurement of the Doppler shifts induced on the SAR image by the line-of-sight component of the sea surface velocity. In our case, the measurement of DCA has been carried out by using the method based on the estimation of the azimuth auto-correlation function proposed in [29]. In particular, to avoid any bias induced by possible filtering of spectral components in the application of the SAR transfer function at the stage of generation of the Single Look Complex products, the method is applied starting from the raw data.

The block diagram that describes the overall processing chain is shown in Figure 1. More specifically, starting from raw data and after the SAR focusing operation, the corresponding Doppler Centroid frequency, referred to in the following as f_{DC}, is estimated over small sliding patches: details on the patch sized choice are provided in Section 7. Subsequently, the DCA, referred to as f_{DCA}, is extracted by subtracting the Doppler centroid induced by the antenna pointing and Earth rotation. The DCA and the radial velocity v_r of the backscattering cells are related by the following simple equation:

$$f_{DCA}(x,r) = f_{DC}(x,r) - f_{DC0}(x,r) = \frac{2}{\lambda}v_r(x,r) \quad (1)$$

where x and r are the azimuth and range, respectively and λ is the wavelength. The term f_{DC0} is the Doppler Centroid corresponding to a "stationary" scene, more precisely to a scene with all scatterers moving at the same velocity of the associate one with the Earth rotation.

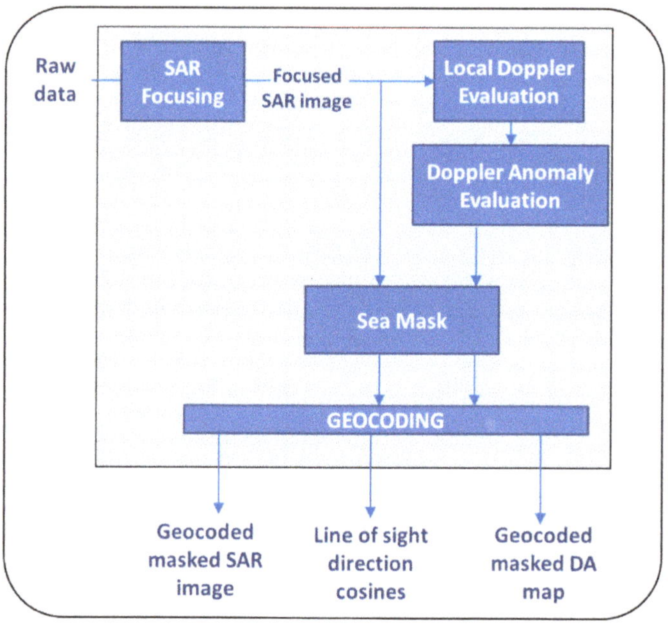

Figure 1. Block diagram of the processing chain.

The radial component of the target velocity v_r is then evaluated by the Doppler Anomaly as follows:

$$v_r = \frac{\lambda f_{DCA}}{2} \quad (2)$$

Under the white Gaussian assumption for the signal and noise, it has been shown that the accuracy of such estimator is given by [30]

$$\sigma_{f_{DC}} = 0.3407\frac{PRF}{\sqrt{M}} \quad (3)$$

where M is the number of independent samples used for the spectral estimation and PRF is the pulse repetition frequency of the illuminating sensor.

It is important to underline that the "background" DC, f_{DC0} in its polynomial approximation form could be extracted directly from the ancillary information in the ENVISAT data. In our case, to account also for additional deviations of the beam that might be present in the acquisition, we preferred to perform an estimation directly from the data by a factorization this term in an azimuth $f_{DC0x}(x)$ and a range $f_{DC0r}(r)$ component:

$$f_{DC0}(x,r) = f_{DC0x}(x) f_{DC0r}(r) \tag{4}$$

Based on the observed experimental patterns for the ENVISAT sensor we modeled $f_{DC0x}(x)$ as a sinusoidal contribution:

$$f_{DC0x}(x) = a\sin(2\pi\xi_0 x) \tag{5}$$

whereas for $f_{DC0r}(r)$ is expressed by the use of a polynomial form as follows:

$$f_{DC0r}(r) = a_0 + a_1 r + a_2 r^2 + \ldots.. = \sum_{n=0}^{L} a_n r^n \tag{6}$$

The azimuth component was modeled as in Equation (5) because such a behavior was observed in the entire scene processed in the experimental results discussed in Section 7 and takes place at a constant frequency of 31.72 Hz. This component, whose origin is assumed to be related to the sensor, likely due to a systematic oscillation of the carrier frequency, was compensated simply via a notch filtering.

As for the range component in f_{DC0}, being the scene located in a coastal region involving both land and sea, we used an approach based on separate estimation in two regions starting from the sea mask. We therefore distinguished f_{DC0r} between a term relative to the land f_{DC0r_L} and the term relative to the sea f_{DC0r_S}:

$$f_{DC0r_L}(r_L) = \sum_{n=0}^{N} a_{L_n} r_L^n \tag{7}$$

$$f_{DC0r_S}(r_S) = \sum_{n=0}^{N} a_{S_n} {r_S}^n \tag{8}$$

where we used the subscript L and R to distinguish the terms related to the land and to the sea, respectively; and N is the polynomial order. Our tests also show that a range expansion to a fourth-order, i.e., $N = 4$, provides a satisfactory trade-off between the removal of the regular pattern related to the background variation of the Doppler Centroid and the preservation of components associated to the sea surface velocity.

The polynomial coefficients are estimated through a classical Least Square approach, and the resulting coefficients are combined afterward as explained in the following.

More specifically, focusing for example on the land component, we assume to have P measurement points on the land:

$$\begin{cases} f_{DC0r_L}(r_{L_1}) = f_1 \\ \cdot \\ \cdot \\ \cdot \\ f_{DC0r_L}(r_{L_P}) = f_P \end{cases} \tag{9}$$

thus allowing the following matrix problem formulation:

$$\begin{bmatrix} 1 & r_{L_1} & \cdots & {r_{L_1}}^N \\ \cdot & \cdot & \cdots & \cdot \\ \cdot & \cdot & \cdots & \cdot \\ \cdot & \cdot & \cdots & \cdot \\ 1 & r_{L_P} & \cdots & {r_{L_P}}^N \end{bmatrix} \begin{bmatrix} a_{L_0} \\ \cdot \\ \cdot \\ \cdot \\ a_{L_N} \end{bmatrix} = \begin{bmatrix} f_1 \\ \cdot \\ \cdot \\ \cdot \\ f_P \end{bmatrix} \tag{10}$$

that is:

$$\mathbf{R}_L \mathbf{a}_L = \mathbf{f}_L \tag{11}$$

where $\mathbf{R}_L$ is the Px $(N + 1)$ range matrix expressed in Equation (10) and $\mathbf{a}_L$ is the $(N+1) \times 1$ vector of coefficients on the land area that we need to find. Hence, now, it is possible to estimate the coefficients $\mathbf{a}_L$ as follows:

$$\mathbf{a}_L = \left(\mathbf{R}_L \mathbf{R}_L^{\mathrm{T}}\right)^{-1} \mathbf{R}_{L}^{\mathrm{T}} \mathbf{f}_L \tag{12}$$

In the same way, it is possible the estimate the coefficients relative to the say Q points located in the sea region:

$$\mathbf{a}_S = \left(\mathbf{R}_S \mathbf{R}_S^{\mathrm{T}}\right)^{-1} \mathbf{R}_{S}^{\mathrm{T}} \mathbf{f}_S \tag{13}$$

The estimated coefficients relevant to the sea and land areas are then combined to achieve the final estimation of the range component: in this process, only the coefficients from order 1 to N are averaged in order to decouple the zero order contribution on the two different regions. Assuming, as an approximation, that the variance of the estimated coefficients is proportional to the inverse of the number of measurements in two areas, i.e,. P and Q, we applied the following combination to find t:

$$[a_1 \ldots\ldots a_N]^{\mathrm{T}} = \left[a_{L_1} \ldots\ldots a_{L_N}\right]^{\mathrm{T}} \frac{P}{P+Q} + \left[a_{S_1} \ldots\ldots a_{S_N}\right]^{\mathrm{T}} \frac{Q}{P+Q} \tag{14}$$

The range component to be subtracted from the data at each azimuth line, i.e., $f_{DC0_r}(r)$, is achieved by substituting $a_0 = a_{L_0}$, i.e., the land offset, and $a_1 \ldots\ldots a_N$ resulting from Equation (14) in Equation (6). The final DCA estimation, is evaluated by subtracting $f_{DC0_r}(r)$ from $f_{DC}(x, r)$ estimated over the patches with the Madsen approach described in [29], and then applying the previously mentioned azimuth notch filter.

The final DCA is geocoded to provide a product loadable in GIS environments. During the geocoding stage, the line of sight (LOS) director cosine is evaluated and provided as an output together with the sea mask.

3. Wind Retrieval from SAR

The dominant contribution to the radar backscatter comes from the wind generated surface gravity-capillary waves through the resonant Bragg wave scattering mechanism [31,32]. Since the caps of these waves tend to align perpendicular to the local wind, the radar backscatter is also wind direction-dependent. Therefore, SAR backscatter can be used in principle to derive sea-surface wind information. Several empirically derived backscatter models, known as Wind Geophysical Model Functions (W-GMF), have been proposed in the literature [33–35] to relate the 10 m wind vector to SAR observations. They are generally defined as

$$\sigma^0 = \mathrm{B0}[1 + \mathrm{B1}\cos\phi + \mathrm{B2}\cos(2\phi)]^n \tag{15}$$

where σ^0 is the backscattering coefficient measured on the SAR image, ϕ is the wind direction relative to the antenna pointing ($\phi = 0$ indicates an upwind, i.e., a wind aligned to the range and blowing towards the radar), and the coefficients B0, B1 and B2 depend on the wind speed w, the local incidence angle, and the polarization and frequency of the radar beam. The Bi parameters were determined statistically after drawing comparisons with wind data from ECMWF atmospheric modeling outputs, or from operational buoy measurements. The exponent n could be 1.6 or 1, depending on the model considered. In this work, the GMF used was developed at IFREMER and is referred to as CMOD-IFREMER [36].

Unfortunately, the inverse relationship of Equation (15) is an inherently underdetermined problem for single-view measurement instruments, as happens for SAR imaging. Indeed, Equation (15) establishes that the backscatter is a monotonically increasing function of the wind speed and that is sinusoidal with ϕ. In particular, fixing the wind speed, σ^0 is maximum for the wind blowing towards (or away from) the radar ($\phi = 0°, 180°$) and minimum for the wind blowing perpendicular to the radar, see Figure 2.

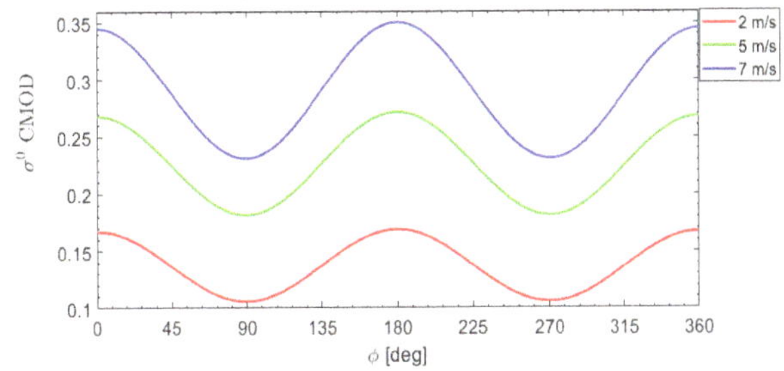

Figure 2. σ^0 computed with CMOD-IFREMER, see Equation (15), as a function of wind speed and direction, for 23° of incidence angle. Results for three different wind speed values are shown: 7 m/s (blue line), 5 m/s (green line), 2 m/s (red line). $\phi = 0°$ indicates a wind blowing towards the antenna.

As a consequence, both wind speed and direction cannot be estimated from a single σ^0 measurement so that some a priori wind information is needed. For this research, a downscaling of the 10 m wind fields obtained from the non-hydrostatic mesoscale model Weather Research and Forecast-Advanced Research (WRF-ARW) version 3.3.1 [37] were considered as a priori wind vector information. The choice fell on this wind database since it has been already assessed for the Mediterranean basin. Indeed, [33] studied the performance of the third-generation wave model WAVEWATCH III (WWIII) in the Mediterranean by using as a wind forcing term the downscaled WRF data and validating their simulation results through buoys data provided by the Rete Ondametrica Nazionale (RON) [38], as better specified in Section 6.

Following [39], the SAR retrieved wind vector components are computed by minimizing the cost function:

$$J = \left(\frac{\sigma^0(u,v) - \sigma^0_m}{\Delta\sigma}\right)^2 + \left(\frac{u_{ww} - u}{\Delta u}\right)^2 + \left(\frac{v_{ww} - v}{\Delta v}\right)^2 \tag{16}$$

where σ^0_m is the SAR measured backscatter, σ^0 is the backscatter value computed by solving Equation (15) with wind vector of components u and v, $\Delta\sigma = 0.078\ \sigma^0_m$, and $\Delta u = \Delta v = \sqrt{3}\ m/s$. Finally, u_{ww} and v_{ww} are the reference wind components provided by the WWIII hindcast. Cost function minimization is performed by looking at the global minimum of Equation (16) computed for velocities from 0 m/s to 25 m/s with a 0.05 m/s spacing and wind direction from 0° to 360° and spacing of 1°.

4. Doppler Modeling of the Wind Component

The topic of evaluating the influence of the wind influence on the Doppler centroid has been already addressed in the literature, see f.i. [4,40]. We focus here on the investigations carried out in [28], which deal with the problem of using the SAR DCA as an additional source of information for wind retrieval. In particular, the goal in [28] was to integrate the DCA measurement in classical approaches, such as the one described in the previous section, for wind inversion of models for the Normalized Radar Cross Section (NRCS) [36]. In doing such an integration, the authors proposed and empirically derived Doppler Geophysical Model Function (D-GMF) relating the wind speed and direction to the Doppler anomaly trough parameters associated with the SAR observation geometry and characteristics. This empirical function, which is summarized in the following, was derived by analyzing a match-up database between the wind speed and direction provided by ECMWF and Doppler measurements derived by ASAR acquisitions for both VV and HH polarizations. The analysis emphasizes that the Doppler shift due to sea surface wind can be described by the wind speed (w), the wind direction (ϕ)

with respect to the antenna look angle in degrees, the incidence angle of the impinging radiation in degree (ϑ) and the radar polarization (*pol*), i.e.,

$$f_{W\_DCA} = CDOP(\phi, w, \vartheta, pol) \tag{17}$$

Expression (17) is developed in [28] as

$$f_{W\_DCA} = \alpha_{pol} F[X(\phi, w, \vartheta, pol)] + \beta_{pol} \tag{18}$$

where α_{pol} and β_{pol} are two coefficients depending on polarization, and

$$F(x) = (1 + e^{-x})^{-1} \tag{19}$$

the argument of the function F is in Equation (18) is defined as

$$X(\phi, w, \vartheta, pol) = \gamma_0^{pol} + \sum_{i=1}^{11} \gamma_i^{pol} F\left[\Gamma_i^{pol}\left(\widetilde{\phi}, w, \vartheta\right)\right] \tag{20}$$

where

$$\Gamma_i^{pol}\left(\widetilde{\phi}, w, \vartheta\right) = \omega_{i,0}^{pol} + \omega_{i,1}^{pol} V_1^{pol}\left(\widetilde{\phi}\right) + \omega_{i,2}^{pol} V_2^{pol}(w) + \omega_{i,3}^{pol} V_3^{pol}(\vartheta) \tag{21}$$

$$V^{pol} = \begin{pmatrix} V_1^{pol} \\ V_2^{pol} \\ V_3^{pol} \end{pmatrix} = \begin{pmatrix} \widetilde{\phi}\lambda_{20}^{pol} + \lambda_{21}^{pol} \\ \vartheta\lambda_{00}^{pol} + \lambda_{01}^{pol} \\ w\lambda_{10}^{pol} + \lambda_{11}^{pol} \end{pmatrix} \tag{22}$$

with $\widetilde{\phi} = 360° - \phi$ if $\phi < 180°$ and $\widetilde{\phi} = \phi$ elsewhere. Notice that ω^{pol}, λ^{pol} and γ^{pol} are polarization-dependent coefficients whose values can be found in [28].

The model described by [28] and here summarized was validated on two cases of complex meteorological situation and it is found that the high sensitivity of the Doppler to the wind direction is useful to retrieve more realistic wind patterns in the case of complex and rapidly changing meteorological situations. More detail about the results and the areas of study can be found in [28].

Figure 3a shows the absolute value of the Dopper Anomaly (f_{W_DCA}), evaluated according to Equations (18)–(22), as a function of wind speed, the latter ranges from 1 m/s to 18 m/s. For the ENVISAT case, the incidence angle is equal to 23° and the polarization is VV. In particular, in this figure are shown the upwind case (green continuous line), which is the case of wind blowing toward the antenna, and the downwind (red dashed line) case, which is the case of wind blowing away from the antenna. In both cases, the measurements shows that a reduction of the wind leads to decrease of the Doppler frequency; however, the DCA does not fall down beyond 8 Hz, corresponding to about 30 cm/s surface radial velocity. Notice that this figure is comparable to Figure 1b reported in [28]. In Figure 3B it is shown the DCA, as a function of wind direction (ϕ), as defined above, for three different values of wind speed: 7 m/s blue line, 5 m/s green line, 2 m/s red line. In this case, the DCA and the surface current both approach zero in the case of orthogonality between the wind direction and the radar LOS.

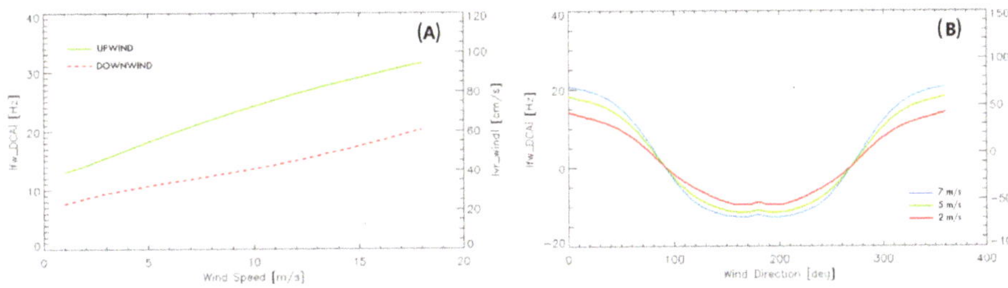

Figure 3. Plots relevant to the Doppler Geophysical Mapping Function describing the Doppler Centroid Anomaly (DCA) in Equations (18)–(22) as a function of the wind speed and direction. (**A**) modulus of the DCA (f_{W_DCA}) as a function of wind speed, for 23° of incidence angle and for VV polarization: upwind condition in green, downwind condition in red dashed. (**B**) DCA (f_{W_DCA}) as a function of the wind direction, same incidence angle polarization as in the left image, for three different wind speed values: 7 m/s (blue line), 5 m/s (green line), 2 m/s (red line).

As our focus was on the analysis of the estimated sea surface currents, we exploited the D-GMF in a direct form, which is the opposite of the method used in [28]. More specifically, given the radar geometry and the wind speed and direction provided estimated by the Bayesian retrieval involving the hindcast of the meteorological ocean model and the measured NRCS a Doppler Centroid shift f_{W_DCA} were evaluated through the D-GMF in Equations (18)–(22) and converted to surface velocity through Equation (2).

5. The Data Analysis Procedure

The former sections addressed the separate description of the different elements considered in this study to investigate the possibility of exploiting the Doppler information derived from the SAR data to achieve products that can provide an indication about the sea currents. As previously noticed, the sea surface current directly extracted from the SAR data can be heavily impacted by meteorological factors, mainly the wind.

The procedure used in this work for the data analysis is depicted in Figure 4: it involves the joint processing of SAR acquisitions and the integration with the external information about the wind achieved by the external atmospheric model.

Figure 4 highlights that SAR data are exploited following two main streams based on the use of the intensity and of the complex (amplitude and phase) information for the (spectral) Doppler analysis. As for the latter, the DCA is generated according to the procedure described in Section 2 and summarized in Figure 1 starting from the complex information associated with the raw data. The choice to start from the Level-0 product is, as already pointed out in Section 2, related to the need for avoiding any possible bias introduced at the data focusing stage for the generation of Level-1 products supplied by the data provider. The latter may in fact incorporate filtering for image enhancement based on ad hoc trade-off of the resolution and the sidelobes suppression. The achieved DCA, one converted to surface current (point 6 left output), is then stored for the subsequent comparison with the response associated with to meteorological factors carried at the final stage of the processing.

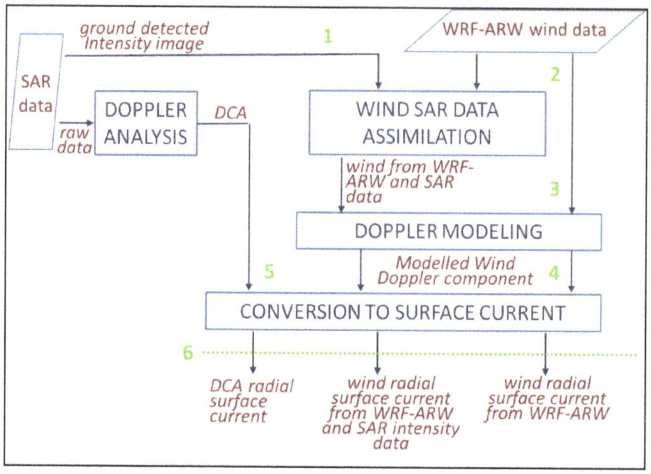

Figure 4. Data analysis scheme explaining the procedure followed for the joint analysis on the Doppler (surface current) derived by the SAR data (complex and detected intensity) and the atmospheric model.

The estimation of the Doppler component induced by the meteorological factors follows the right side flow in Figure 4 and is composed of two sequential steps. In the first step, the wind information provided by the weather forecast model is enriched via the assimilation, based on the Bayesian strategy described in Section 3, of the High Resolution SAR intensity. A High-Resolution wind map is thus also produced (level 3).

The second step exploits the Doppler GMF described in Section 4 to derive the wind related Doppler information from the wind intensity and direction achieved after from both the original wind data by the model and the HR wind results after integration with the intensity SAR data. The retrieved wind Doppler components are converted to surface current, see level 6 at middle and right, according to Equation (2). At this stage, a final subtraction between the DCA derived surface current and the one derived by the wind information could be implemented to extract the residual radial surface current component. As long as wind data derived from the weather forecast are given with a high confidence level, the output produced by the procedure shown in Figure 4 should therefore contain only the information relevant to the current associated with the ocean current and possible effects of the sea state. This subtraction is, however, not included because of the aim of this work was to evaluate the influence of the wind on the surface currents derived from the Doppler analysis.

The above described procedure was adopted in the experimental section for the data analysis of the case studies. The green numbers shown in Figure 4 identify the results of the different processing steps, which are shown and described in the experimental analysis carried out in Section 7. To summarize, they are

(1) The HR detected ground image (SAR intensity).

(2) The Wind vector information from the external atmospheric model.

(3) The Wind vector retrieval from SAR intensity image (point 1 in Figure 4) using as a-priori guess the external atmospheric model (point 2 in in Figure 4).

(4) The Estimated wind driven Doppler component.

(5) The Doppler Centroid Anomaly.

(6) The radial surface current computed from the DCA (point 5 in Figure 4) and the Doppler modeled starting from the two different wind information (point 4 in Figure 4).

6. Test Site and Data Set

The test site is situated in the south part of Italy, see Figure 5 on the left, and covers a coastal area of about 90 km × 100 km (see Figure 5 on the right). More specifically, it starts in the south of the Gulf of Gaeta, includes the Gulf of Naples with the Ischia, Procida and Capri Islands, and part of the Gulf of Salerno.

Figure 5. The area of interest is located in the South part of Italy. The red placemarks show the corners of the validation area of WRF-ARW wind model (left side). On the right side, there is a zoom on the test site, delimitated by red line, which includes the south of the Gulf of Gaeta, the Gulf of Naples with the Ischia, Procida and Capri Islands, and part of the Gulf of Salerno.

In Figure 5 left, red placemarks were highlighted to delimitate the corners of the validation area of WRF-ARW wind model. The validation was performed at the sites that are vertices of a quadrangular containing the study area of this paper: Ponza, Cetraro, Palermo and Capo Comino. Moreover, [41] developed hindcast covering 32 years (from 1979 to 2010) and containing, among the other, the wind downscaled data used in the present study.

In this work, we focused our study on the coast described above because it is an area densely urbanized and characterized by heavy maritime traffic. Hence, the analysis carried out could be functional for the study of the sea state and have an interesting outcome for the activities oriented toward the definition of a long-term strategy in support of sustainable growth in the marine and maritime sectors, i.e., those known as Blue Economy or Blue Growth. Moreover, the results on the surface current obtained in this work could also be useful for a meteorological analysis.

Among the several available SAR sensors, in this work, we decided to use ENVISAT ASAR for its characteristics relative to the swath. As a matter of fact, ENVISAT, in case of the (standard) stripmap acquisition, provides seamless and continuous coverage over a single swath width up to 100 km.

The swath analyzed in this work is shown on the right side of Figure 5 and delimitated by the yellow placemarks and red line. For the case of COSMO-SkyMed the swath width, still considering the stripmap acquisition, would have been approximatively halved in both directions. A more recent SAR sensor, which offers a very wide swath and operates in C-Band, is Sentinel-1 (S1), which can boast swath of about 250 km. S1 captures three sub-swaths using Terrain Observation with Progressive Scans SAR (TOPSAR) mode; this mode allows covering wide coastal areas for each single orbit. The operation mode provides, however, additional complexity in the estimation of Doppler centroid for the azimuth beam steering reasons, and would lead from burst to burst to the presence of discontinuities in the estimation Doppler Anomaly due to the change on the borders of the antenna pointing: the discontinuities impacts also the final radial velocity product [28].

Following the above considerations, as the best choice for this area, we found 47 images acquired by ENVISAT over ascending orbits between 2002 and 2010 (i.e., almost the whole operative period of the sensor). The ENVISAT sensor operates at C-band (wavelength equal to 5.6 cm), in VV polarization. Such a sensor is characterized by a spatial resolution of 5 m by 20 m in azimuth and ground range, respectively.

The dataset was screened for the presence of potential amplitude patterns related to the sea-air interface, in particular from the wind or from mass displacements of marine currents. Among them, a total of five images were selected for the subsequent numerical analysis on the Doppler characteristics, in particular, they span from 2004 to 2010 and are listed in Table 1 together with the indication of the available model/models.

Table 1. Acquisitions dates and used models.

Acquisition Date	WRF-ARW Model	Oceanografic Numerical Model
17 November 2004	✔	
22 December 2004	✔	
02 November 2005	✔	
11 January 2006	✔	
22 September 2010	✔	✔

It is worth noting that the dataset presented had the aim of discussing the open issues, especially concerning the influence of wind on surface currents and hence, represents a case study, rather than a statistically significant analysis. We underline that the choice of the period is related to the availability of ENVISAT data that, as discussed above, are characterized by an operational mode more suitable to the scope of the work.

7. Results

In this section, we discuss the results of the numerical analysis carried out on the selected images of Table 1.

DCA was evaluated, in particular, according to the method described in Section 2 with patches sliding along the whole SAR image. We considered patches of width 512 azimuth by 128 range pixels corresponding to a 2.5 km- by 2.5 km-wide window. The size of the patch was set in such a way to guarantee a good balancing between spatial and spectral resolutions. In particular, this latter is equal to 2.85 Hz. The stationary component $f_{DC0}(r)$ was evaluated by removing azimuth independent range polynomial components of the fourth order estimated on land and sea regions. The resulting radial velocity was estimated from (1): the notch filter for the removal of the undulations of the Doppler Centroid modeled as in Equation (5) and present on all the processed images.

On the other hand, the wind vector retrieval through CMOD model was performed by multi-looking the SAR data, i.e., averaging the number of cells, in order to reduce the speckle noise. As a trade-off between spatial resolution and speckle noise, the removal a multi-looked pixel of 1 km $\times$ 1 km was considered. The wind a-priori guess, provided by the 10 m wind fields obtained using the non-hydrostatic mesoscale model WRF-ARW, is instead downscaled to a computational domain model, covering the whole Mediterranean, with a ~10 km resolution Lambert conformal grid. Initial and boundary conditions for the atmospheric simulations with the WRF-ARW model were provided from the CFSR (Climate Forecast System Reanalysis) database [42].

Figures 6–10 show the results obtained for the five dates listed in the previous section, organized in a panel comprising eight images. Specifically, for each date are shown:

- the σ^0 image (point 1 in Figure 4);

- the estimated Doppler Anomaly maps converted to the sea surface (line-of-sight component) velocity (point 6 left arrow in Figure 4);
- the wind direction obtained from the (Bayesian) CMOD and from the WRF-ARW model (point 3 in Figure 4);
- the wind speed relative to the CMOD model and to the WRF-ARW model (point 3 in Figure 4)
- the surface radial velocity due to the wind component, evaluated by the D-GMF as described in Section 4 for both winds estimated by WRF-ARW and refined by the CMOD-based Bayesian model (point 6 in Figure 4, middle and right products).

It can be noticed that the estimated velocity ranges between ± 100 cm/s, red and blue colors (redshift and blue shift) of the estimated sea surface velocity component indicate a motion toward and away from the radar antenna, respectively.

Figure 6 refers to the acquisition of 17 November 2004; in this case, the amplitude image (Figure 6A) shows an evident pattern in the north of the Ischia Island area. This pattern is observed also in the image corresponding to the surface currents (Figure 6B) with a redshift located in the same position. The image in Figure 6B also shows another pattern with a component of the surface current toward the sensor in the south-most region of the swath: no significant variation, however, is observed in the same area in Figure 6A. Both patterns can be explained, although partially, by looking at the wind data of the WRF-ARW model wind: the arrows' direction (Figure 6D) indicates in fact the presence of a wind blowing towards the sensor (upwind condition). However, the north pattern has a sharper transition than the southern one in terms of wind direction. In the most westerly part of both Figure 6C,D, a wind blowing almost parallel to the azimuth direction, i.e., in the direction orthogonal in the LOS, can be noticed: a feature that seems to be correctly mapped in the DCA measurement on the left part of Figure 6B. Such a transition is even more emphasized in the refined wind measurements following the MAP based on CMOD estimator for the wind described in Section 3 (see Figure 6C,E,F for the intensities). The surface radial velocity obtained via the Doppler model described in Section 4 shows a rather good match with the Doppler measurement achieved by processing the complex SAR data (Figure 6B) for both the WRF-ARW (Figure 6H) and the CMOD wind (Figure 6G). However, the CMOD derived surface velocity (Figure 6G) shows a higher spatial match with Figure 6B when compared to the WRF-ARW derived surface velocity in Figure 6H. Finally, in Figure 6G it is noticeable only a partial mismatch on the northern pattern, likely due to a bias of the WRF-ARW wind source on the estimation of the wind direction in CMOD MAP based estimator.

In Figure 7 are presented the results relative to the acquisition on 22 December 2004, also in this case, in the amplitude image (Figure 7A), there is a pattern in the northern part of Ischia Island (less evident than the case of Figure 6A), which is also present in the surface radial velocity of Figure 7B. In the south-east area of the DCA map (Figure 7B) a redshift is again observed; it represents a surface current that moves toward the sensor. Wind directions of Figure 7C,D can explain such a behavior because they show, also in this case, the presence of an upwind: the wind magnitude (see Figure 7E,F) confirms this evidence too. Finally, Figure 7G,H highlight that in the upwind condition the radial velocity obtained via the D-GMF described in Section 4 tends to overestimate the radial velocity, i.e., shows a redshift also in areas of with low wind magnitude. This behavior is in agreement with the plots in Figure 3, which show that the in the upwind condition the D-GMF provides radial velocities different from zero even in the case of very low wind magnitude (Figure 3 left), whereas with wind directions orthogonal to the LOS (Figure 3 right), it does not provide contributions. Despite the bias for low wind, there are many congruencies between the patterns shown in Figure 7B,C, especially in areas with substantial wind conditions.

Figure 8 shows the results for the acquisition on 2 November 2005. Moreover, in this case, the presence of a strong backscattering pattern in the northern area and in the south part of Figure 8A it is evident and the surface radial velocity shows a similar pattern, confirming a good correspondence between the amplitude and the product derived by the DCA and hence, likely an effect of the wind. Observing Figure 8C, and similarly Figure 8D, it is possible to notice a change of the wind direction:

it starts blowing from land to sea then it rotates orthogonally to the LOS. Thiss change is mapped the red to green shift present in Figure 8B. Wind intensities evaluated from the WRF-ARW model (Figure 8F) and CMOD MAP based estimator (Figure 8E) look slightly different. Anyway, the surface radial velocity due to wind component (Figure 8G,H) shows that the redshift in Figure 8B can be due to the presence of wind, which is in a quasi-upwind condition. It is also noticed that, in agreement with the plot in Figure 3 right, the change of wind direction in the western part is mapped in a decrease of the surface current signal. In this case the CMOD derived surface velocity (Figure 8G) is in a better agreement with the behavior of Figure 8B than the WRF-ARW derived surface velocity (Figure 8H).

The results for the acquisition of 11 January 2006 are presented in Figure 9: they again involve specific patterns present in the amplitude image (Figure 9A) and in the surface radial velocity (Figure 9B) in the Gulf of Naples, as well as in the north and south regions. It can be noticed that CMOD wind and WRF-ARW show differences mainly in the wind intensity (Figure 9E,F), which also impacts the direction (Figure 9C,D) especially in the region in the north of the Gulf of Naples. Here, it is rather evident in both the amplitude (Figure 9A) and surface velocity (Figure 9B) images in the north region that the decrease of the surface current in the north-east area, with reduced wind intensity, is impacted by the change of direction. The intensity increase of surface velocity visible in the south part of Figure 9B is on the other hand due to the quasi-upwind condition, particularly evident in Figure 9D and to the high wind intensity shown in Figure 9F. The CMOD derived surface velocity (Figure 9G) is in agreement with the surface radial velocity (Figure 9B), while for the WRF-ARW derived surface velocity (Figure 9H), a distributed overestimation of the model essentially due to the alignment of the wind to the LOS is present. The comparison of Figure 9B,G shows also a decrease of the surface velocity estimated with the DCA: this reduction could be originated by the subtraction polynomial range component discussed in Section 2 at the stage of generation of the DCA.

Finally, in Figure 10 are shown the results for the acquisition of 22 September 2010. This case study is interesting because, as in some other cases, there is a change of the wind direction going offshore, but the wind is rather intense all over the image, see Figure 10D. In the amplitude image (Figure 10A), a pattern in the intensity image in the northern region is evident, which is also mapped by the surface velocity derived by the DCA in Figure 10B with a redshift corresponding to a current toward the sensor. This pattern, however, is not mapped in the intensity of the WRF-ARW model (Figure 10F). Apart from the area shown in the south-east part of the image, the surface current derived by the WRF-ARW model (Figure 10H) does not show a satisfactory match with the measured DCA shown in Figure 10B. The pattern in the north of the Gulf of Naples in Figure 10B, is only roughly mapped in Figure 10H. The situation drastically changes with the wind derived by CMOD. The pattern with large intensity in the north of the Gulf of Naples in Figure 10A provides, in fact, a refinement mainly affecting the wind intensity, see Figure 10E. This refinement originates from a better matching of the surface current estimated by the wind in Figure 10G and the surface current derived by the DCA in Figure 10B with respect to Figure 10H.

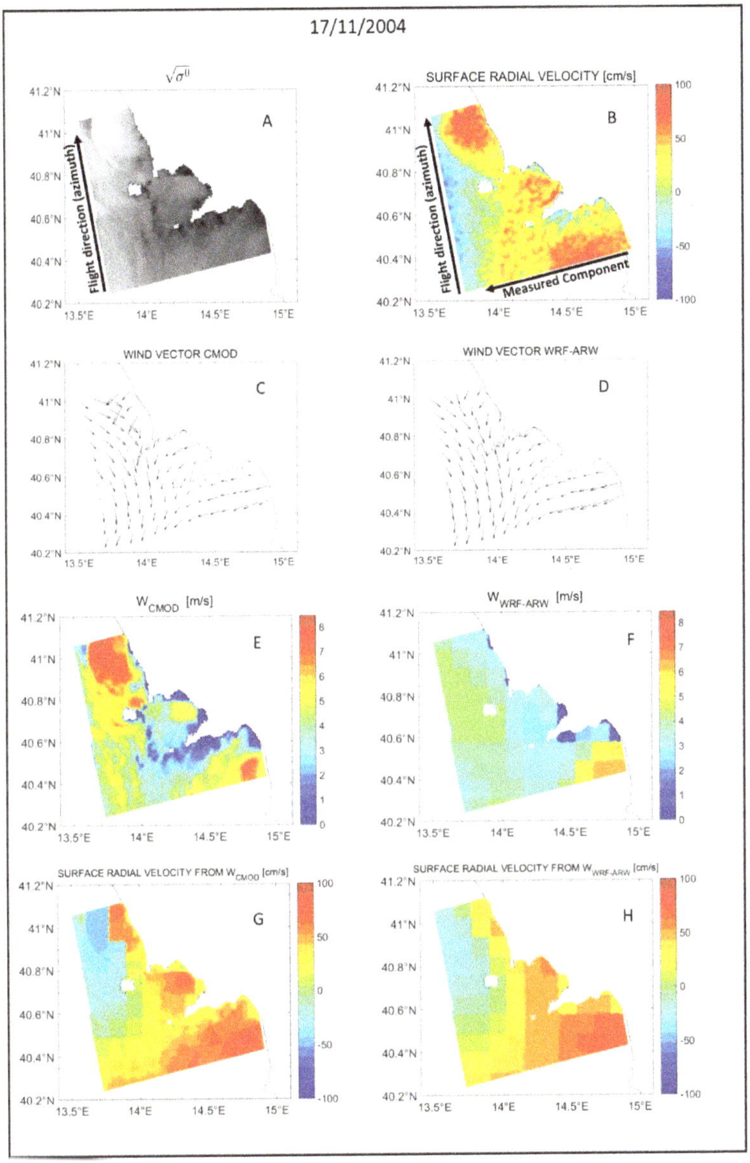

Figure 6. Results related to the acquisition of 17/11/2004.The image is divided as follow: (**A**) is the σ^0; (**B**) is the estimated Doppler Anomaly maps of the marine area converted into the line-of-sight component of the sea surface velocity; (**C**) is the wind direction obtained from the CMOD model; (**D**) is the wind direction obtained from the WRF-ARW model; (**E**) is the wind speed relative to the CMOD model; (**F**) is the wind speed relative to the WRF-ARW model; (**G**) and (**H**) is the surface radial velocity due to the wind component, evaluated as descried in Section 4, for the CMOD model and for the WRF-ARW model, respectively.

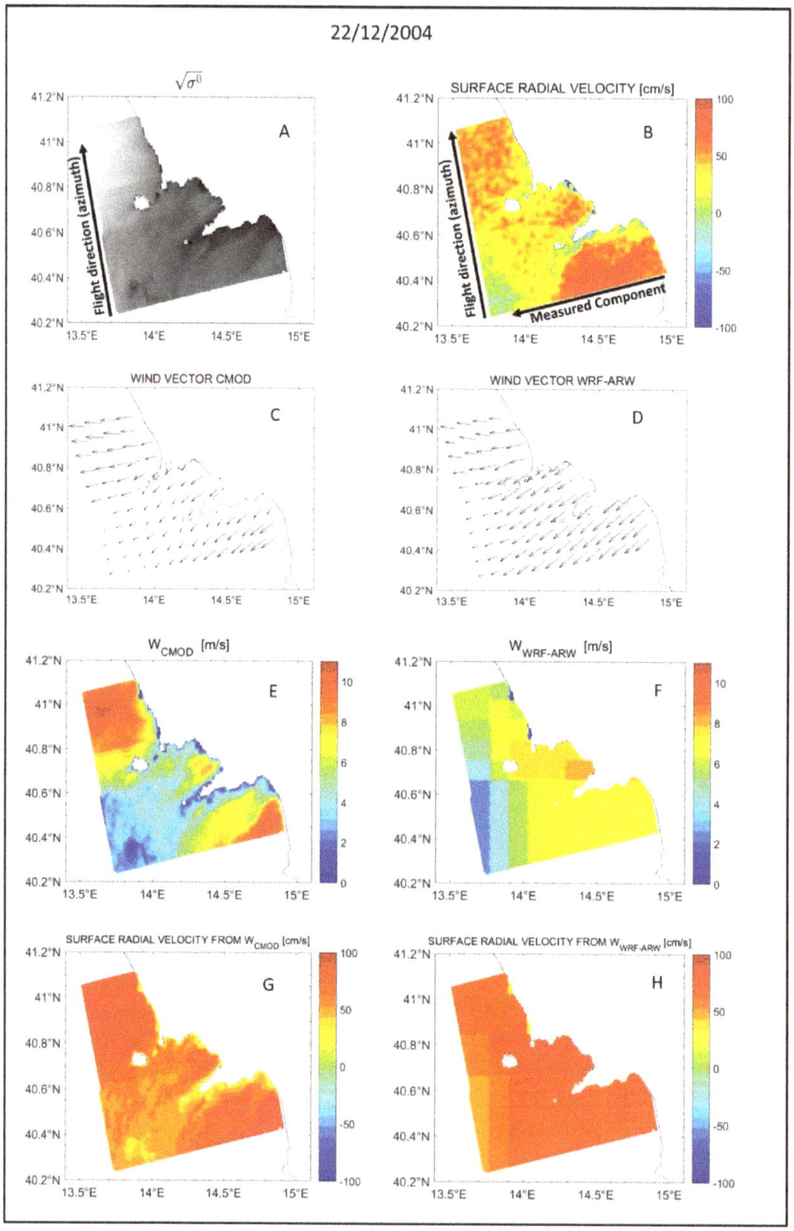

Figure 7. Results related to the acquisition of 22/12/2004. The image is divided as follow: (**A**) is the σ^0; (**B**) is the estimated Doppler Anomaly maps of the marine area converted into the line-of-sight component of the sea surface velocity; (**C**) is the wind direction obtained from the CMOD model; (**D**) is the wind direction obtained from the WRF-ARW model; (**E**) is the wind speed relative to the CMOD model; (**F**) is the wind speed relative to the WRF-ARW model; (**G**) and (**H**) is the surface radial velocity due to the wind component, evaluated as descried in Section 4, for the CMOD model and for the WRF-ARW model, respectively.

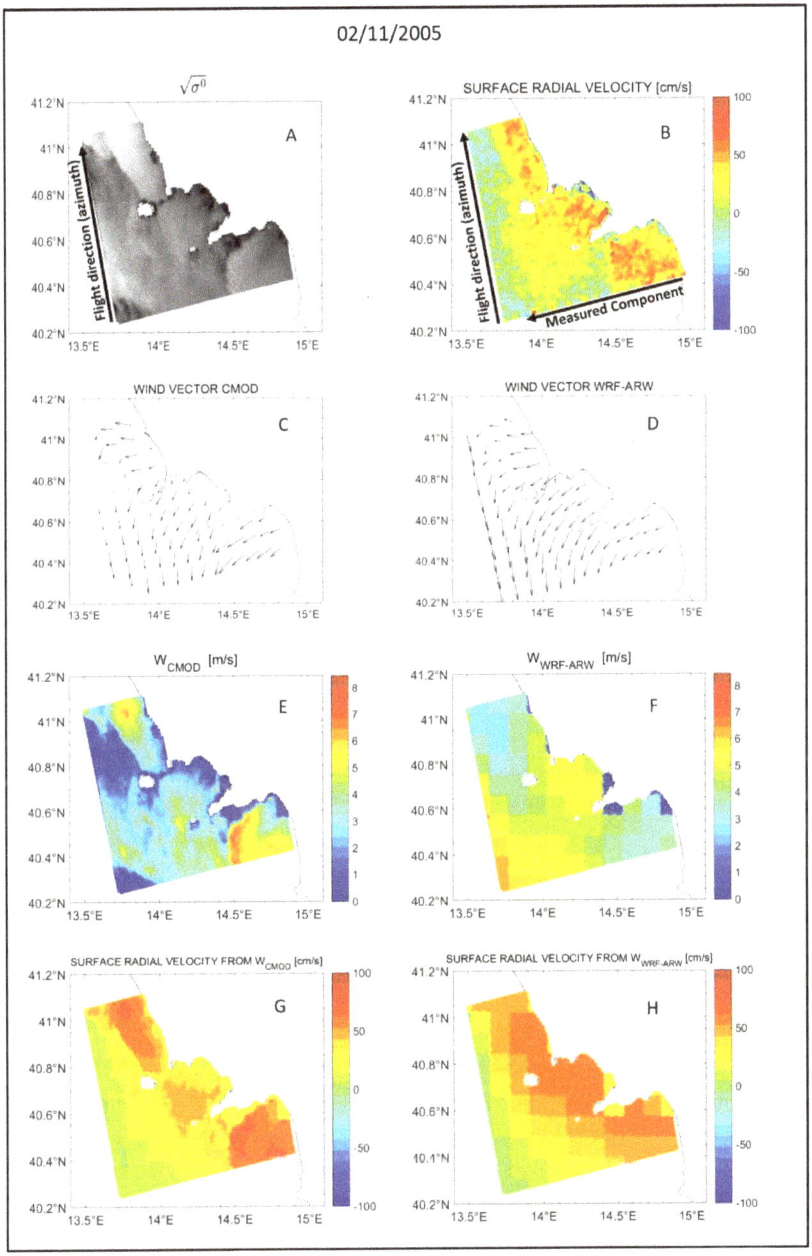

Figure 8. Results related to the acquisition of 02/11/2005. The image is divided as follow: (**A**) is the σ^0; (**B**) is the estimated Doppler Anomaly maps of the marine area converted into the line-of-sight component of the sea surface velocity; (**C**) is the wind direction obtained from the CMOD model; (**D**) is the wind direction obtained from the WRF-ARW model; (**E**) is the wind speed relative to the CMOD model; (**F**) is the wind speed relative to the WRF-ARW model; (**G**) and (**H**) is the surface radial velocity due to the wind component, evaluated as descried in Section 4, for the CMOD model and for the WRF-ARW model, respectively.

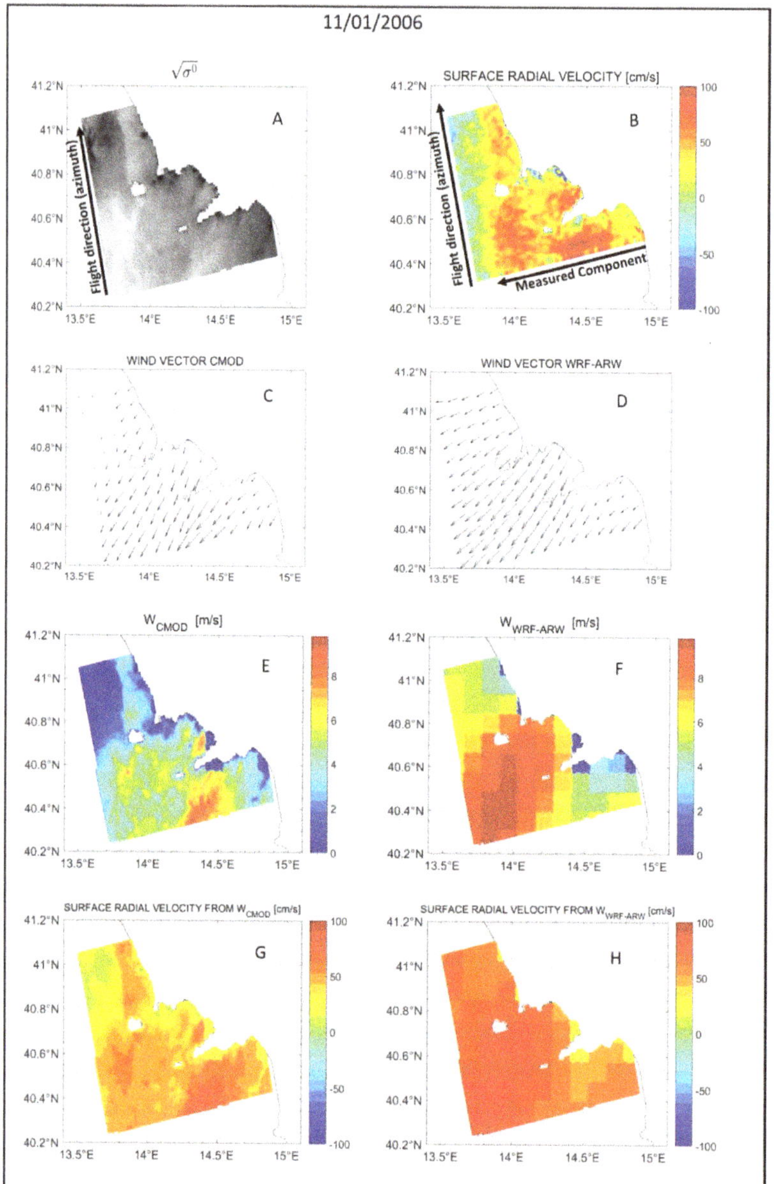

Figure 9. Results related to the acquisition of 11/01/2006. The image is divided as follow: (**A**) is the σ^0; (**B**) is the estimated Doppler Anomaly maps of the marine area converted into the line-of-sight component of the sea surface velocity; (**C**) is the wind direction obtained from the CMOD model; (**D**) is the wind direction obtained from the WRF-ARW model; (**E**) is the wind speed relative to the CMOD model; (**F**) is the wind speed relative to the WRF-ARW model; (**G**) and (**H**) is the surface radial velocity due to the wind component, evaluated as descried in Section 4, for the CMOD model and for the WRF-ARW model, respectively.

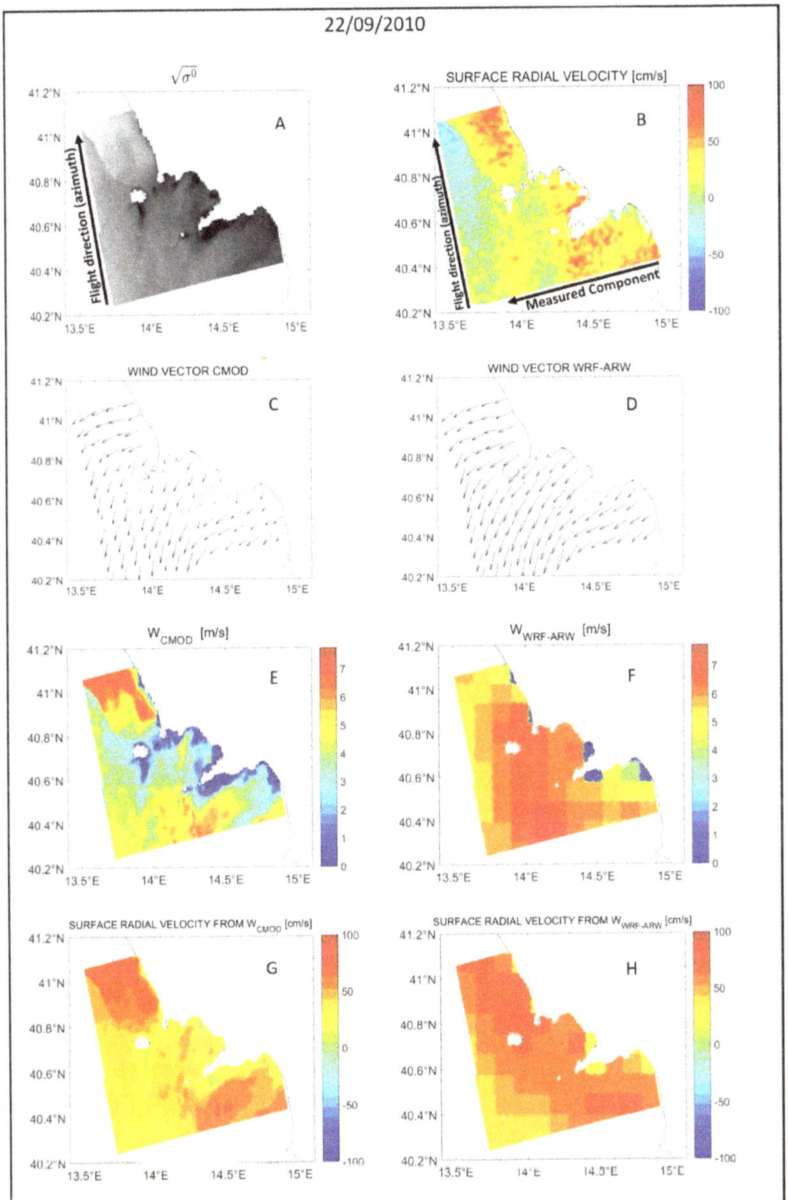

Figure 10. Results related to the acquisition of 22/09/2010. The image is divided as follow: (**A**) is the σ^0; (**B**) is the estimated Doppler Anomaly maps of the marine area converted into the line-of-sight component of the sea surface velocity; (**C**) is the wind direction obtained from the CMOD model; (**D**) is the wind direction obtained from the WRF-ARW model; (**E**) is the wind speed relative to the CMOD model; (**F**) is the wind speed relative to the WRF-ARW model; (**G**) and (**H**) is the surface radial velocity due to the wind component, evaluated as descried in Section 4, for the CMOD model and for the WRF-ARW model, respectively.

In order to a comprehensive interpretation of the results with respect to the circulation currents, the surface current in the area has been investigated by using an oceanographic numerical model for the date of 22 September 2010. Specifically, a state-of-the-art ocean model (SHYFEM, [43]) was applied to compute the circulation patterns in the area during the whole of September 2010. The model has been widely used to reproduce with and predict with success the three-dimensional circulation patterns induced by wind, tides and thermohaline forcing, at both coastal and shelf scale in the Mediterranean basin [44–49]. A simulation run was carried out considering the main meteo-marine forcing affecting the local circulation as the heat flows to the surface, the direct wind action and thermohaline forcing. Atmospheric and oceanographic data needed as model surface and boundary conditions were provided by the operational meteorological prediction system SKIRON [50] and by the European Earth Observation Programme Copernicus, respectively. The same model parameters and simulation setup used in [43] was adopted, unless the model domain that, in this application, included the Naples Gulf and the surrounding coastal and shelf areas.

In Figure 11, upper panels, the surface current fields obtained from simulation results between the 0 and 3 m depths, corresponding to the model first vertical layer, are reported for the interested domain at the time between the 20:00 and 21:00 UTC on 22 September 2010. In both panels the flow pattern is characterized by the presence of an intense coastal flow directed northwestward with maximum current speeds higher than 0.3 m/s.

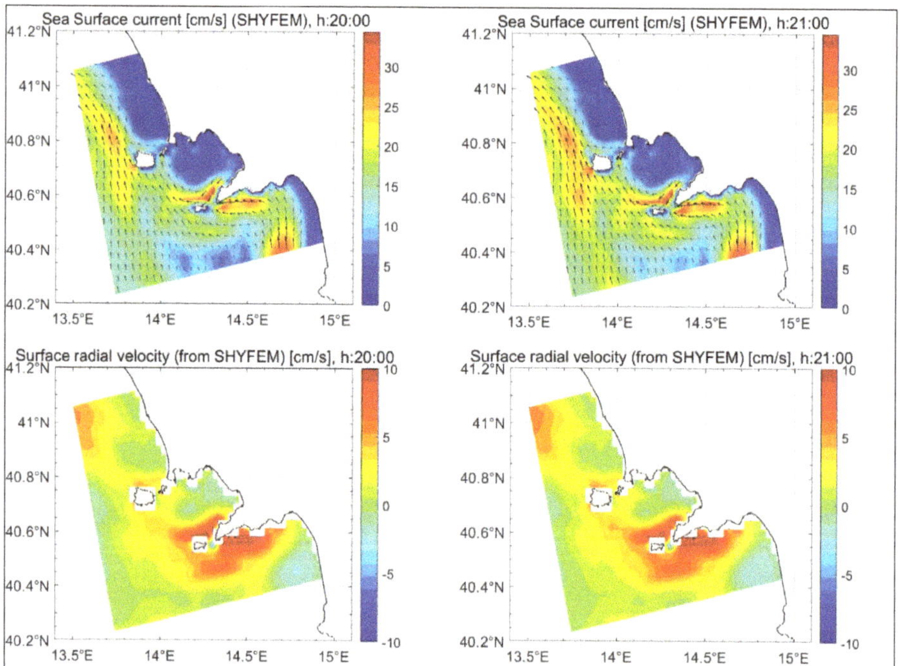

Figure 11. Ocean model results: surface current fields (upper panels) and intensities of the radial components (lower panels), computed for the 0 to 3 m surface water layer depth for the interested domain at the time 22/09/2010, 20:00 and 21:00 UTC.

In the nearby Gulf of Naples, an HF radar system is operationally providing the water current at the surface but limited to the bay area [23,51]. A comparison between the measurements derived with the Doppler Centroid Anomaly from SAR and the measurements of the HF radar system available in the bay is reported in [52], and shows reasonable agreement although the limitation of covered area.

On the other hand, several studies based on the use of oceanographic models have been carried out to reproduce the water circulation in both the Gulf of Naples and the surrounding coastal zone [23,27]. In particular, in [27] a similar modeling application was carried out applying an ocean model to investigate the water circulation in both the shelf and coastal area. Even if in the two applications the temporal coverages differ, there is an evidence that in [27] results obtained for both Spring and Autumn periods reproduce exactly the same hydrodynamic patterns reported in Figure 11. In particular in [27], both the model results and the analysis of the geostrophic currents depicted a northward flow leaning on the shelf break that interacting with the coast geometry generates cyclonic circulation patterns on the adjacent coastal area (see Figure 3, 13 and 14 in [27]). The obtained results are in line with previous studies in the area indicating the northward coastal flow as a typical circulation structure characterizing the area [27].

Changes in the flow directions are detected near the coastlines and on the leeside of the main Capes and Islands. Specifically, the interaction between the coastal flow, the coastline geometry and the wind action give rise to an edge separating the main external flow from the circulation cells in the shallow water part of the domain, north to the Gulf.

Similarly, in Figure 10B, inversions of the signal direction are detected for the same areas, thus confirming the presence of potential oceanographic heterogeneities as fronts or edges. Considering the wind role in modulating the pattern flow in this area, in the off-shore part of the domain, due to the higher water depths, the Coriolis force allows a direct contribution of the wind action to the main northwestward flow. In the shallow part of the domain, the off-shore wind direction and the complicated coastal geometry generate highly variable oceanographic structures that cannot be accurately solved by the adopted numerical model application. Within this context, thought the model results are very helpful in detecting the edge between open ocean and coastal circulation patterns, they cannot provide accurate information on the wind induced transport processes characterizing the skin layers of the coastal waters. Consequently, in the lower panels of Figure 11, the decomposition of the modeled flow field into the radial component is not favoring the detection of a direct correspondence between the wind speed and direction and the effects on the computed surface circulation.

8. Discussion and Conclusions

SAR sensors have significant applications in the marine sector due to the high sensitivity to different phenomena occurring under and above the sea surface.

SAR data can be used to extract the sea radial surface velocity, a key variable in maritime scenario. In the present work, we investigated the potentialities and the limitation in the extraction from ENVISAT data of information about the sea currents in the coastal area of the Gulf of Naples (South Italy) in the Mediterranean Sea.

In the cases examined in the previous section, rather typical of the area considered, we have seen that the wind shaped the Doppler Anomaly derived from the analysis of the SAR images. One of the reasons of that is the weakness of the surface currents, which, according to the ocean model hindcast (Figure 11), roughly range between ± 5 cm/s, which is similar to the resolution of the Doppler Anomaly derived from SAR. Despite the inability to quantify the effect of circulation currents on the Doppler measurements, the study does provide evidence of the effects in terms of the observed Doppler patterns.

Doppler Anomaly measured form SAR image shows some differences with the model-derived one, a fact that may be ascribed to the coarse spatial resolution of the wind's model, which was adopted both to force the Doppler Geophysical Model Functions [28] (D-GMF) and as a first guess to the retrieve wind field from SAR (D-GMF).

The results presented in this paper suggest that this well known issue [53,54] could be partly overcome. The coarse spatial resolution of external data could indeed be refined by wind/Doppler approach that highlights changes in wind direction, as shown in this paper. By combining D-GMF and W-GMF with an atmospheric wind model, a wind vector fine map could be thus obtained.

Once all the above problems are solved, the Doppler contribution produced by the D-GMF should be subtracted to that derived from SAR, to perform a validation of the residual Doppler Anomaly. Future investigations of the effects of wind and currents on SAR derived surface currents should thus consider coastal regions for which current data (e.g., from moored Acoustic Doppler Current Profiler velocity data [55], High-Frequency Radar currents (e.g., [56]), are available for a direct quantitative comparison. It should be remarked that over large areas (as those studied here) that cannot be sensed by the coastal HF radars, there is the need of a more detailed ocean surface modelling assimilating the available in-situ data, possible nowadays with the enhanced computer facilities and the data assimilation techniques. Along this line, altimetry-derived surface currents (e.g., [57,58]) can be exploited also in coastal regions [59] as an additional source of information.

The work along these lines would provide a contribution to the exploitation of SAR derived currents products to the advance of coastal ocean observing systems [60].

Author Contributions: For research articles with several authors, a short paragraph specifying their individual contributions must be provided. The following statements should be used "Conceptualization, V.Z. and G.F.; methodology, V.Z., F.D.S., G.D.C. and G.F.; software, V.Z., F.D.S., A.C., G.F.; validation, V.Z., F.D.S. and A.C.; data curation, V.Z. and F.D.S.; writing—original draft preparation, V.Z., F.D.S., G.D.C., G.F.; writing—review and editing, V.Z., F.D.S., A.C., S.Z., G.D.C., and G.F.; supervision, G.D.C and G.F.; funding acquisition, S.Z., G.D.C., G.F. All authors have read and agreed to the published version of the manuscript.

Funding: This research started with funding of the Italian Space Agency (ASI), under the project agreement I/055/09/0, and of the Italian Ministry of Education, University and Research Italian (MIUR) under the RITMARE flagship project; it achieved recent funding under the Contract ASI n° 217-I-E.0: "Progetto Premiale "Rischi Naturali indotti dalle Attività Umana—COSTE".

Acknowledgments: The authors wish to thank Dr. Alexis Mouche for providing the source code implementing the wind Doppler model discussed in [28] and summarized in Section 4, exploited for the generation of the images shown in the lower rows in the figure panels from Figures 6–10. The authors wish also to thank Consiglia Rasulo for proofreading the English.

Conflicts of Interest: The authors declare no conflict of interest.

References

1. Covello, F.; Battazza, F.; Coletta, A.; Manoni, G.; Valentini, G. COSMO-SkyMed Mission Status: Three out of Four Satellites in Orbit. In Proceedings of the International Geoscience and Remote Sensing Symposium '09 (IGARSS '09), Cape Town, South Africa, 12–17 July 2009.
2. Schulze, D.; Zink, M.; Krieger, G.; Böer, J.; Moreira, A. TANDEM-X Mission Concept and Status. In Proceedings of the FRINGE'09, Frascati, Italy, 30 November–4 December 2009.
3. Berger, M.; Moreno, J.; Johannessen, J.A.; Levelt, P.F.; Hanssen, R.F. ESA's sentinel missions in support of Earth system science. *Remote Sens. Environ.* **2012**, *120*, 84–90. [CrossRef]
4. Chapron, B.; Collard, F.; Ardhuin, F. Direct measurements of ocean surface velocity from space: Interpretation and validation. *J. Geophys. Res.* **2005**, *110*, C07008. [CrossRef]
5. Gerling, T. Structure of the surface wind field from Seasat SAR. *J. Geophys. Res.* **1986**, *91*, 2308–2320. [CrossRef]
6. Thompson, T.W.; Liu, W.T.; Weissman, D.E. Synthetic Aperture Radar observation of ocean roughness from rolls in an unstable marine boundary layer. *Geophys. Res. Lett.* **1983**, *10*, 1172–1175. [CrossRef]
7. Etling, D.; Brown, R.A. Roll vortices in the planetary boundary layer: A review. *Bound. Layer Meteorol.* **1993**, *65*, 215–248. [CrossRef]
8. Tsai, W.T. On the formation of streaks on wind-driven water surfaces. *Geophys. Res. Lett.* **2001**, *28*, 3959–3962. [CrossRef]
9. Tsai, U.C.; Huang, Y.F.; Yang, J.T. Strategies for the development of offshore wind technology for far-east countries–A point of view from patent analysis. *Renew. Sust. Energ. Rev.* **2016**, *60*, 182–194. [CrossRef]
10. Pierson, W.J. The measurement of the synoptic scale wind over the ocean. *J. Geophys. Res.* **1983**, *88*, 1683–1708. [CrossRef]
11. Panofsky, H.A.; Dutton, J.A. *Atmospheric Turbulence*; John Wiles & Sons: Hoboken, NJ, USA, 1984.

12. Romeiser, R.; Alpers, W. An improved composite surface model for the radar backscattering cross-section of the ocean surface, Model response to surface roughness variations and the radar imaging of the underlying topography. *J. Geophys. Res.* **1997**, *102*, 25251–25267. [CrossRef]
13. Jackson, G.; Fornaro, G.; Berardino, P.; Esposito, C.; Lanari, R.; Pauciullo, A.; Reale, D.; Zamparelli, V.; Perna, S. Experiments of Sea Surface Currents Estimation with Space and Airborne Sar Systems. In Proceedings of the International Geoscience and Remote Sensing Symposium '15 (IGARSS'15), Milano, Italy, 26–31 July 2015.
14. Zamparelli, V.; Jackson, G.; Cucco, A.; Fornaro, G.; Zecchetto, S. Sar Based Sea Current Estimation In The Naples Coastal Area. In Proceedings of the International Geoscience and Remote Sensing Symposium '16 (IGARSS'16), Beijing, China, 10–15 July 2016.
15. Serafino, F.; Lugni, C.; Nieto Borge, J.C.; Zamparelli, V.; Soldovieri, F. Bathymetry Determination via X-Band Radar Data: A New Strategy and Numerical Results. *Sensors* **2010**, *10*, 6522–6534. [CrossRef]
16. Fu, L.L.; Holt, B. Some example of detection of oceanic mesoscale eddies by Seasat synthetic aperture radar. *J. Geophys. Res.* **1983**, *88*, 1844–1852. [CrossRef]
17. Johannessen, J.A.; Suchman, R.A.; Digranes, G.; Lyzenga, D.R.; Wackerman, C.; Johannessen, O.M.; Vachon, P. Coastal ocean fronts and eddies imaged with ERS-1. *J. Geophys. Res.* **1996**, *101*, 6651–6667. [CrossRef]
18. Romeiser, R.; Runge, H.; Suchandt, S.; Kahle, R.; Rossi, C.; Bell, P. Quality assessment of surface current fields from TerraSAR-X and TanDEM-X along-track interferometry and Doppler centroid analysis. *IEEE Trans. Geosci. Rem. Sens.* **2014**, *52*, 2759–2772. [CrossRef]
19. Biron, K.; Van Wychen, W.; Vachon, P. Gulf Stream detection from SAR Doppler Anomaly. *Can. J. Remote Sens.* **2018**, *44*, 311–320. [CrossRef]
20. Rouault, M.J.; Mouche, A.; Collard, F.; Johannessen, J.A.; Chapron, B. Mapping the Agulhas Current from space: An assessment of ASAR surface current velocities. *J. Geoph. Res.* **2010**, *115*, C10026. [CrossRef]
21. Kerbaol, V.; Collard, F. SAR-Derived Coastal and Marine Applications: From Research to Operational Products. *IEEE J. Ocean. Eng.* **2005**, *30*, 482–486. [CrossRef]
22. Ardhuin, F.; Chapron, B.; Maes, C.; Romeiser, R.; Gommenginger, C.; Cravatte, S.; Morrow, R.; Donlon, C.; Bourassa, M. Satellite Doppler observations for the motions of the oceans. *Bull. Am. Meteorol. Soc.* **2019**, *100*, ES215–ES219. [CrossRef]
23. Iermano, I.; Moore, A.M.; Zambianchi, E. Impacts of a 4-Dimensional Variational Data Assimilation in a Coastal Ocean Model of Southern Tyrrhenian Sea. *J. Marine Syst.* **2016**, *154*, 157–171. [CrossRef]
24. Cianelli, D.; Uttieri, M.; Buonocore, B.; Falco, P.; Zambardino, G.; Zambianchi, E. Dynamics of a very special Mediterranean coastal area: The Gulf of Naples. In *Mediterranean Ecosystems: Dynamics, Management and Conservation*; Williams, G.S., Ed.; Nova Science Publishers, Inc.: New York, NY, USA, 2012; pp. 129–150.
25. Iacono, R.; Napolitano, E.; Marullo, S.; Artale, V.; Vetrano, A. Seasonal Variability of the Tyrrhenian Sea Surface Geostrophic Circulation as Assessed by Altimeter Data. *J. Phys. Oceanogr.* **2013**, *43*, 1710–1732. [CrossRef]
26. Bakun, A.; Agostini, V.N. Seasonal patterns of wind-induced upwelling/downwelling in the Mediterranean Sea. *Sci. Mar.* **2001**, *65*, 243–257. [CrossRef]
27. De Ruggiero, P.; Napolitano, E.; Iacono, R.; Pierini, S. A high-resolution modelling study of the circulation along the Campania coastal system, with a special focus on the Gulf of Naples. *Cont. Shelf Res.* **2016**, *122*, 85–101. [CrossRef]
28. Mouche, A.A.; Collard, F.; Chapron, B.; Dagestad, K.F.; Guitton, G.; Johannessen, J.A.; Kerbaol, V.; Hansen, M.W. On the Use of Doppler Shift for Sea Surface Wind Retrieval from SAR. *IEEE Trans. Geosci. Rem. Sens.* **2012**, *50*, 2901–2909. [CrossRef]
29. Madsen, S.N. Estimating the Doppler centroid of SAR data. *IEEE Trans. Aerosp. Electron. Syst.* **1989**, *25*, 134–140. [CrossRef]
30. Bamler, R. Doppler Frequency Estimation and the Cramer-Rao Bound. *IEEE Trans. Geosci. Rem. Sens.* **1996**, *29*, 385–390. [CrossRef]
31. Hasselmann, K.; Raney, R.K.; Plant, W.J.; Alpers, W.; Shuchman, R.A.; Lyzenga, D.R.; Rufenach, C.L.; Tucker, M.J. Theory of synthetic aperture radar ocean imaging: A MARSEN view. *J. Geophys. Res.* **1985**, *90*, 4659–4686. [CrossRef]
32. Plant, W.J. Relationship between stress wind and wave slope. *J. Geophys. Res.* **1982**, *87*, 1961–1967. [CrossRef]
33. Mentaschi, L.; Besio, G.; Cassola, F.; Mazzino, A. Performance evaluation of WavewatchIII in the Mediterranean Sea. *Ocean Model.* **2015**, *90*, 82–94. [CrossRef]

34. Hersbach, H. *CMOD5. N: A C-Band Geophysical Model Function for Equivalent Neutral Wind*; European Centre for Medium-Range Weather Forecasts: Reading, UK, 2008.
35. Hersbach, H.; Stoffelen, A.; de Haan, S. An improved C-band scatterometer ocean geophysical model function: CMOD5. *J. Geophys. Res. Oceans* **2007**, *112*, C03006. [CrossRef]
36. Quilfen, Y. *ERS-1 off-line Wind Scatterometer Products*; Technical Report IFREMER-SCAT/IOA/DOS-01; Institut Français de Recherche pour l'Exploitation de la Mer: Brest, France, 1993.
37. Skamarock, W.C.; Klemp, J.B.; Dudhia, J.; Gill, D.O.; Barker, D.; Duda, M.G.; Huang, X.Y.; Wang, W.; Powers, J.G. A description of the Advanced Research WRF version 3. NCAR Technical note-475+ STR. 2008.
38. Mentaschi, L.; Besio, G.; Cassola, F.; Mazzino, A. Problems in RMSE-based wave model validations. *Ocean Model.* **2013**, *72*, 53–58. [CrossRef]
39. Portabella, M.; Stoffelen, A.; Johannessen, J.A. Toward an optimal inversion method for synthetic aperture radar wind retrieval. *J. Geophys. Res.* **2002**, 107. [CrossRef]
40. Johannessen, J.A.; Chapron, B.; Collard, F.; Kudryavtsev, V.; Mouche, A.; Akimov, D.; Dagestad, K.F. Direct ocean surface velocity measurements from space: Improved quantitative interpretation of Envisat ASAR observations. *Geophys. Res. Lett.* **2008**, *35*, L22608. [CrossRef]
41. Mentaschi, L.; Besio, G.; Cassola, F.; Mazzino, A. Developing and validating a forecast/hindcast system for the Mediterranean Sea. *J. Coast. Res.* **2013**, *2*, 1551–1556. [CrossRef]
42. Saha, S.; Moorthi, S.; Pan, H.; Wu, X.; Wang, J.; Nadiga, S.; Tripp, P.; Kistler, R.; Woollen, J.; Behringer, D.; et al. The NCEP Climate Forecast System Reanalysis. *Bull. Amer. Meteor. Soc.* **2010**, *91*, 1015–1058. [CrossRef]
43. Umgiesser, G.; Melaku Canu, D.; Cucco, A.; Solidoro, C. A Finite Element Model for the Venice Lagoon. Development, Set Up, Calibration and Validation. *J. Mar. Syst.* **2004**, *51*, 123–145. [CrossRef]
44. Cucco, A.; Sinerchia, M.; Ribotti, A.; Olita, A.; Fazioli, L.; Perilli, A.; Sorgente, B.; Borghini, M.; Schroeder, K.; Sorgente, R. A high-resolution real-time forecasting system for predicting the fate of oil spills in the Strait of Bonifacio (western Mediterranean Sea). *Mar. Pollut. Bull.* **2012**, *64*, 1186–1200. [CrossRef]
45. Canu, D.M.; Solidoro, C.; Bandelj, V.; Quattrocchi, G.; Sorgente, R.; Olita, A.; Fazioli, L.; Cucco, A. Assessment of oil slick hazard and risk at vulnerable coastal sites. *Mar. Pollut. Bull.* **2015**, *94*, 84–95. [CrossRef]
46. Federico, I.; Pinardi, N.; Coppini, G.; Oddo, P.; Lecci, R.; Mossa, M. Coastal ocean forecasting with an unstructured grid model in the southern Adriatic and northern Ionian seas. *Nat. Hazards Earth Syst Sci.* **2017**, *17*, 45–59. [CrossRef]
47. Cucco, A.; Quattrocchi, G.; Olita, A.; Fazioli, L.; Ribotti, A.; Sinerchia, M.; Tedesco, C.; Sorgente, R. Hydrodynamic modelling of coastal seas: The role of tidal dynamics in the Messina Strait, Western Mediterranean Sea. *Nat. Hazards Earth Syst. Sci.* **2016**, *16*, 1553–1569. [CrossRef]
48. Ferrarin, C.; Davolio, S.; Bellafiore, D.; Ghezzo, M.; Maicu, F.; Mc Kiver, W.; Drofa, O.; Umgiesser, G.; Bajo, M.; De Pascalis, F.; et al. Cross-scale operational oceanography in the Adriatic Sea. *J. Oper. Oceanogr.* **2019**, *12*, 86–103. [CrossRef]
49. Ribotti, A.; Antognarelli, F.; Cucco, A.; Falcieri, M.; Fazioli, L.; Ferrarin, C.; Olita, A.; Oliva, G.; Pes, A.; Quattrocchi, G.; et al. An Operational Marine Oil Spill Forecasting Tool for the Management of Emergencies in the Italian Seas. *J. Mar. Sci. Eng.* **2019**, *7*, 1. [CrossRef]
50. Kallos, G.; Nickovic, S.; Jovic, D.; Kakaliagou, O.; Papadopoulos, A.; Misirlis, N.; Boukas, L.; Mimikou, M. The ETA model operational forecasting system and its parallel implementation. Presented at 1st Workshop on Large-Scale Scientific Computations, Varna, Bulgaria, 7–11 June 1997.
51. Cianelli, D.; Falco, P.; Iermano, I.; Mozzillo, P.; Uttieri, M.; Buonocore, B.; Zambardino, G.; Zambianchi, E. Inshore/offshore Water Exchange in the Gulf of Naples. *J. Marine Syst.* **2015**, *145*, 37–52. [CrossRef]
52. Fornaro, G.; Zamparelli, V.; Jackson, G.; Zecchetto, S.; Bovenga, F.; Refice, A.; Belmonte, A.; Pasquariello, G.; Adamo, P.; Rana, F.M.; et al. Validazione sui Test Bed, RITMARE Project Deliverable, SP5_WP2_AZ2_D05 2016, in italian.
53. Zecchetto, S.; Accadia, C. Diagnostics of T1279 ECMWF analysis winds in the Mediterranean Basin by comparison with ASCAT 12.5 km winds, Q.J.R. *Meteorol. Soc.* **2014**, *140*, 2506–2514. [CrossRef]
54. Zecchetto, S. Wind Direction Extraction from SAR in Coastal Areas. *Remote Sens.* **2018**, *10*, 261. [CrossRef]
55. Weisberg, R.H.; Liu, Y.; Mayer, D.A. Mean circulation on the west Florida continental shelf observed with long-term moorings. *Geophys. Res. Lett.* **2009**, *36*, L19610. [CrossRef]
56. Liu, Y.; Weisberg, R.H.; Merz, C.R. Assessment of CODAR and WERA HF radars in mapping currents on the West Florida Shelf. *J. Atmos. Ocean. Technol.* **2014**, *31*, 1363–1382. [CrossRef]

57. Johnson, E.S.; Bonjean, F.; Lagerloef, G.S.E.; Gunn, J.T.; Mitchum, G.T. Validation and error analysis of OSCAR sea surface currents. *J. Atmos. Oceanic Technol.* **2007**, *24*, 688–701. [CrossRef]
58. Sudre, J.; Maes, C.; Garcon, V. On the global estimates of geostrophic and Ekman surface currents. *Limnol. Oceanogr. Methods* **2013**, *3*, 1–20. [CrossRef]
59. Liu, Y.; Weisberg, R.H.; Vignudelli, S.; Roblou, L.; Merz, C.R. Comparison of the X-TRACK altimetry estimated currents with moored ADCP and HF radar observations on the West Florida Shelf. *Adv. Space Res.* **2012**, *50*, 1085–1098. [CrossRef]
60. Liu, Y.; Kerkering, H.; Weisberg, R.H. *Coastal Ocean Observing Systems*; Elsevier (Academic Press): London, UK, 2015; 461p, ISBN 978-0-12-802022-7.

Journal of
*Marine Science
and Engineering*

MDPI

Article

An Integrated Reconstruction of the Multiannual Wave Pattern in the Gulf of Naples (South-Eastern Tyrrhenian Sea, Western Mediterranean Sea)

Simona Saviano [1,2,*], Daniela Cianelli [2,3], Enrico Zambianchi [1,2], Fabio Conversano [3] and Marco Uttieri [2,4]

1 Department of Science and Technologies, Parthenope University of Naples, Centro Direzionale Is.C4, 80143 Naples, Italy; enrico.zambianchi@uniparthenope.it
2 CoNISMa (National Inter-University Consortium for Marine Sciences), Piazzale Flaminio, 00196 Rome, Italy; daniela.cianelli@szn.it (D.C.); marco.uttieri@szn.it (M.U.)
3 Department of Research Infrastructures for MArine Biological Resources, Stazione Zoologica Anton Dohrn, Villa Comunale, 80121 Naples, Italy; fabio.conversano@szn.it
4 Department of Integrative Marine Ecology, Stazione Zoologica Anton Dohrn, Villa Comunale, 80121 Naples, Italy
* Correspondence: simona.saviano@uniparthenope.it

Received: 15 April 2020; Accepted: 20 May 2020; Published: 23 May 2020

Abstract: Surface gravity waves retrieved by a network of HF (High Frequency) radars and measured in situ by an ADCP (Acoustic Doppler Current Profiler) current meter connected to an elastic beacon were used to carry out a multiple-year characterization of the wave field of the Gulf of Naples (south-eastern Tyrrhenian Sea, western Mediterranean). The aim of the work was to create a climatology of the study area and to demonstrate the potential of an integrated platform for coastal studies. The patterns recorded by the different instruments were in agreement with the wave climatology of the southern Tyrrhenian Sea as well as with previous scores for the same area. The results presented in this work also highlight seasonal and interannual consistency in the wave patterns for each site. In a wider context, this study demonstrates the potential of HF radars as long-term monitoring tools of the wave field in coastal basins, and supports the development of integrated observatories to address large-scale scientific challenges such as coastal ocean dynamics and the impact of global change on the local dynamics.

Keywords: Gulf of Naples; HF radar; ADCP; augmented observatory; wave field

1. Introduction

The monitoring of the wave field over several years is a fundamental prerequisite for the knowledge and definition of the wave climate of a given basin. Wave climatology has direct consequences on the design of offshore and coastal structures and their management as well as on renewable energy assessment [1]. Wave climate can be defined as the distribution of significant wave height (Hs), period (T), and direction (θ) averaged over a defined period for a given site [2,3]. While for the open sea, the wave climate can be assumed as rather invariant and may be similar over large distances [3], it might be profoundly modified in coastal areas (i.e., sea level rise, climate change) [4,5].

The correct assessment of wave climatology is strongly dependent on the number of years included in the analysis [3]. Wave climatology has been traditionally based on moored buoys, returning accurate but spatially limited measurements [6,7]. More recently, the availability of long-term satellite altimetry data and the implementation of third-generation wave models have paved the way to wave climate studies from global [8] to regional scales [9], allowing for the recreation of the wave field not directly covered by in situ buoys. Satellite altimetry data allow for the estimation of wave climate statistics with a resolution of 1° to 2°, depending on the satellite spatial and time coverage [10,11]. Wave models, instead, have enabled the development of wave climatology based on more extensive wind data (i.e., wind reanalysis), permitting the creation of global to regional scale retesting [12,13].

In this context, HF (High Frequency) radars can provide high-resolution (both spatial and temporal) information and near real-time monitoring of the wave field in coastal areas, with virtually continuous functioning over several years, thus potentially representing a valid alternative to wave buoys. HF radars employ the echo backscattered by the rough sea surface, according to the physics first described by Crombie [14]. In the presence of Bragg scattering, the signal is reflected off waves with a wavelength exactly half that emitted by the radar and travelling radially with respect to the antenna itself (either approaching or moving away from it). The spectral analysis of the electromagnetic signal reveals a Doppler-shifted peak, resulting from the current field within which the wave is propagating and from the theoretical wave speed [15]. The first-order echoes generally produce strong signals, used for robust measurements of surface currents [16,17]. Additional contributions to the Doppler signature are due to nonlinear wave interactions and double-scattering processes leading to second-order echoes [18], which, in turn, can be used to extract information on the wave field itself [17,19,20]. To date, studies using HF radars to evaluate wave features have been mostly limited to short periods (up to a few months; e.g., [18,21–27], or to specific events (e.g., tsunamis [28]; storms [29]). Only a few works have so far delved into the assessment of wave retrieval by HF radars over long periods, in the UK [30], in Central California (USA) [31], in Galway Bay (IE) [32], in the Gulf of Naples (IT) [33], and along the Galician coast (ES) [34], creating a new benchmark for augmented observatories.

To date, however, no study has yet made use of HF radar-derived wave measurements to build the wave climatology of a given area. In the present contribution, we intend to raise the bar for HF operators by integrating multiannual HF radar wave measurements with in situ recordings collected through an ADCP (Acoustic Doppler Current Profiler) to build the climatology of a coastal basin, the Gulf of Naples (GoN) (south-eastern Tyrrhenian Sea, western Mediterranean). HF radar derived measurements in the GoN have been previously compared with those recorded by a fixed buoy installed outside the coverage area of the radars [33] and with model simulations [35] returning good agreement in the seasonal patterns, reconstructed by the different platforms and in the statistical comparison of the wave parameters themselves (e.g., with normalized root mean square error in the order of few centimeters for Hs). The HF system operating in the GoN is the longest running in Europe [36], being operative since 2004 (Figure 1). Wave recording and analysis was initialized in May 2008 and ended in December 2012, permitting the collection of data over several consecutive years, providing the unique opportunity of depicting typical seasonal dynamics and highlighting characteristic features during storm conditions. Complimentary to the HF data, in this contribution, the wave recordings collected by means of a bottom-mounting ADPC interlocked with a MEDA elastic beacon operating in the GoN were included in the analysis, with reference to the period November 2015–July 2018. This integrated approach allowed the creation of a dataset spanning more than seven years, here used to construct a multiple-year pattern of wave conditions in the coastal area of the GoN. In particular, thanks to the geometry of the HF system and to the positioning of the MEDA elastic beacon, the dataset was analyzed to pinpoint the specific wave patterns of three different sectors of the basin and to reconstruct their seasonal patterns.

The results confirm previous records from the same basin [33,37–40], evidencing specific features depending on the season and on the sub-basin investigated. The findings presented in this work are additionally in line with the general wave climate of the southern Tyrrhenian Sea [41], indicating a commonality with the large-scale processes. As a wider-breath implication, the results shown here emphasize the possibility of resorting to HF radars as a long-term platform to monitor the wave field in coastal areas, and support the adoption of an integrated, multi-platform approach to build a robust observation system, retrieving near real-time data for the proper management and monitoring of coastal areas.

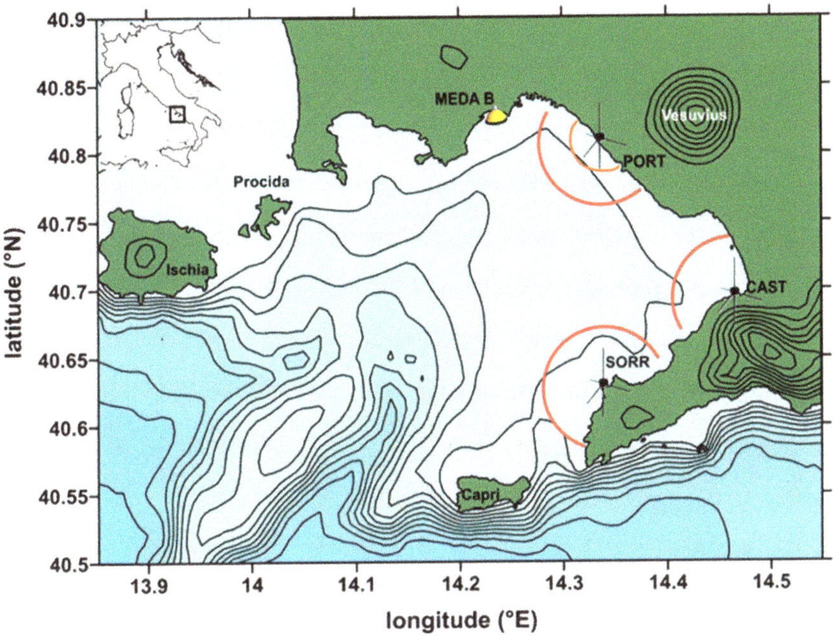

Figure 1. Map of the Gulf of Naples (south-eastern Tyrrhenian Sea, western Mediterranean Sea; see inset) with the locations of the three HF radar sites (antenna icons) and of ADCP connected to the MEDA B elastic beacon (buoy icon). The annuli centered on HF antennas represent RC5 (red) and RC2 (orange) (see text for details). The bathymetric and orographic contours were spaced every 100 m; elevation isolines were plotted starting from 300 m for graphical reasons.

2. Study Area

The GoN is a marginal sub-basin of the south-eastern Tyrrhenian Sea, characterized by an intricate bottom topography (including two deep canyons) and orography (Vesuvius volcano, hills in the urban area of Naples city, Lattari mountains on the south) [42–44]. The area is subject to intense anthropic pressure [45–48], but at the same time it is particularly rich in terms of biodiversity [49–53] and punctuated by numerous coastal archeological sites [54–56], with the presence of habitats with a high value of natural capital stock [57].

The surface circulation of the GoN is mainly driven by the local wind field, with the creation of specific patterns affecting both the physical and biological processes developing in the area [42,58–63]. During the winter months, the predominant and most intense winds come from the NNE and NE directions; in spring and autumn, the principal directions are NE and SW, while in summer, winds are generally weaker than the rest of the year [64]. This general wind regime is associated with numerous low pressure systems passing over the basin, with frequent stormy and windy events principally during winter and autumn. In contrast, in spring and summer, the setup and reinforcement of the Azores anticyclone, and in the latest years of the African anticyclone, determine the onset of a stable, moderate breeze system [65,66].

Previous investigations on the wave characteristics in the GoN, using non-directional wave buoys [37–40] and HF radars [33], have shown a strong seasonality related to the meteorological features of the study area, resulting in severe sea storms in winter and autumn, and lower wave heights due to stable and persistent high-pressure field in spring–summer [42,58,64]. A year-long analysis using HF radars [33] showed that, on average, wave heights in the inner sub-basins of the GoN ranged from 1.19 to 1.70 m, while areas closer to the Tyrrhenian Sea scored higher values (1.77–2.38 m) with typical wave periods between 6 and 8 s. Storms impacting the GoN occur mostly in winter and autumn, and are typically of the frequent extratropical type (i.e., linked to the oscillation of polar front), covering large areas and remaining in the offshore sectors for a significant time, so that large waves can build up [67]. In some cases, storms can also be driven by the Mediterranean low pressure systems [67]. In such extreme events, wave heights can be significantly higher than the average records, often getting values greater than 3 m.

It is worth underlining here that the previous investigations for the GoN were focused on short time windows (typically from a few weeks to months) and included the use of one instrument only. The results thus collected provide information uniquely for one site. Only in [33], was a year-long investigation with three HF radars performed, retrieving the first description of the wave field in three different areas of the GoN.

3. Materials and Methods

The HF radar network installed in the GoN, operated by the Department of Science and Technology at the 'Parthenope' University of Naples, is a system composed of three SeaSonde 25 MHz transceiving stations (CODAR Ocean Sensors, Mountain View, CA, USA) located in Portici (PORT), Castellammare di Stabia (CAST), and Massa Lubrense (SORR) (Figure 1; for its characteristics, data validation, and use in numerical models see [58,63,68]).

In SeaSonde systems, a Pierson–Moskowitz model [20] is used to extract wave parameters (significant wave height Hs_{HF}, direction θ_{HF}, and centroid period T_{HF}) from the second-order spectrum. In the present work, wave parameters were measured and averaged along annuli (range cells, RCs) centered on the antenna and regularly spaced every 1 km. For a 25 MHz system, the operational range for Hs_{HF} is typically bounded between 0.50 and 4.00 m. Hs_{HF} below 0.50 m are characteristic of weak second-order echoes [26], whereas at $Hs_{HF} > 4.00$ m the wave spectrum saturates and retrieval of wave parameters is unachievable [20]. As a QA/QC (Quality Assurance/Quality Control) procedure, Hs_{HF} values over the theoretical upper threshold were checked for consistency with the wave pattern recorded in the same time window, following the protocol designed in [33]. Moreover, waves should have a period of 5 s at least, or longer if the noise floor increases, which is to be interpreted by the radar [20]. Wave measurements were performed over 12 RCs at the PORT and SORR sites, and over 15 RCs in CAST, by using the proprietary CODAR software (Seasonde Radial Suite R7u2), and were provided every 10 min by averaging four spectra falling within the given time window. Based on previous analysis in [33], wave measurements from RC5 (i.e., the RC representative of the annulus 5–6 km far from the antenna) were selected as representative of an operational trade-off between distance from the coast (sufficient echo intensity) and depth (deep water condition; mean depths over the annuli: PORT: 117 m; CAST: 65 m; SORR: 98 m). To perform a more accurate comparison between the measurements collected by PORT and the ADCP installed within the MEDA framework (see below for details), for this specific site RC2 (i.e., representative of the annulus covering the area between 2 and 3 km for the antenna, mean depth over the annulus: 70 m) was also taken into account. In the present study, wave data collected from all three sites over the period between May 2008 and December 2012 were used. This time window corresponded to an optimal multiannual functioning of the network (i.e., system calibration and validation, minimal data gaps), ensuring the construction of a robust dataset.

The elastic beacon MEDA B managed by the Stazione Zoologica Anton Dohrn is located at a distance of about 450 m from the coast of Naples on a seabed 17.5 m deep (Figure 1). MEDA B has been operating since November 2015. An ADCP current meter is connected to the elastic beacon on the bottom, not far from the anchor of the MEDA. It is equipped with a sensor for directional wave measurements indicating how much wave energy exists (Hs_B), from which direction (θ_B), and at what frequency (T_B, peak period) (Teledyne RD Instruments—WHSW 600 kHz model with NEMO^{TM} real-time wave processing module). The module is configured to sample wave data every 20 min by recording spectra, which are then averaged every hour (2100 pings at 2 Hz). Data referring to the time period considered were QA/QC checked using the same protocol adopted for HF radar data. This ensured obtaining a coherent dataset, without spikes or incongruent values. Characteristics of the measurement systems and the dataset are shown in Table 1.

As a first step in the data analysis, an overview of the records was produced as Hs time series, conveying an immediate, qualitative description of the interannual patterns for each platform. In addition, wave recordings were seasonally grouped, divided in three-month seasons (January–March: winter; April–June: spring; July–September: summer; October–December: autumn) based on the specific evidences discussed in [61], showing the onset of typical spring–summer climatologic features (e.g., stable high pressure systems, breeze regimes) in April. This categorization differs from the typical classification given by [69], more realistically mirroring the climatology of the GoN.

The wave field characterization for all the selected periods was carried out over the dataset to distinguish different basin behavior consequent to different meteorological forcing.

Table 1. Characteristics of the measurement systems and the dataset.

	SeaSonde Systems	**ADCP**
Parameters		
Significant wave height (Hs)	x	x
Mean Direction (θ)	x	x
Centroid period (T_{HF})	x	
Peak period (T_B)		x
Time series	May 2008–December 2012	November 2015–December 2018
Number of observations	81,888	21,590
Acquisition Time	10 min	1 h

Descriptive and numerical statistics were then employed to analyze the seasonally grouped results (Hs, T, and θ). Data were interpreted using standard descriptors (mean, median, standard deviation, minimum and maximum values). A more detailed comparison was additionally performed by means of seasonal boxplots, a standardized way of displaying the distribution of data based on five descriptors: 5th percentile, 25th percentile, median, 75th percentile, and 95th percentile. In the simplest boxplot, the central rectangle spans the first quartile to the third quartile (interquartile range, IQR). A segment inside the rectangle shows the median and "whiskers" above and below the box show the locations of the 5th and 95th percentiles. Data outside this range are displayed as single outliers. In the present analysis, the HF radar outliers were checked for consistency based on the results obtained in previous studies [33,35] and retained in the analysis without being discarded as aberrant values. Boxplots visually overlapping in terms of IQR identified statistically similar records.

Polar rose diagrams were employed to visualize the directional distribution of waves for each site. Additionally, in the plots, a radial scale reported the frequency associated to each directional class, while spokes were color-coded according to Hs ranges. The multiple-year study was complemented by the joint frequency distribution. Data were grouped in eight Hs classes of 0.50 m each, and in direction sectors with an angle of 15°. The number of occurrences for each class was calculated by dividing the number of events falling within the considered class and the total number of data. The statistical indicator taken into account was consequently the joint frequency distribution of significant Hs with respect to θ.

A storm can be defined as a strong atmospheric perturbation accompanied, among other elements, by a low pressure system and intense winds. At sea, the most direct effects are the increase in Hs, and under certain conditions, in sea level (storm surge) [70]. In the GoN, such events are more frequently recorded in autumn and winter, but also occasionally in spring. In the literature, different methods for the identification of sea storms can be found, generally considering a parameter exceeding a threshold over a certain period (e.g., Hs [71], energy flux [72], wind [73]). In the framework of the present analysis, to investigate extreme events the storm energy content E (m^2 h) [71] was calculated according to:

$$E = \int_{t1}^{t2} Hs^2 \mathrm{dt} \tag{1}$$

where t_1 and t_2 define the storm duration. In line with the available literature, in order to identify storm events in the dataset analyzed, observations of Hs > 1.25 m over a period longer than six hours were taken into consideration [71,74].

4. Results

4.1. Interannual Patterns

The interannual Hs time series recorded by the HF radars and by the ADCP are shown in Figure 2. The time series clearly indicated seasonal patterns, with higher Hs in autumn/winter due to the higher occurrence of depression systems, and lower scores in spring/summer in association with the establishment of stable high pressures [33,64]. With reference to HF radars measurements, as expected from previous evidences [33], PORT and CAST recorded Hs_{HF} values lower than the concomitant ones by the SORR site, which is closer to the open Tyrrhenian waters. The visual comparison of the time series recorded at RC5 and RC2 at the PORT site showed no significant difference in Hs_{HF} as the waves moved shoreward, as long as the deep water condition was respected. As expected, Hs_B recorded by the ADCP were lower than those obtained by the HF radars. This evidence can be explained by different factors, namely, the sheltered position of the ADCP with respect to HF antennas, determining the refraction of incoming waves and reduction of Hs [75] and close vicinity of the gauge to the coastline.

The variability in Hs was further investigated at seasonal scale through the characterization of measurements by means of statistical descriptors, as summarized in Table 2. This was coupled with the comparison of the boxplots reported in Figure 3. Overall, the integration of the different approaches confirmed what qualitatively emerged in the visual analysis of the time series shown in Figure 2, namely: more intense wave conditions at SORR; no significant differences between PORT and CAST, which both covered the internal part of the basin, nor between RC5 and RC2 at PORT; and lower Hs at the MEDA B location. The extreme maximum Hs_{HF} recorded by the HF radars were compatible with previous investigations for the same basin [33,35], although falling outside the theoretical radar detection limits (see [33] for an extensive discussion). The discordance in the mean value of the seasonal data can be partly attributed to HF radar low sea-state limitations [33] and to environmental noise, which decreases the accuracy of raw measurements. It is evident from Figure 2 that Hs_B went below the 0.5 m threshold, instead the HF radars could not detect the waves efficiently.

The seasonal polar rose diagrams shown in Figure 4 enabled the immediate visual identification of the most frequent wave directions θ. Site-specific features depended upon the local bathymetry and orography, pointing to spatially variable characteristics of the wave field for the different sub-basins of the GoN. The ADCP recordings showed a strong directionality in winter, spring and fall with waves approaching the shore from the SW, while during the summer a high percentage of waves with low height (<0.5) was observed from a broader sector (95°–265°). All HF radar observations had typically unimodal seasonal directional climates. Using the color-coding of the spokes and the radial axis of the roses, it was possible to identify the frequency of occurrence for each site in terms of Hs along each angular sector considered. The color labelling of the spokes confirmed that the ADCP typically recorded lower Hs compared to the HF radars, which instead monitored more offshore sectors. At the same time, rose diagrams visually communicated the occurrence of higher Hs_{HF} at SORR than at PORT and CAST.

The site-specific relationship between Hs and θ was further confirmed by the joint frequency distributions of the multiannual datasets (Table 3). The ADCP recorded lower Hs, with a predominance of occurrences (>75%) from 165°–195°. On the other hand, as already deduced from the previous analyses, the HF radars recorded higher Hs. The directional spreading of PORT (195°–225°, >80%) and CAST (270°–300°, >96%) was similar to the ADCP one, whereas SORR displayed a wider θ range (285°–330°, 79%). The percentage of captured data with Hs < 0.5 m was low for all HF sites due to instrumental intrinsic limitations, as discussed by [76] and the references therein.

The analysis carried out in the present work also focused on the assessment of the wave period (T) estimates from the HF radars (T_{HF}) and the ADCP (T_B). T_{HF} returned a period representing the centroid of the model being fitted to the second-order Doppler spectrum, while T_B provided the peak period measured by the ADCP. As discussed in [31], T_{HF} can be assumed as a stable estimator just like the mean period retrieved by wave buoys. The boxplots for T_{HF} and T_B (Figure 5) show that the HF radars returned statistically comparable values, grouped in the 6–8 s window, in line with the invariance of wave period. T_B, on the other hand, always returned smaller values (4–6 s), but typically with higher statistical dispersion around the median compared to T_{HF}. For each site and platform, consistent values were recorded for the different seasons.

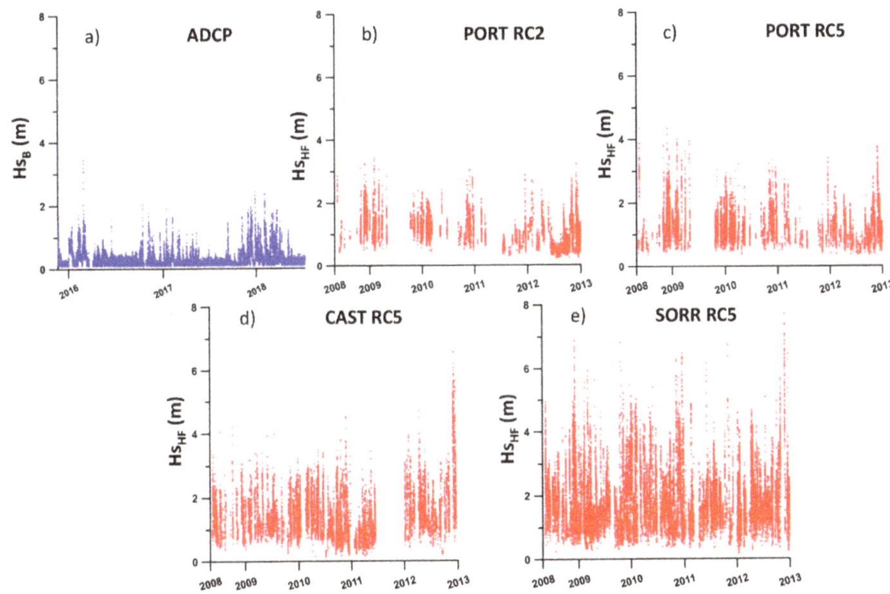

Figure 2. Hs time series for the ADCP (Hs_B; blue, panel **a**) and the three HF radar sites (Hs_{HF}; red, panels **b–e**). For the PORT site, the Hs_{HF} values recorded both at RC2 and RC5 are plotted.

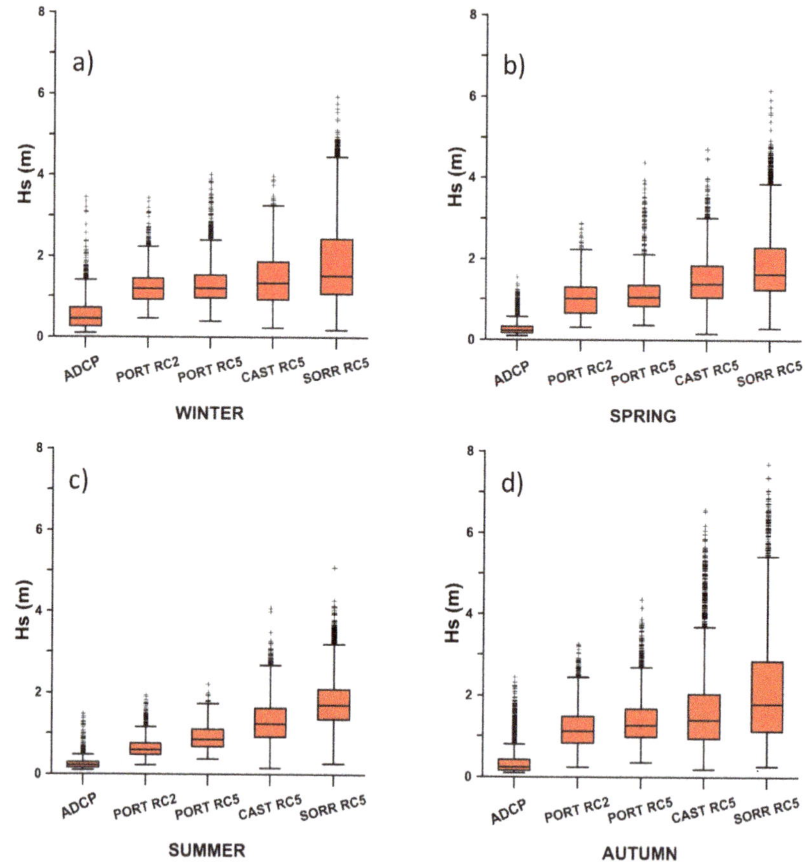

Figure 3. Cumulative seasonal Hs boxplots recorded by the ADCP (T_B) and the HF radars (T_{HF}). For the PORT site, both RC2 and RC5 data are shown. The panels show the different seasons: (**a**) winter; (**b**) spring; (**c**) summer; and (**d**) autumn.

Table 2. Statistical descriptors (min: minimum value; max: maximum value; mean: mean value; std: standard deviation; median: median value) of the seasonal Hs values recorded by the HF radars and the ADCP. For the PORT site, both RC2 and RC5 are included.

	PORT-RC2	PORT-RC5	CAST-RC5	SORR-RC5	ADCP	PORT-RC2	PORT-RC5	CAST-RC5	SORR-RC5	ADCP
Hs (m)			WINTER					SPRING		
min	0.46	0.39	0.23	0.18	0.10	0.31	0.37	0.16	0.29	0.10
max	3.43	3.95	3.90	5.90	3.45	2.87	4.30	4.70	6.10	1.55
mean	1.24	1.3	1.43	1.84	0.54	1.06	1.18	1.49	1.85	0.28
std	0.43	0.52	0.63	1.02	0.34	0.48	0.55	0.61	0.83	0.17
median	1.19	1.20	1.34	1.52	0.45	1.01	1.05	1.38	1.63	0.23
	PORT-RC2	PORT-RC5	CAST-RC5	SORR-RC5	ADCP	PORT-RC2	PORT-RC5	CAST-RC5	SORR-RC5	ADCP
Hs (m)			SUMMER					AUTUMN		
min	0.22	0.37	0.43	0.26	0.1	0.24	0.35	0.19	0.26	0.10
max	1.94	2.20	1.23	5.00	1.46	3.25	4.30	6.50	7.20	2.44
mean	0.65	0.92	0.74	1.76	0.24	1.19	1.38	1.63	2.08	0.35
std	0.27	0.33	0.56	0.63	0.13	0.49	0.56	0.99	1.20	0.32
median	0.59	0.85	1.23	1.70	0.22	1.11	1.26	1.39	1.78	0.24

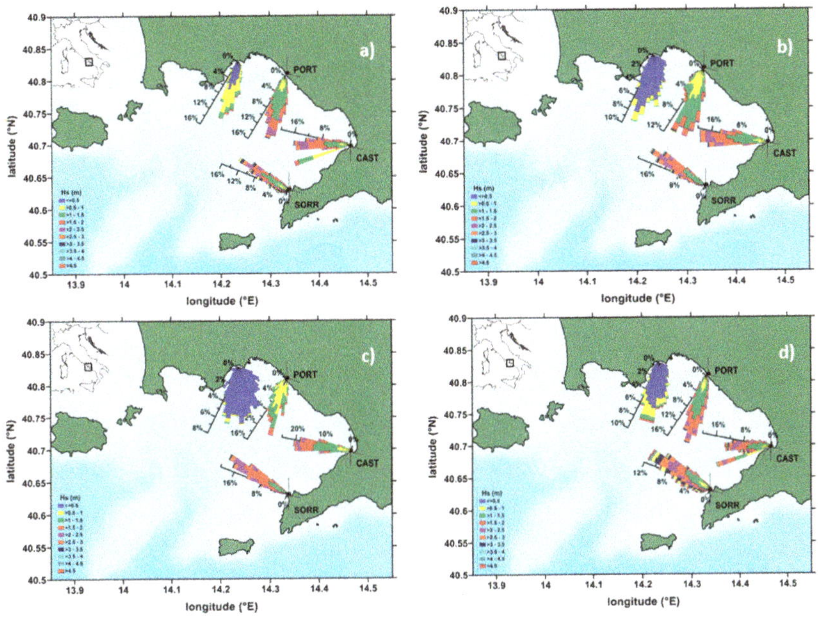

Figure 4. Seasonal wave rose diagrams for the HF radar and the ADCP entire dataset. The panels show the different seasons: (**a**) winter; (**b**) spring; (**c**) summer; and (**d**) autumn. The radial scale reports wave frequency, while each spoke is color-divided according to wave height ranges (see legend).

Table 3. Joint frequency distributions of the wave measurements (Hs and θ) recorded by the ADCP and the HF radars. Data were grouped in eight Hs classes of 0.50 m each, and in direction sectors with an angle of 15°. For each combination of Hs and θ values, the number of occurrences was calculated.

ADCP (November 2015–July 2018)																	
Wave Direction (deg N)	3454	2484	564	125	22	2	6	0	0	0	0	0	0	0	0	0	6655
360	-	-	-	-	-	-	-	-	-	-	-	-	-	-	-	-	-
345	-	-	-	-	-	-	-	-	-	-	-	-	-	-	-	-	-
330	-	-	-	-	-	-	-	-	-	-	-	-	-	-	-	-	-
315	-	-	-	-	-	-	-	-	-	-	-	-	-	-	-	-	-
300	-	-	-	-	-	-	-	-	-	-	-	-	-	-	-	-	-
285	-	-	-	-	-	-	-	-	-	-	-	-	-	-	-	-	-
270	-	-	-	-	-	-	-	-	-	-	-	-	-	-	-	-	-
255	-	-	-	-	-	-	-	-	-	-	-	-	-	-	-	-	-
240	168	41	9	1	1	1	-	-	-	-	-	-	-	-	-	-	221
225	190	53	13	-	-	-	3	-	-	-	-	-	-	-	-	-	259
210	288	139	40	3	-	1	-	-	-	-	-	-	-	-	-	-	471
195	541	395	94	27	3	-	3	-	-	-	-	-	-	-	-	-	1063
180	857	794	209	60	13	-	-	-	-	-	-	-	-	-	-	-	1933
165	958	843	172	33	4	-	-	-	-	-	-	-	-	-	-	-	2010
150	288	157	22	1	1	-	-	-	-	-	-	-	-	-	-	-	469
135	96	31	2	-	-	-	-	-	-	-	-	-	-	-	-	-	129
120	35	21	1	-	-	-	-	-	-	-	-	-	-	-	-	-	57
105	32	9	2	-	-	-	-	-	-	-	-	-	-	-	-	-	43
90	-	-	-	-	-	-	-	-	-	-	-	-	-	-	-	-	-

Table 3. *Cont.*

75	-	-	-	-	-	-	-	-	-	-	-	-	-	-	-	-	-
60	-	-	-	-	-	-	-	-	-	-	-	-	-	-	-	-	-
45	-	-	-	-	-	-	-	-	-	-	-	-	-	-	-	-	-
30	-	-	-	-	-	-	-	-	-	-	-	-	-	-	-	-	-
15	-	-	-	-	-	-	-	-	-	-	-	-	-	-	-	-	-
	0.5	**1**	**1.5**	**2**	**2.5**	**3**	**3.5**	**4**	**4.5**	**5**	**5.5**	**6**	**6.5**	**7**	**7.5**	**8**	
								Significant Wave Height (m)									
PORT HF radar (May 2008–December 2012)																	
Wave Direction (deg N)	78	2907	3739	1538	617	275	90	29	4	0	0	0	0	0	0	0	9277
360	-	-	-	-	-	-	-	-	-	-	-	-	-	-	-	-	-
345	-	-	-	-	-	-	-	-	-	-	-	-	-	-	-	-	-
330	-	-	-	-	-	-	-	-	-	-	-	-	-	-	-	-	-
315	-	-	-	-	-	-	-	-	-	-	-	-	-	-	-	-	-
300	-	-	-	-	-	-	-	-	-	-	-	-	-	-	-	-	-
285	-	1	1	-	-	-	-	-	-	-	-	-	-	-	-	-	**2**
270	1	27	31	5	6	1	2	-	-	-	-	-	-	-	-	-	**73**
255	1	126	94	25	16	9	14	1	-	-	-	-	-	-	-	-	**286**
240	6	258	187	40	22	13	5	1	-	-	-	-	-	-	-	-	**532**
225	22	522	538	134	45	35	9	8	1	-	-	-	-	-	-	-	**1314**
210	26	992	1347	604	236	108	37	10	2	-	-	-	-	-	-	-	**3362**
195	16	746	1164	536	221	84	13	5	-	-	-	-	-	-	-	-	**2785**
180	6	225	368	188	69	24	10	4	1	-	-	-	-	-	-	-	**895**
165	-	10	9	6	2	1	-	-	-	-	-	-	-	-	-	-	**28**
150	-	-	-	-	-	-	-	-	-	-	-	-	-	-	-	-	-
135	-	-	-	-	-	-	-	-	-	-	-	-	-	-	-	-	-
120	-	-	-	-	-	-	-	-	-	-	-	-	-	-	-	-	-
105	-	-	-	-	-	-	-	-	-	-	-	-	-	-	-	-	-
90	-	-	-	-	-	-	-	-	-	-	-	-	-	-	-	-	-
75	-	-	-	-	-	-	-	-	-	-	-	-	-	-	-	-	-
60	-	-	-	-	-	-	-	-	-	-	-	-	-	-	-	-	-
45	-	-	-	-	-	-	-	-	-	-	-	-	-	-	-	-	-
30	-	-	-	-	-	-	-	-	-	-	-	-	-	-	-	-	-
15	-	-	-	-	-	-	-	-	-	-	-	-	-	-	-	-	-
	0.5	**1**	**1.5**	**2**	**2.5**	**3**	**3.5**	**4**	**4.5**	**5**	**5.5**	**6**	**6.5**	**7**	**7.5**	**8**	
								Significant Wave Height (m)									
CAST HF radar (May 2008–December 2012)																	
Wave Direction (deg N)	365	2878	3880	2623	1405	639	174	75	62	42	32	12	2	2	0	0	12.191
360	-	-	-	-	-	-	-	-	-	-	-	-	-	-	-	-	-
345	-	-	-	-	-	-	-	-	-	-	-	-	-	-	-	-	-
330	-	-	-	-	-	-	-	-	-	-	-	-	-	-	-	-	-
315	17	135	98	54	89	23	9	3	-	-	-	-	-	-	-	-	**428**
300	71	526	650	368	213	93	19	18	2	-	-	-	-	-	-	-	**1960**
285	66	1120	1882	1281	657	290	79	16	16	13	8	7	-	2	-	-	**5437**
270	52	615	979	804	387	195	63	29	38	28	22	4	2	-	-	-	**3218**
255	159	482	271	116	59	38	4	9	6	1	2	1	-	-	-	-	**1148**
240	-	-	-	-	-	-	-	-	-	-	-	-	-	-	-	-	-
225	-	-	-	-	-	-	-	-	-	-	-	-	-	-	-	-	-
210	-	-	-	-	-	-	-	-	-	-	-	-	-	-	-	-	-

Table 3. *Cont.*

195	-	-	-	-	-	-	-	-	-	-	-	-	-	-	-	-	-
180	-	-	-	-	-	-	-	-	-	-	-	-	-	-	-	-	-
165	-	-	-	-	-	-	-	-	-	-	-	-	-	-	-	-	-
150	-	-	-	-	-	-	-	-	-	-	-	-	-	-	-	-	-
135	-	-	-	-	-	-	-	-	-	-	-	-	-	-	-	-	-
120	-	-	-	-	-	-	-	-	-	-	-	-	-	-	-	-	-
105	-	-	-	-	-	-	-	-	-	-	-	-	-	-	-	-	-
90	-	-	-	-	-	-	-	-	-	-	-	-	-	-	-	-	-
75	-	-	-	-	-	-	-	-	-	-	-	-	-	-	-	-	-
60	-	-	-	-	-	-	-	-	-	-	-	-	-	-	-	-	-
45	-	-	-	-	-	-	-	-	-	-	-	-	-	-	-	-	-
30	-	-	-	-	-	-	-	-	-	-	-	-	-	-	-	-	-
15	-	-	-	-	-	-	-	-	-	-	-	-	-	-	-	-	-
	0.5	1	1.5	2	2.5	3	3.5	4	4.5	5	5.5	6	6.5	7	7.5	8	
	Significant Wave Height (m)																
SORR HF radar (May 2008–December 2012)																	
Wave Direction (deg N)	**158**	**2531**	**5488**	**4568**	**2795**	**1795**	**1364**	**755**	**437**	**167**	**56**	**31**	**28**	**16**	**4**	**0**	**20.193**
360	5	34	49	38	13	3	6	1	1	1	-	-	-	-	-	-	**151**
345	6	121	167	84	45	15	12	2	-	-	-	-	-	-	-	-	**454**
330	15	329	449	247	141	71	46	29	10	5	1	-	-	-	-	-	**1343**
315	55	921	2224	1867	952	509	260	109	67	18	4	6	6	2	-	-	**7000**
300	45	609	1729	1699	1038	617	376	206	104	35	6	7	3	6	-	-	**6480**
285	7	239	476	421	374	330	331	178	105	50	20	9	7	5	2	-	**2554**
270	4	89	161	123	162	177	227	159	92	29	19	3	9	3	2	-	**1259**
255	3	66	94	63	46	59	92	58	46	25	5	5	2	-	-	-	**564**
240	3	62	93	19	14	13	13	11	11	3	1	-	-	-	-	-	**243**
225	7	22	22	4	3	-	1	-	1	1	-	-	-	-	-	-	**61**
210	6	32	19	1	4	1	-	2	-	-	-	-	-	-	-	-	**65**
195	2	6	5	2	2	-	-	-	-	-	-	-	-	-	-	-	**17**
180	-	1	-	-	1	-	-	-	-	-	-	-	-	-	-	-	**2**
165	-	-	-	-	-	-	-	-	-	-	-	-	-	-	-	-	-
150	-	-	-	-	-	-	-	-	-	-	-	-	-	-	-	-	-
135	-	-	-	-	-	-	-	-	-	-	-	-	-	-	-	-	-
120	-	-	-	-	-	-	-	-	-	-	-	-	-	-	-	-	-
105	-	-	-	-	-	-	-	-	-	-	-	-	-	-	-	-	-
90	-	-	-	-	-	-	-	-	-	-	-	-	-	-	-	-	-
75	-	-	-	-	-	-	-	-	-	-	-	-	-	-	-	-	-
60	-	-	-	-	-	-	-	-	-	-	-	-	-	-	-	-	-
45	-	-	-	-	-	-	-	-	-	-	-	-	-	-	-	-	-
30	-	-	-	-	-	-	-	-	-	-	-	-	-	-	-	-	-
15	-	-	-	-	-	-	-	-	-	-	-	-	-	-	-	-	-
	0.5	**1**	**1.5**	**2**	**2.5**	**3**	**3.5**	**4**	**4.5**	**5**	**5.5**	**6**	**6.5**	**7**	**7.5**	**8**	
	Significant Wave Height (m)																

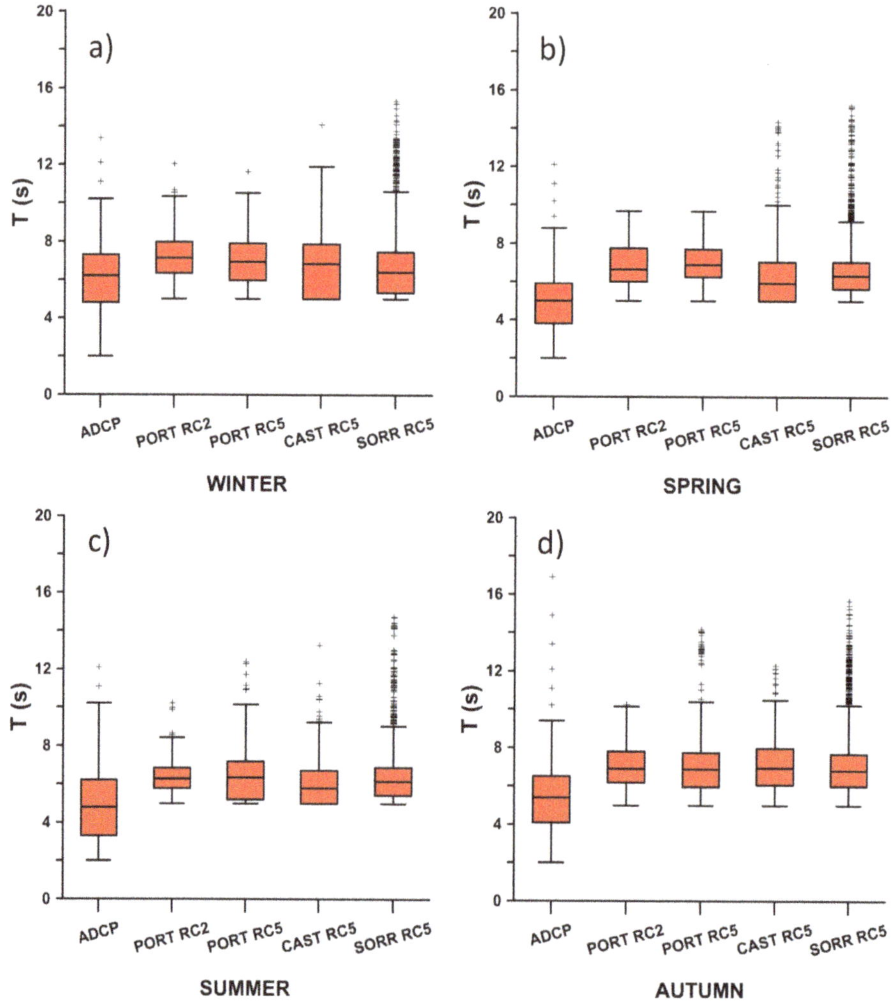

Figure 5. Cumulative seasonal T boxplots recorded by the ADCP and the HF radars. For the PORT site, both RC2 and RC5 data are shown. The panels show the different seasons: (**a**) winter; (**b**) spring; (**c**) summer; and (**d**) autumn.

4.2. Analysis of Extreme Events

During the sampled years, the HF radars and the ADCP recorded several storm events. Among the study cases selected, four events were selected and considered for the present investigation (Table 4): one recorded by the PORT and SORR HF radars (24–25 November 2008, total duration: 24 h); one recorded by all three radar sites (10–11 January 2010, total duration: 36 h); two recorded by the ADCP (28–29 February 2016, total duration: 32 h; 27–29 December 2017, total duration 45 h). Out of the possible events, the attention was focused on storms with a minimum duration of 24 h, and which were considered particularly severe (i.e., determining the suspension of ferry boat service to the isles of the GoN).

The analysis of these events allowed us to investigate the wave energy content (E) in different coastal sectors of the GoN. Unfortunately, the lack of temporal overlap between the ADCP and the HF radars prevented a direct comparison of the energy content measured by the two platforms. Despite this, some preliminary conclusions can be drawn from the analyzed events. First, as expected, the E values at SORR were always higher than the concomitant ones recorded at the innermost sites, in line with the site specificities outlined in the previous section. As a second outcome, the calculation based on ADCP data demonstrated that, even in the innermost sector of the GoN, the E associated with prolonged storms could reach scores comparable to those on farther offshore sectors during shorter events.

Table 4. Duration (h), Hs mean (m), and energy content E (m^2 h) of extreme events recorded by HF radars and ADCP.

Event	**1 (24–25/11/08) Duration 24 h**		**2 (10–11/1/10) Duration 36 h**	
Parameters	Hs mean (m)	E (m^2 h)	Hs mean (m)	E (m^2 h)
PORT-RC2	1.79	85.29	1.47	77.29
PORT-RC5	2.24	118.29	1.59	92.80
CAST-RC5	-	-	1.75	116.23
SORR-RC5	3.52	316.93	2.32	234.40
Event	**3 (28–29/02/16) Duration 32 h**		**4 (27–29/12/17) Duration 45 h**	
Parameters	Hs mean (m)	E (m^2 h)	Hs mean (m)	E (m^2 h)
ADCP	1.85	122.85	1.49	109.48

5. Discussion

An accurate knowledge of wave conditions in coastal basins is essential for a broad range of scientific as well as applied purposes (e.g., the study of coastal shoreline stability and the seasonal longshore migration of sediment, engineering activities, renewable energy and climate studies centered on air–sea interaction [77,78]).

A previous study on the wave climate in Italian seas showed different behaviors depending on the investigated areas, and classified the wave climate into two main regimes: the first characterized by bi- or multi-modal waves arriving from different sectors (as happens in the Ionian and Adriatic seas), while the second regime is characterized by unimodal patterns (Tyrrhenian Sea) [41]. In addition, the wave climate in the Tyrrhenian Sea varies strongly during the seasons, with low Hs (in the order of a few tens of centimeters) in summer replaced by waves up to 3 m during winter and autumn [41]. In this context, the results presented in this study show that the GoN adheres to this larger-scale regime. The multiple year analysis of the wave dataset recorded by the HF radar network (2008–2012) and by in situ measurements gathered by the ADCP (2015–2018) suggest that the wave field is predominantly locally wind-driven and, for each site, the wave data are coherent among the years, with specific patterns depending on the sub-basin of the GoN under analysis. The results presented in this contribution align with previous studies focused on shorter temporal windows [33,37–40,63], pointing out a strong seasonality of Hs related to the meteorological features of the study area. In addition, covering different sectors of the basin (each with peculiar geomorphological features and bathymetry), the integrated observation platform used here was able to recognize the different dynamics characterizing the internal and the external sub-regions of the GoN.

In terms of directionality, θ is strongly affected by the morphology of the GoN. The area closer to the Naples urban site (monitored by PORT HF radar and ADCP) is aligned with the main aperture of the GoN ("Bocca Grande", the channel separating Ischia and Capri islands), while the presence of Capri Island creates a shadow zone shielding the coastline covered by SORR and, to a lesser degree, the CAST antennas, as confirmed by recent comparisons between HF radar data and wave models [35].

For each site, the directional pattern was strongly unimodal, validating for the different sectors of the GoN the larger scale behavior discussed in [41]. In all cases, due to the relative closeness of the measurements to the coast, refraction appears to contribute substantially in letting the wave front approach the coast almost perpendicularly.

The comparison of T brings challenging results in the evaluation of the inter-platform performance. Previous literature indicates that T_{HF} typically sits between the mean and the peak period retrieved by wave buoys [29,31], being a suitable indicator of the mean period T_m [29,31,34] as demonstrated also specifically for the GoN with both wave buoys [33] and models [35]. In the present analysis, the peak period T_B returned by the ADCP showed values lower than T_{HF} and displayed a greater statistical dispersion, as typical for peak periods that are noisier than mean ones [31]. This inconsistency may be attributed to specific instrument data retrieval and analysis processes, but to date has no previous literature background reference. It should also be considered that HF radar cannot detect waves with a period lower than 5 s [20,34], thus introducing another element of disalignment between the two platforms. As such, this specific topic deserves further studies to be investigated in detail.

Taken together, the results presented in this contribution depict specific and recurrent features in the wave field of the different sub-basins of the GoN. Due to its proximity to the outer sector of the GoN, the SORR site recorded the highest E values, while at CAST and PORT the strength of the wave field was reduced. Interestingly, the comparison of the measurements recorded by PORT and by the ADCP revealed differences despite their close vicinity. These differences were only in part mitigated when considering RC2 (i.e., the cell closer to the coast which could be analyzed in the HF radar dataset). As discussed in a previous section, this result is likely to be due to the reduction of Hs induced by refraction, mirrored in a 30° rotation of the main lobe compared to PORT as well as by a more shielded position of the gauge. The robustness of HF radar measurements at RC5 has been recently validated through the comparison with wave models [35], supporting the reliability of these records. Future implementations will also address the validation of ADCP records with wave models, setting up a specific numerical benchmark. In addition, prospective developments of the physical infrastructure will permit the contemporary functioning of both platforms, thus allowing for a direct comparison of the measurements. These next steps will shed light on the details ruling the nearshore propagation of the field and the associated changes in wave parameters.

The Hs observations during the sea storms selected in this work were in agreement with previous studies for the same area [33,38–40]. According to the storm classification presented in [71], all selected events can be categorized as class I (weak), which are the most probable in terms of occurrence frequency, except for the event 1 at the SORR location that was categorized as class II (moderate) based on its E value. The differences observed among the HF observations as well as those between the HF and ADCP records, mirror the site-specific patterns highlighted in the time series analysis, but also reflect the importance of storm duration. As a second outcome, the calculation based on ADCP data demonstrated that, even in the innermost sector of the GoN, the E associated with prolonged storms could reach scores comparable to those on farther offshore sectors during shorter events. These aspects, here preliminarily addressed by resorting to a small subset of events, will be deepened in a future work by including a higher number of stormy conditions and by complementing the analysis with numerical simulations to better resolve the spatial variability of the wave field in the GoN.

HF radars typically present an optimal window of functioning in terms of detectable Hs. For a 25 MHz system like the one installed in the GoN, this window ranges from 0.50 m to 4.00 m. The results discussed in [33,35], complemented with those presented here, show that HF measurements can be considered as sufficiently robust indicators of the wave field pattern even when Hs > 4.00 m in the presence of storm events, upon application of appropriate QA/QC procedures. This aspect, however, needs further investigation, as demonstrated by comparative analysis with wave model simulations [35]. This operational window overlaps with the one guaranteeing the maximum efficiency of wave energy converters (Hs > 2.00 m, as discussed in [78]). As such, HF radars candidate as eligible tools for the preliminary assessment of sites of interest for the installation of wave energy farms [79,80].

The adoption of the integrated approach tested in this contribution is compliant with the indications provided by [81] in relation to the development of the next-generation of marine observatories, favoring the integration of different sampling strategies. Further implementations may encompass the inclusion of additional platforms (e.g., X-band radars [82]) to retrieve an even more detailed description of the wave field in coastal areas. Aside from the specific results focusing on the dynamics of the GoN, the present study also provides wider-breath elements in relation to the potential use of HF radars as a long-term monitoring platform for wave fields. ADCPs have a well-documented track of records in surface wave measurements [83], whereas HF radars have a proven ability to measure surface currents accurately even over long temporal scales (e.g., [84,85]). In contrast, only a few works have used HF radar data to characterize the wave field over periods as long as almost two years [30–34]. The results presented here represent the first report of a multiannual analysis of wave data retrieved through a network of HF radars, setting a new standard for the operational applications of this technology.

In addition, by monitoring the wave field over long periods, HF radars demonstrate their eligibility as source of information for hindcast analyses as well as input for model projections. Implementations in the functioning of HF systems and in the methods of data retrieval and analysis [86–89] will enhance the robustness of the measurements, providing increasingly more precise information on ocean surface parameters.

6. Conclusions

Multiannual surface gravity wave data retrieved by HF radar measurements and in situ measurements by an ADCP were employed to describe and analyze the wave field of the GoN. The integrated acquisition revealed clear seasonal patterns and pointed to specific features, depending on the sub-basin analyzed. The approach used in the present paper represents, to the best of present knowledge, the first use of HF radars combined with ADCP measurements to reconstruct the wave climatology (at least partially, due to the limitation in the number of years of recording) of a coastal basin. The results demonstrate the ability of an integrated system to provide a detailed view of surface wave dynamics over a decade, and represent a spur to further integrations and implementations of current ocean observatories. The multiannual wave patterns depicted can provide useful information for stakeholders and policy managers for several applications (e.g., the design of port and marine infrastructures and the identification of sites suitable for wave energy farms.

Author Contributions: Methodology: S.S., D.C., F.C., and M.U.; Software: S.S.; Validation: S.S., D.C., and M.U.; Data curation: F.C. and S.S.; Writing review and editing: S.S., D.C., and M.U. and F.C.; Project administration: F.C., M.U., and E.Z. All authors have read and agreed to the published version of the manuscript.

Funding: This work was funded by the Flagship Project RITMARE (The Italian Research for the Sea) coordinated by the Italian National Research Council and funded by the Italian Ministry of Education, University and Research within the National Research Program 2011–2013 and the AQUILA project (Parthenope University internal grant) (CUP:I63C17000060005). EZ was partly supported by the Parthenope University internal individual research grant. MU was supported by an internal SZN grant. This work was partly supported by DORA, Deployable Optics for Remote sensing Applications (ARS01-00653), a project funded by MIUR-PON Research and Innovation/PNR 2015-2020.

Acknowledgments: The authors thank the Editor and anonymous Reviewers for critical discussion on an earlier version of the study. The Department of Science and Technology (formerly the Department of Environmental Sciences) of the Parthenope University of Naples operates the HF radar system on behalf of the AMRA consortium (formerly CRdC AMRA), a regional competence center for the analysis and monitoring of environmental risks. Our radar remote sites are hosted by the ENEA Centre of Portici, the Villa Angelina Village of High Education and Professional Training and La Villanella resort in Massa Lubrense, and the Fincantieri shipyard in Castellammare di Stabia, whose hospitality is gratefully acknowledged. Mention of trade names or commercial products does not constitute endorsement or recommendation. Technical support by Giovanni Zambardino provided invaluable support to the proper functioning of the HF radar system. Exchanges with Berardino Buonocore and Pierpaolo Falco were greatly appreciated.

Conflicts of Interest: The authors declare no conflicts of interest.

References

1. Appendini, C.M.; Torres-Freyermuth, A.; Salles, P.; López-González, J.; Mendoza, E.T. Wave climate and trends for the Gulf of Mexico: A 30-yr wave hindcast. *J. Clim.* **2013**, *27*, 1619–1632. [CrossRef]
2. Wiegel, R.K. *Oceanographical Engineering*; Prentice Hall: Englewood Cliffs, NJ, USA, 1964; p. 532.
3. Herbich, J.B.; Walters, T. Wave climate. In *Beaches and Coastal Geology*; Schwartz, M.L., Ed.; Springer: Boston, MA, USA, 1982; pp. 855–856.
4. Chini, N.; Stansby, P.; Leake, J.; Wolf, J.; Roberts-Jones, J.; Lowe, J. The impact of sea level rise and climate change on inshore wave climate: A case study for East Anglia (UK). *Coast. Eng.* **2010**, *57*, 973–984. [CrossRef]
5. Semedo, A.; Weisse, R.; Behrens, A.; Sterl, A.; Bengtsson, L.; Günther, H. Projection of global wave climate change toward the end of the twenty-first century. *J. Clim.* **2013**, *26*, 8269–8288. [CrossRef]
6. Menéndez, M.; Méndez, F.J.; Izaguirre, C.; Luceño, A.; Losada, I.J. The influence of seasonality on estimating return values of significant wave height. *Coast. Eng.* **2009**, *56*, 211–219. [CrossRef]
7. Stefanakos, C.N.; Athanassoulis, G.A.; Cavaleri, L.; Bertotti, L.; Lefèvre, J.-M. Wind and wave climatology of the Mediterranean Sea. Part II: Wave statistics. In Proceedings of the 14th International Offshore and Polar Engineering Conference (ISOPE 2004), Toulon, France, 23–28 May 2004.
8. Izaguirre, C.; Méndez, F.J.; Menéndez, M.; Losada, I.J. Global extreme wave height variability based on satellite data. *Geophys. Res. Lett.* **2011**, *38*, L10607. [CrossRef]
9. Inghilesi, R.; Orasi, A.; Catini, F. The ISPRA Mediterranean coastal wave forecasting system: Evaluation and perspectives. *J. Oper. Oceanogr.* **2016**, *9*, 89–98. [CrossRef]
10. Collard, F.; Ardhuin, F.; Chapron, B. Extraction of coastal ocean wave fields from SAR images. *IEEE J. Ocean. Eng.* **2005**, *30*, 526–533. [CrossRef]
11. Quilfen, Y.; Chapron, B. Ocean surface wave-current signatures from satellite altimeter measurements. *Geophys. Res. Lett.* **2019**, *46*, 253–261. [CrossRef]
12. Stopa, J.E.; Cheung, K.F.; Tolman, H.L.; Chawla, A. Patterns and cycles in the climate forecast system reanalysis wind and wave data. *Ocean Model.* **2013**, *70*, 207–220. [CrossRef]
13. Galanis, G.; Hayes, D.; Zodiatis, G.; Chu, P.C.; Kuo, Y.H.; Kallos, G. Wave height characteristics in the Mediterranean Sea by means of numerical modeling, satellite data, statistical and geometrical techniques. *Mar. Geophys. Res.* **2012**, *33*, 1–15. [CrossRef]
14. Crombie, D. Doppler spectrum of sea echo at 13.56 mc/s. *Nature* **1955**, 681–682. [CrossRef]
15. Barrick, D.E. The ocean waveheight non directional spectrum from inversion of the HF sea echo Doppler spectrum. *Remote Sens. Environ.* **1977**, *6*, 201–227. [CrossRef]
16. Paduan, J.D.; Graber, H.C. Introduction to high-frequency radar: Reality and myth. *Oceanography* **1997**, 10. [CrossRef]
17. Gurgel, K.-W.; Antonischki, G.; Essen, H.-H.; Schlick, T. Wellen Radar WERA: A new ground-wave HF radar for ocean remote sensing. *Coast. Eng.* **1999**, *37*, 219–234. [CrossRef]
18. Lopez, G.; Conley, D.C.; Greaves, D. Calibration, validation, and analysis of an empirical algorithm for the retrieval of wave spectra from HF radar sea echo. *J. Atmos. Ocean. Technol.* **2016**, *33*, 245–261. [CrossRef]
19. Barrick, D.E. The role of the gravity-wave dispersion relation in HF radar measurements of the sea surface. *IEEE J. Ocean. Eng.* **1986**, *11*, 286–292. [CrossRef]
20. Lipa, B.J.; Nyden, B. Directional wave information from the SeaSonde. *IEEE J. Ocean Eng.* **2005**, *30*, 221–231. [CrossRef]
21. Lipa, B.J.; Barrick, D.; Isaacson, J.; Lilleboe, P.M. CODAR wave measurements from a North Sea semisubmersible. *IEEE J. Ocean Eng.* **1990**, *15*, 119–125. [CrossRef]
22. Wyatt, L.R. The ocean wave directional spectrum. *Oceanography* **1997**, *10*. [CrossRef]
23. Wyatt, L.R.; Thompson, S.P.; Burton, R.R. Evaluation of HF radar wave measurement. *Coast. Eng.* **1999**, *37*, 259–282. [CrossRef]
24. Wyatt, L.R.; Green, J.J.; Middleditch, A. HF radar data quality requirements for wave measurement. *Coast. Eng.* **2011**, *58*, 327–336. [CrossRef]
25. Siddons, L.; Wyatt, L.; Wolf, J. Assimilation of HF radar data into the SWAN wave model. *J. Mar. Syst.* **2009**, *77*, 312–324. [CrossRef]
26. Wyatt, L.R.; Green, J.J. Measuring high and low waves with HF radar. In Proceedings of the OCEANS 2009-EUROPE, Bremen, Germany, 11–14 May 2009; pp. 1–5.

27. Jin, L.; Wen, B.; Zhou, H. A new method of wave mapping with HF radar. *Int. J. Antennas Propag.* **2016**, 1–7. [CrossRef]
28. Lipa, B.J.; Barrick, D.E.; Bourg, J.; Nyden, B.B. HF radar detection of tsunamis. *J. Oceanogr.* **2006**, *62*, 705–716. [CrossRef]
29. Lipa, B.J.; Barrick, D.E.; Alonso-Martirena, A.; Fernandes, M.; Ferrer, M.I.; Nyden, B. Brahan project high frequency radar ocean measurements: Currents, winds, waves and their interactions. *Remote Sens.* **2014**, *6*, 12094–12117. [CrossRef]
30. Wyatt, L.R.; Green, J.J.; Middleditch, A.; Moorhead, M.D.; Howarth, J.; Holt, M.; Keogh, S. Operational wave, current and wind measurements with the Pisces HF radar. *IEEE J. Ocean. Eng.* **2006**, *31*, 819–834. [CrossRef]
31. Long, R.M.; Barrick, D.; Largier, J.L.; Garfield, N. Wave observations from central California: SeaSonde systems and in situ wave buoys. *J. Sens.* **2011**, 1–18. [CrossRef]
32. Atan, R.; Goggins, J.; Harnett, M.; Agostinho, P.; Nash, S. Assessment of wave characteristics and resource variability at a 1/4-scale wave energy test site in Galway Bay using waverider and high frequency radar (CODAR) data. *Ocean Eng.* **2016**, *117*, 272–291. [CrossRef]
33. Saviano, S.; Kalampokis, A.; Zambianchi, E.; Uttieri, M. A year-long assessment of wave measurements retrieved from an HF radar network in the Gulf of Naples (Tyrrhenian Sea, Western Mediterranean Sea). *J. Oper. Oceanogr.* **2019**, *12*, 1–15. [CrossRef]
34. Basañez, A.; Lorente, P.; Montero, P.; Álvarez-Fanjul, E.; Pérez-Muñuzuri, V. Quality Assessment and practical interpretation of the wave parameters estimated by HF Radars in NW Spain. *Remote Sens.* **2020**, *12*, 598. [CrossRef]
35. de Leo, F.; Besio, G.; Saviano, S.; Zambianchi, E.; Uttieri, M. Evaluation of HF-radar wave measures in the Gulf of Naples. In Proceedings of the IMEKO TC19 International Workshop on Metrology for the Sea: Learning to Measure Sea Health Parameters (MetroSea 2019), Genova, Italy, 3–5 October 2019.
36. Rubio, A.; Mader, J.; Corgnati, L.; Mantovani, C.; Griffa, A.; Novellino, A.; Quentin, C.; Wyatt, L.; Schulz-Stellenfleth, J.; Horstmann, J. HF radar activity in European coastal seas: Next steps toward a pan-European HF radar network. *Front. Mar. Sci.* **2017**, *4*, 8. [CrossRef]
37. Pugliese Carratelli, E.; Sansone, E. Rilievi ondametrici nel Golfo di Napoli. *Ann. IUN* **1987**, *50*, 87–97.
38. Benassai, G.; de Maio, A.; Mainolfi, R.; Sansone, E. Previsione delle onde estreme e persistenza delle mareggiate nel Golfo di Napoli. *Ann. IUN* **1994**, *61*, 47–60.
39. Benassai, G.; de Maio, A.; Sansone, E. Altezze e periodi delle onde significative nel Golfo di Napoli dall'aprile 1986 al giugno 1987. *Ann. IUN* **1994**, *61*, 3–9.
40. Buonocore, B.; Sansone, E.; Zambardino, G. Rilievi ondametrici nel Golfo di Napoli. *Ann. IUN* **2003**, *67*, 203–211.
41. Morucci, S.; Picone, M.; Nardone, G.; Arena, G. Tides and waves in the Central Mediterranean Sea. *J. Oper. Oceanogr.* **2016**, *9*, 10–17. [CrossRef]
42. Cianelli, D.; Uttieri, M.; Buonocore, B.; Falco, P.; Zambardino, G.; Zambianchi, E. Dynamics of a very special Mediterranean coastal area: The Gulf of Naples. In *Mediterranean Ecosystems*; Williams, G.S., Ed.; Nova Science Publishers Inc.: New York, NY, USA, 2012; pp. 129–150.
43. De Maio, A.; Moretti, M.; Sansone, E.; Spezie, G.; Vultaggio, M. Outline of marine currents in the Bay of Naples and some considerations on pollutant transport. *Nuovo Cim. C* **1985**, *8*, 955–969. [CrossRef]
44. Moretti, M.; Sansone, E.; Spezie, G.; Vultaggio, M.; De Maio, A. Alcuni aspetti del movimento delle acque del Golfo di Napoli. *Ann. IUN* **1976**, *45–46*, 207–217.
45. Arienzo, M.; Donadio, C.; Mangoni, O.; Bolinesi, F.; Stanislao, C.; Trifuoggi, M.; Toscanesi, M.; Di Natale, G.; Ferrara, L. Characterization and source apportionment of polycyclic aromatic hydrocarbons (pahs) in the sediments of gulf of Pozzuoli (Campania, Italy). *Mar. Pollut. Bull.* **2017**, *124*, 480–487. [CrossRef]
46. Qu, C.; Albanese, S.; Lima, A.; Hope, D.; Pond, P.; Fortelli, A.; Romano, N.; Cerino, P.; Pizzolante, A.; De Vivo, B. The occurrence of OCPs, PCBs, and PAHs in the soil, air, and bulk deposition of the Naples metropolitan area, southern Italy: Implications for sources and environmental processes. *Environ. Int.* **2019**, *124*, 89–97. [CrossRef]
47. Mercogliano, R.; Santonicola, S.; De Felice, A.; Anastasio, A.; Murru, N.; Ferrante, M.C.; Cortesi, M.L. Occurrence and distribution of polycyclic aromatic hydrocarbons in mussels from the gulf of Naples, Tyrrhenian Sea, Italy. *Mar. Pollut. Bull.* **2016**, *104*, 386–390. [CrossRef] [PubMed]

48. Trifuoggi, M.; Donadio, C.; Mangoni, O.; Ferrara, L.; Bolinesi, F.; Nastro, R.A.; Stanislao, C.; Toscanesi, M.; Di Natale, G.; Arienzo, M. Distribution and enrichment of trace metals in surface marine sediments in the Gulf of Pozzuoli and off the coast of the brownfield metallurgical site of Ilva of Bagnoli (Campania, Italy). *Mar. Pollut. Bull.* **2017**, *124*, 502–511. [CrossRef] [PubMed]
49. Bavestrello, G.; Bo, M.; Canese, S.; Sandulli, R.; Cattaneo-Vietti, R. The red coral populations of the gulfs of Naples and Salerno: Human impact and deep mass mortalities. *Ital. J. Zool.* **2014**, *81*, 552–563. [CrossRef]
50. Ferrigno, F.; Appolloni, L.; Russo, G.F.; Sandulli, R. Impact of fishing activities on different coralligenous assemblages of Gulf of Naples (Italy). *J. Mar. Biol. Assoc. UK* **2018**, *98*, 41–50. [CrossRef]
51. Donnarumma, L.; Appolloni, L.; Chianese, E.; Bruno, R.; Baldrighi, E.; Guglielmo, R.; Russo, G.F.; Zeppilli, D.; Sandulli, R. Environmental and benthic community patterns of the shallow hydrothermal area of Secca delle Fumose (Baia, Naples, Italy). *Front. Mar. Sci.* **2019**, *6*, 685. [CrossRef]
52. Polese, G.; Semprucci, F.; Campoli, L.; Maselli, V.; Ribera d'Alcalà, M.; Balsamo, M.; Di Cosmo, A. Meiofaunal assemblages of the bay of Nisida and the environmental status of the Phlegraean area (Naples, Southern Italy). *Mar. Biodiv.* **2018**, *48*, 127. [CrossRef]
53. Crocetta, F.; Riginella, E.; Lezzi, M.; Tanduo, V.; Balestrieri, L.; Rizzo, L. Bottom-trawl catch composition in a highly polluted coastal area reveals multifaceted native biodiversity and complex communities of fouling organisms on litter discharge. *Mar. Environ. Res.* **2020**, *155*, 104875. [CrossRef]
54. Mattei, G.; Rizzo, A.; Anfuso, G.; Aucelli, P.P.C.; Gracia, F.J. A tool for evaluating the archaeological heritage vulnerability to coastal processes: The case study of Naples Gulf (southern Italy). *Ocean Coast. Manag.* **2019**, *179*, 104876. [CrossRef]
55. Mattei, G.; Rizzo, A.; Anfuso, G.; Aucelli, P.P.C.; Gracia, F.J. Enhancing the protection of archaeological sites as an integrated coastal management strategy: The case of the Posillipo Hill (Naples, Italy). *Rend. Fis. Acc. Lincei* **2020**, *31*, 139–152. [CrossRef]
56. Aucelli, P.P.C.; Mattei, G.; Caporizzo, C.; Cinque, A.; Troisi, S.; Peluso, F.; Stefanile, M.; Pappone, G. Ancient coastal changes due to ground movements and human interventions in the Roman Portus Julius (Pozzuoli Gulf, Italy): Results from photogrammetric and direct surveys. *Water* **2020**, *12*, 658. [CrossRef]
57. Buonocore, E.; Donnarumma, L.; Appolloni, L.; Miccio, A.; Russo, G.F.; Franzese, P.P. Marine natural capital and ecosystem services: An environmental accounting model. *Ecol. Model.* **2020**, *424*. [CrossRef]
58. Uttieri, M.; Cianelli, D.; Buongiorno Nardelli, B.; Buonocore, B.; Falco, P.; Colella, S.; Zambianchi, E. Multiplatform observation of the surface circulation in the Gulf of Naples (Southern Tyrrhenian Sea). *Ocean Dyn.* **2011**, *61*, 779–796. [CrossRef]
59. de Ruggiero, P. A high-resolution ocean circulation model of the Gulf of Naples and adjacent areas. *Nuovo Cim. C* **2013**, *36*, 143–150. [CrossRef]
60. de Ruggiero, P.; Napolitano, E.; Iacono, R.; Pierini, S. A high-resolution modelling study of the circulation along the Campania coastal system, with a special focus on the Gulf of Naples. *Cont. Shelf Res.* **2016**, *122*, 85–101. [CrossRef]
61. Cianelli, D.; Falco, P.; Iermano, I.; Mozzillo, P.; Uttieri, M.; Buonocore, B.; Zambardino, G.; Zambianchi, E. Inshore/offshore water exchange in the Gulf of Naples. *J. Mar. Syst.* **2015**, *145*, 37–52. [CrossRef]
62. Cianelli, D.; D'Alelio, D.; Uttieri, M.; Sarno, D.; Zingone, A.; Zambianchi, E.; Ribera d'Alcalà, M. Disentangling physical and biological drivers of phytoplankton dynamics in a coastal system. *Sci. Rep.* **2017**, *7*, 15868. [CrossRef]
63. Falco, P.; Buonocore, B.; Cianelli, D.; De Luca, L.; Giordano, A.; Iermano, I.; Kalampokis, A.; Saviano, S.; Uttieri, M.; Zambardino, G. Dynamics and sea state in the Gulf of Naples: Potential use of high-frequency radar data in an operational oceanographic context. *J. Oper. Oceanogr.* **2016**, *9*, 33–45. [CrossRef]
64. Menna, M.; Mercatini, A.; Uttieri, M.; Buonocore, B.; Zambianchi, E. Wintertime transport processes in the Gulf of Naples investigated by HF radar measurements of surface currents. *Nuovo Cim.* **2007**, *30*, 605–622.
65. Bartzokas, A. Annual variation of pressure over the Mediterranean area. *Theor. Appl. Climatol.* **1989**, *40*, 135. [CrossRef]
66. Hatzaki, M.; Flocas, H.A.; Simmonds, I.; Kouroutzoglou, J.; Keay, K.; Rudeva, I. Seasonal aspects of an objective climatology of anticyclones Affecting the Mediterranean. *J. Clim.* **2014**, *27*, 9272–9289. [CrossRef]
67. Mazzarella, V.; Maiello, I.; Ferretti, R.; Capozzi, V.; Picciotti, E.; Alberoni, P.P.; Marzano, F.S.; Budillon, G. Reflectivity and velocity radar data assimilation for two flash flood events in central Italy: A comparison between 3D and 4D variational methods. *Q. J. R. Meteorol. Soc.* **2019**, 1–19. [CrossRef]

68. Iermano, I.; Moore, A.M.; Zambianchi, E. Impacts of a 4-dimensional variational data assimilation in a coastal ocean model of southern Tyrrhenian Sea. *J. Mar. Syst.* **2016**, *154*, 157–171. [CrossRef]
69. Glickman, T.S. *Glossary of Meteorology*, 2nd ed.; American Meteorological Society: Boston, MA, USA, 2000.
70. Demirel, H.; Kompil, M.; Nemry, F. A framework to analyze the vulnerability of European road networks due to Sea-Level Rise (SLR) and sea storm surges. *Transp. Res. A Policy Pract.* **2015**, *81*, 62–76. [CrossRef]
71. Mendoza, E.T.; Jimenez, J.A.; Mateo, J. A coastal storms intensity scale for the Catalan sea (NW Mediterranean). *Nat. Hazards Earth Syst. Sci.* **2011**, *11*, 2453–2462. [CrossRef]
72. Molina, R.; Manno, G.; Lo Re, C.; Anfuso, G.; Ciraolo, G. Storm energy flux characterization along the Mediterranean coast of Andalusia (Spain). *Water* **2019**, *11*, 509. [CrossRef]
73. Hudak, D.R.; Young, J.M.C. Storm climatology of the Southern Beaufort sea. *Atmos. Ocean* **2002**, *40*, 145–158. [CrossRef]
74. Atan, R.; Goggins, J.; Harnett, M.; Nash, S.; Agostinho, P. Assessment of extreme wave height events in Galway Bay using high frequency radar (CODAR) data. In *Renewable Energies Offshore*; CRC Press: Boca Raton, FL, USA, 2015; pp. 49–56. [CrossRef]
75. Dally, W.R. Surf Zone Processes. Encyclopedia of Earth Science Series; In *Encyclopedia of Coastal Science*; Schwartz, M.L., Ed.; Springer: Dordrecht, The Netherlands, 2005.
76. Laws, K.; Paduan, J.D.; Vesecky, J. Estimation and assessment of errors related to antenna pattern distortion in CODAR SeaSonde high-frequency radar ocean current measurements. *J. Atmos. Ocean. Technol.* **2010**, *27*, 1029–1043. [CrossRef]
77. Wolf, J.; Woolf, D.K. Waves and climate change in the north-east Atlantic. *Geophys. Res. Lett.* **2006**, *33*, L06604. [CrossRef]
78. Vicinanza, D.; Contestabile, P.; Ferrante, V. Wave energy potential in the north-west of Sardinia (Italy). *Renew. Energy* **2013**, *50*, 506–521. [CrossRef]
79. Wyatt, L.R. Wave power measurements in the Celtic Sea using HF radar. In Proceedings of the OCEANS 2017, Aberdeen, UK, 19–22 June 2017; pp. 1–4. [CrossRef]
80. Wyatt, L.R. Wave and tidal power measurement using HF radar. *Int. Mar. Energy J.* **2018**, *1*, 123–127. [CrossRef]
81. Crise, A.; Ribera d'Alcalà, M.; Mariani, P.; Petihakis, G.; Robidart, J.; Iudicone, D.; Bachmayer, R.; Malfatti, F. A Conceptual framework for developing the next generation of Marine OBservatories (MOBs) for science and society. *Front. Mar. Sci.* **2018**, *5*, 318. [CrossRef]
82. Ludeno, G.; Serafino, F. Estimation of the significant wave height from marine radar images without external reference. *J. Mar. Sci. Eng.* **2019**, *7*, 432. [CrossRef]
83. Rao, L.; Wang, C.; Deng, K.; Zhang, Z.; Lu, S. Surface wave measurement with ADCP: A review of publications from 1979 to 2017. In Proceedings of the OCEANS 2017, Anchorage, AK, USA, 18–21 September 2017; pp. 1–7.
84. Ren, L.; Miao, J.; Li, Y.; Luo, X.; Li, J.; Hartnett, M. Estimation of coastal currents using a soft computing method: A case study in Galway Bay, Ireland. *J. Mar. Sci. Eng.* **2019**, *7*, 157. [CrossRef]
85. Lorente, P.; Basañez Mercader, A.; Piedracoba, S.; Pérez-Muñuzuri, V.; Montero, P.; Sotillo, M.G.; Álvarez-Fanjul, E. Long-term skill assessment of SeaSonde radar-derived wave parameters in the Galician coast (NW Spain). *Int. J. Remote Sens.* **2019**, *40*. [CrossRef]
86. Cosoli, S. Implementation of the listen-before-talk mode for SeaSonde High-Frequency ocean radars. *J. Mar. Sci. Eng.* **2020**, *8*, 57. [CrossRef]
87. Lopez, G.; Conley, D.C. Comparison of HF radar fields of directional wave spectra against in situ measurements at multiple locations. *J. Mar. Sci. Eng.* **2019**, *7*, 271. [CrossRef]
88. Lipa, B.; Barrick, D.; Whelan, C. A quality control method for broad-beam HF radar current velocity measurements. *J. Mar. Sci. Eng.* **2019**, *7*, 112. [CrossRef]
89. Hardman, R.L.; Wyatt, L.R. Inversion of HF radar Doppler spectra using a neural network. *J. Mar. Sci. Eng.* **2019**, *7*, 255. [CrossRef]

Journal of
Marine Science and Engineering

MDPI

Article

Wave Orbital Velocity Effects on Radar Doppler Altimeter for Sea Monitoring

Ferdinando Reale *, Eugenio Pugliese Carratelli, Angela Di Leo and Fabio Dentale

Department of Civil Engineering, University of Salerno, 84084 Fisciano, Italy; epc@unisa.it (E.P.C.); andileo@unisa.it (A.D.L.); fdentale@unisa.it (F.D.)

* Correspondence: freale@unisa.it

Received: 22 May 2020; Accepted: 14 June 2020; Published: 19 June 2020

Abstract: The orbital velocity of sea wave particles affects the value of sea surface parameters as measured by radar Doppler altimeters (also known as delay Doppler altimeter (DDA)). In DDA systems, the along-track resolution is attained by algorithms that take into account the Doppler shift induced by the component along the Earth/antenna direction of the satellite velocity, V_S. Since the vertical component of the wave particle orbital velocity also induces an additional Doppler effect (in the following R-effect), an error arises on the positioning of the target on the sea surface. A numerical investigation shows that when the wavelength of sea waves is of the same order of magnitude of the altimeter resolution, the shape of the waveform might be significantly influenced by the R-effect. The phenomenon can be particularly important for the monitoring of long swells, such as those that often take place in the oceans.

Keywords: radar Doppler altimeter; significant wave height; orbital velocities; waveforms; swell

1. Introduction

Radar altimetry is a widely acknowledged and largely diffused tool for open sea monitoring. Its basic principles are well known, and need not be described here, since they have been the object of extensive literature [1–3].

Conventional radar altimeters (low resolution mode (LRM)), with their along-track resolution on the order of a few kilometres, have indeed proven to be essential in monitoring sea parameters such as significant wave height (SWH) and sea surface height (SSH) from satellites such as ERS-1, ERS-2, Envisat, Jason-1, etc. They have thus become an essential tool in ocean forecasting, climate analysis, and sea level monitoring [4].

Recent developments have led to a widespread adoption of radar Doppler altimeters (also known as delay Doppler altimeter (DDA)), which provide a much higher along-track resolution (on the order of a few hundred metres); while this potentially improves the possibility of monitoring the ocean parameters, it also enhances some problems related to sea surface data analysis.

Over the years, a number of difficulties have indeed arisen in assessing how various physical effects may impact the reliability of the measurements. For instance, a well-known problem—common to both conventional and Doppler altimetry—is the so-called electromagnetic bias [5], through which the SWH influences the SSH; also, the presence of foam on the sea surface may influence both SWH and SSH measurements [6,7].

The possible influence of sea state on the oscillation of measured SSH was investigated by different authors [8–10], who showed that 10 Hz root mean square SSH is statistically linked to the value of SWH itself.

An aspect that recently was shown to be of considerable importance for DDA is the influence of very long waves upon the monitoring of SWH and SSH; the along-track resolution of DDA is indeed on the same order of, or even lower than, wavelength of the ocean swell.

Moreau et al. [11–13] compared eight months of CryoSat-2 data over long ocean swell and found that, on the one hand, the waveforms derived from DDA are dissimilar from those obtained through conventional altimetry, and on the other hand, the shape of such waveforms is considerably different from the usual shape. By investigating such shapes, they found that sometimes—but not always—the presence of swell is highlighted by a double peak in DDA measurements (Figure 1). It is worth noting that most SWH retrieving algorithms assume a Gaussian distribution of the instantaneous water height—which is not true in a swell dominated sea.

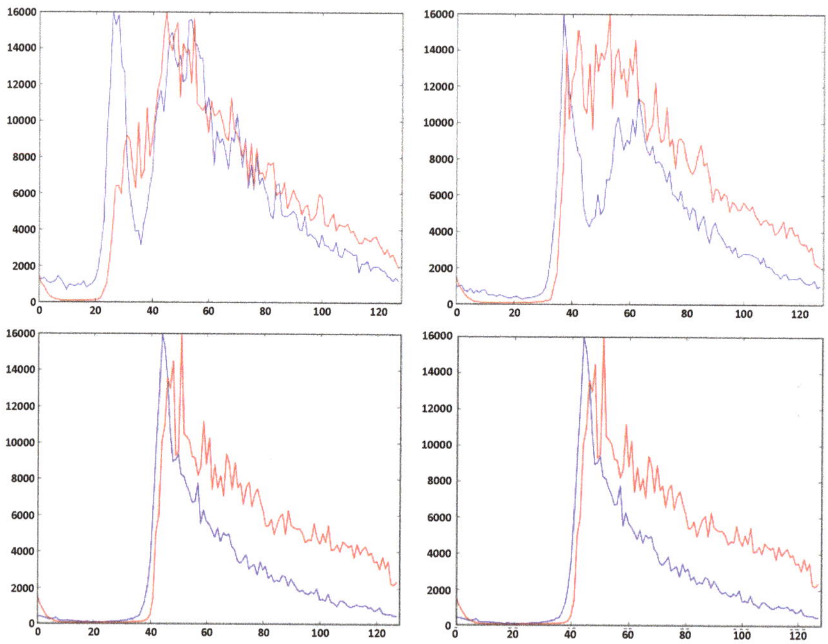

Figure 1. Experimental waveforms both in delay Doppler altimeter (DDA) mode (blue line) and conventional altimeter mode (red line) for various swell events detected by CryoSat-2 altimeter (track 681, cycle 030) as reported in [11].

These results raise concerns about the potential impact of such ocean wave effects on the monitoring of sea state conditions.

The present paper deals with a related problem, i.e., the interference of wave orbital velocity on the DDA algorithm. Such an effect, first reported in 2016 by Reale et al. [14–16], was recently (2019) also tackled in [17–20]. In the following, after a brief recall of the principles of DDA, a numerical procedure is implemented to show the influence of such an effect on the DDA response, and, where possible, to confirm its presence on the basis of published results.

2. Materials and Methods

DDA is similar in concept to conventional radar altimetry, in that a power/time curve (in the following: "waveform") of a reflected radio wave pulse is produced, and by analyzing such a curve, an accurate monitoring can be carried out of both SSH and SWH. In order to clarify the subject of the present paper, it is necessary to briefly recall main principles of DDA and its basic definitions, according to current literature [21–24].

DDA provides a decisive improvement of the resolution along the direction of the flight track (also known as along-track or azimuth direction) by taking into account the Doppler shift caused by the relative velocity of the satellite—or aircraft—borne antenna. The waveform is formed, in the same way as in conventional altimetry, by the variation in time of the illuminated sea surface (see Figure 2), but in DDA, the return signal is split into Δ_{dy} wide strips perpendicular to the along-track direction Y.

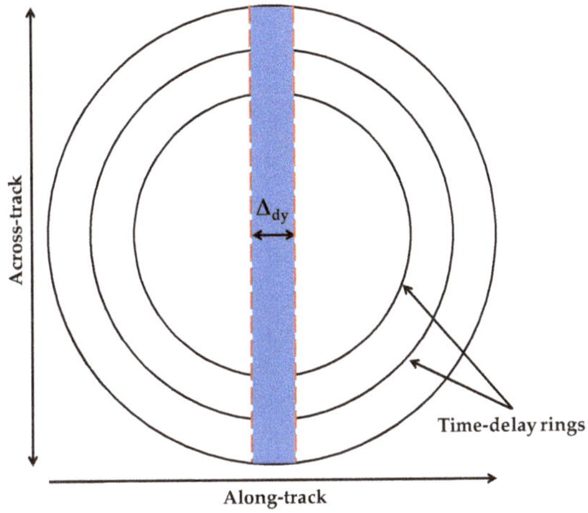

Figure 2. Doppler resolution strip along-track (from ESA CryoSat-2 Product Handbook-Baseline D 1.0 -C2-LI-ACS-ESL-5319).

The resolution along the Y direction is given by the Doppler shift resolution f_{res}, i.e., by the accuracy with which the altimeter can separate the return beam frequencies. Following [21,22], the Doppler frequency f_d and the distance along Y from the Nadir are linked by the following two equations:

$$f_d = \frac{2V_S}{\lambda}\frac{Y}{h}, \tag{1}$$

$$Y = \frac{\lambda h}{2V_S} f_d, \tag{2}$$

where Vs is the Earth/satellite relative velocity, h is the distance from the ground to the antenna, and λ is the electromagnetic wavelength. It is easy to see that the effect is governed by the projection Vc of Vs along the connection from the sea surface to the antenna, which can be taken to be approximately equal to $V_S Y/h$. A point (or a surface element on the sea) gives a contribution to the reflected intensity (and is therefore considered in the measurement procedure) only if its Doppler shift frequency f_d is comprised between $f_{res}/2$ and $-f_{res}/2$, and therefore its velocity component Vc is comprised between $Vs\Delta_{dy}/(2h)$ and $-Vs\Delta_{dy}/(2h)$, Δ_{dy} being the nominal width of the resolution along Y (see Figure 3).

We have thus:

$$\Delta d_y = \frac{\lambda h f_{res}}{2V_S}. \tag{3}$$

In a typical DDA altimeter, the footprint corresponds to an elongated strip on the surface, with an across-track resolution of 5–6 km and an along-track resolution of around 300 m. In presence of waves, the water surface is not still, and each water particle is affected—among others—by a vertical orbital velocity component V_Z; such a component algebraically adds to the Vc velocity due to the antenna relative movement, thus introducing a further Doppler shift f_R, indicated in the following as "R-effect" [14,16].

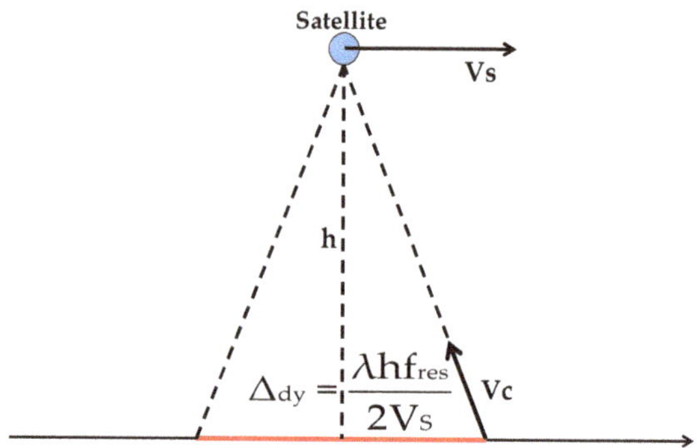

Figure 3. Along-track resolution as determined by antenna velocity relative to the Earth.

It is worth noting that a similar disturbance effect, sometimes known as "velocity bunching", on imaging SAR radars has been discovered as far back as in the 1980s [25], and has been considered and studied ever since [26]. It constitutes a limit to the applications of SAR, but it was recently found that it can been used to extract further information about the wind and the sea state (cut-off effect, [27,28]).

Since the radial direction from the water surface to the antenna is nearly vertical, f_R is given by following Equation (4) (Figure 4):

$$f_R = \frac{2V_Z}{\lambda}. \tag{4}$$

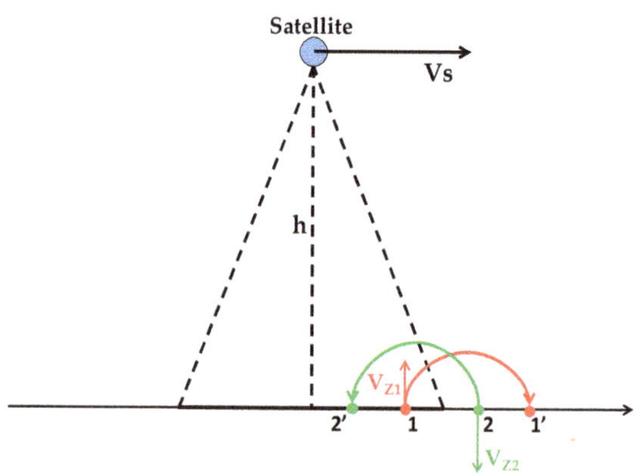

Figure 4. Effect of particle vertical velocity on the apparent position of an elementary surface area. The red point, positioned in 1, has a positive vertical velocity component V_{Z1}, which adds to its Vc; it therefore appears to be farther (position 1') from the nadir; the green point, positioned in 2, has a negative V_{Z2}; it therefore looks closer (position 2').

The nearly vertical apparent velocity of a water particle with respect to the antenna is therefore V_C+V_SY/h, and its total Doppler frequency shift f_t is given by Equation (5):

$$f_t = f_d + f_R = \frac{2V_SY}{\lambda h} + \frac{2V_Z}{\lambda} = \frac{2}{\lambda}\left(\frac{V_SY}{h} + V_Z\right). \tag{5}$$

If f_t for a given point (red point in Figure 4) is greater than f_{res} /2 or less than $-f_{res}/2$, the point will be not taken into account, and the value of its instantaneous elevation η will not be considered by the algorithms, even if its geometrical position (position 1 in Figure 4) is located within the nominal width Δ_{dy}; conversely, a point located outside the nominal position (green point in Figure 4) will appear, due to the R-effect, to be located inside (position 2′ in Figure 4), and will be considered in all the statistics.

It is easy to see that a potential bias does exist, and it has to be evaluated; it remains to be seen whether the phenomenon is significant enough, i.e., if the order of magnitude of the R-effect is the same as that of the Vs induced Doppler, and if the error averages out in practical application.

Considering, for instance, the parameters of the CryoSat-2 satellite operating in SAR mode (see Table 1), and assuming, e.g., a 2 meter-high sinusoidal 225 meter-long wave in deep water ($T \approx 12$ s), the vertical orbital velocity component V_Z varies between −0.52 m/s and +0.52 m/s, while for a shorter and higher wave ($H = 4$ m; $L = 125$ m; $T \approx 9$ s) the effect is more important (V_Z varies between −1.41 m/s and +1.41 m/s).

Table 1. Main parameters of CryoSat-2 altimeter in DDA mode.

Parameter	Value
Satellite Height h	700,000 m
Satellite Velocity V_S	7000 m/s
Electromagnetic Wavelength λ	0.0221 m
Frequency Resolution f_{res}	312.5 Hz
Nominal Space Resolution Δ_{dy}	360.1 m

Figure 5 shows, graphically, some comparisons between the orbital velocity induced Doppler frequency shift f_R and the satellite velocity Doppler effect f_d.

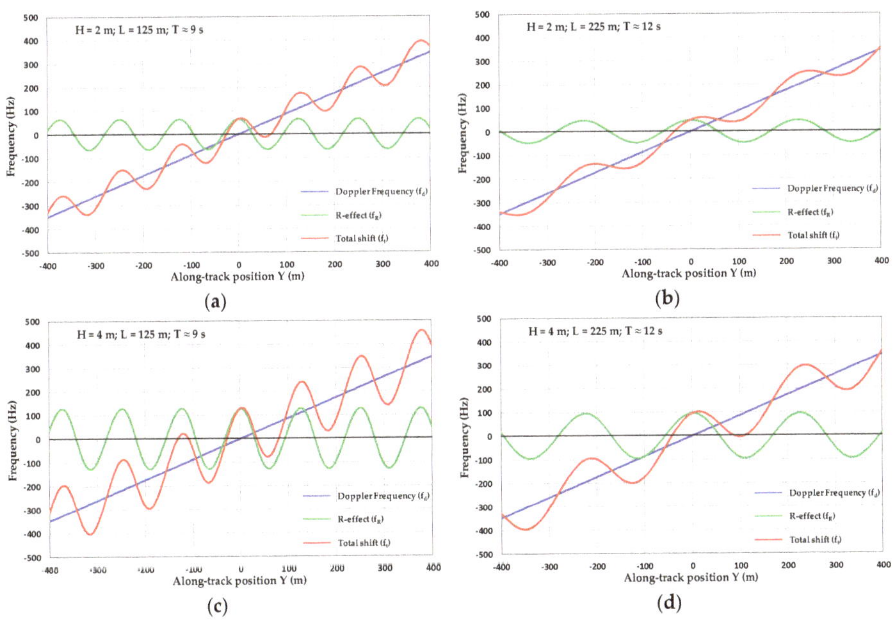

Figure 5. Along-track Doppler frequency shift values for different sinusoidal waves: (**a**) wave height $H = 2$ m, wavelength $L = 125$ m, wave period $T \approx 9$ s; (**b**) $H = 2$ m, $L = 225$ m, $T \approx 12$ s; (**c**) $H = 4$ m, $L = 125$ m, $T \approx 9$ s; (**d**) $H = 2$, $L = 225$ m, $T \approx 12$ s.

A more accurate assessment is thus necessary, and some numerical experiments described in the next paragraph help clarify the issue.

3. Simulation and Results

The consequences of R-effect on DDA measurement can only be assessed numerically; for this work, a conceptual model was used for the reconstruction of sea surface height from DDA. The examples provided here are relative to the CryoSat-2 satellite altimeter, with the parameters as reported in Table 1 above.

A computational procedure was thus implemented on an XY grid, whose side Lx is equal to the across-track resolution, and Ly is sufficiently long to include all the points whose apparent positions are located within the along-track resolution Δ_{dy}. The computational steps are called DX and DY, respectively.

The sea water instantaneous height η(X,Y), as well as its vertical velocity component V_Z(X,Y), was computed on all the points in the grid. With reference to Figure 4, the Doppler frequency shift f_t of a given point is given by Equation (5), so that its apparent position Y_{ai} is:

$$Y_{ai} = Y_i + (f_R + f_d)\frac{\lambda h}{2V_S}. \tag{6}$$

The strip will thus belong, or not, to the Doppler resolution strip Δ_{dy}, according to whether $\Delta_{dy}/2 < Y_{ai} < \Delta_{dy}/2$ or not.

The influence of wave induced Doppler shift can be first observed by considering the formation of waveforms; as stated above, R-effect changes the apparent position of reflected spots of the sea surface, so that the effective borders of the resolution zone are no longer simply given by Equation (2); this in turn will affect the waveform shape. In order to investigate this aspect, the points on the sea geometrically positioned in and around the nominal beam Δ_{dy} were relocated by taking the R-effect into account. By numerical computation, the distribution of distances between the antenna and the instantaneous simulated waveform (in the following: WF) can thus be constructed. This algorithm is simplified, since it neglects purely electromagnetic (e/m) effects, such as the antenna miss-pointing and the measurements noise, and it does not take into accounts the effects of the curvature of the Earth; it is therefore only meant to verify the effect that is being discussed here, which is purely geometrical and hydrodynamical.

The procedure was repeated twice, once by activating the R-effect, and once by ignoring it so that the simulated waveforms could be compared.

3.1. Single Waveforms

In the following, a few single WF results are reported for monochromatic waves, in order to highlight the basic aspects of the problem. The sea state is assumed to be made up of sine wave-trains in deep water, directed along the flight direction Y. The geometric (size and spatial resolution of the test area) computational parameters are reported in Table 2.

Table 2. Main parameters of the single measurement numerical simulations

Parameters of the Simulations	Value
Size of test area in across-track (X) direction Lx	6000 m
Size of test area in along-track (Y) direction Ly	1000 m
Computational grid along the across-track (X) direction DX	1 m
Computational grid along the track (Y) direction DY	1 m
Resolution of distance between antenna and sea surface	0.47 m

Figure 6 provides a comparison between the waveforms for a single swell wave-train aligned with satellite track (Y direction), computed with, and without, R-effect.

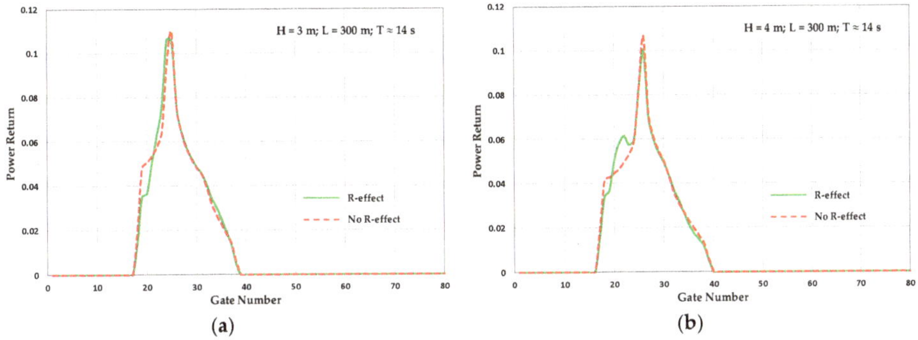

Figure 6. Simulated waveform with (solid green line), and without (dashed red line), R-effect for two different single sine wave trains: (**a**) wave height H = 3 m, wavelength L = 300 m, wave period T ≈ 14 s; (**b**) H = 4 m, L = 300 m, T ≈ 14 s.

The difference between the two reconstructions is very small, to the point of being negligible: it is most likely that the positive effect (points geometrically outside the nominal Doppler strip moving into the resolution area) is balanced by the negative effect (points exiting the nominal strip).

Things change, sometimes significantly, when the interaction between two wave trains is considered. A common situation is the superimposition of two wave trains, as represented in the following, by adding a wave train of height H_1 and a length L_1 of a few hundred metres—such as can be expected from a long ocean swell—to a shorter one with height H_2 and wavelength L_2. Figure 7 shows some results.

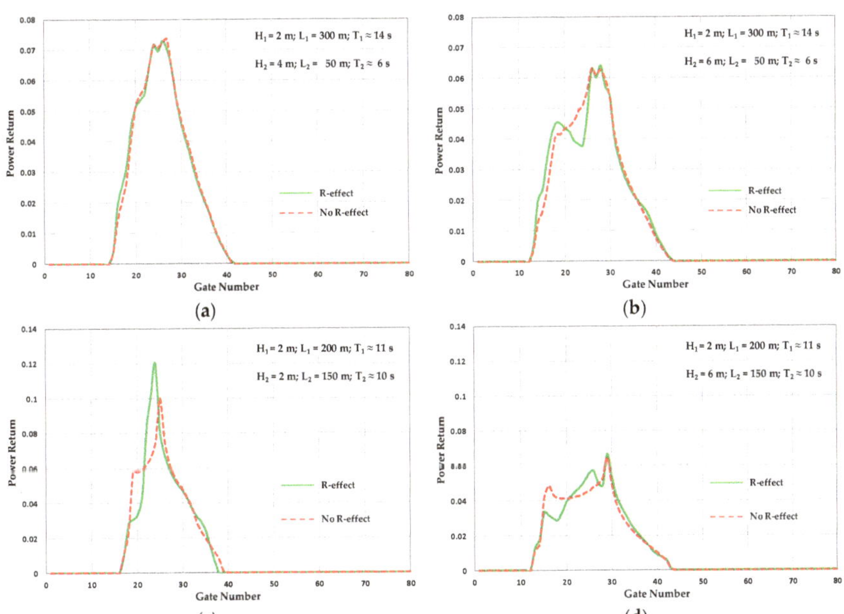

Figure 7. Simulated waveform with (solid green line), and without (dashed red line), R-effect when a longer sinusoidal wave train (wave height H, wavelength L_1) is superimposed to a shorter one (wave height H_2 and wavelength L_2): (**a**) H_2 = 4 m, L_2 = 50 m; (**b**) H_2 = 6 m, L_2 = 50 m. (**c**) H_2 = 2 m, L_2 = 150 m; (**d**) H_2 = 6 m, L_2 = 150 m.

The influence of the R-effect is clearly negligible in the case represented in Figure 7a, while it is relevant in the other examples. It is also useful to notice that some of the resulting waveforms bear a similarity to the shapes reported in [11,13] and represented in Figure 1. Similar tests carried out by varying the wavelength and the height of the two components consistently provide a pseudo waveform well different from the classical shape of a Gaussian sea, thus opening new possibilities to use high frequency DDA data to monitor the presence of very long swells, as already suggested in [11,12]. While it appears that the R-effect can, in certain circumstances, affect the formation of DDA response, it is however necessary, in order to get a more realistic appraisal, to take into account what happens when successive measurements are taken from different positions of the satellites, as well as when more realistic sea state is considered.

3.2. 20 Hz Data—Wind Sea/Swell Interaction

A common procedure in altimetric sea monitoring is to consider successive measurements at regular intervals, with 20 Hz being the usual standard. Since the satellite velocity is about 7 km/s, a 20 Hz sampling yields a 350 m interval, which is roughly equivalent to the nominal Doppler resolution. A computational algorithm with the following parameters (Table 3) was therefore set up to simulate this procedure.

Table 3. Main parameters of the multi-look 20 Hz numerical simulations.

Parameters of the Simulations	Value
Size of test area in across-track (X) direction Lx	6000 m
Size of test area in along-track (Y) direction Ly	6000 m
Computational grid along the across-track (X) direction DX	5 m
Computational grid along the track (Y) direction DY	5 m
Resolution of distance between antenna and sea surface	0.35 m
Number of measurements in a second	20 Hz

In order to take into account the interaction between a long swell and a shorter wind sea, a composite sea state was fed into the simulation, based on a JONSWAP spectrum (significant wave height Hs, spreading parameter Sp = 80, and peak wavelength Lp) and a sinusoidal long wave (H, T). Typical results are shown in Figure 8.

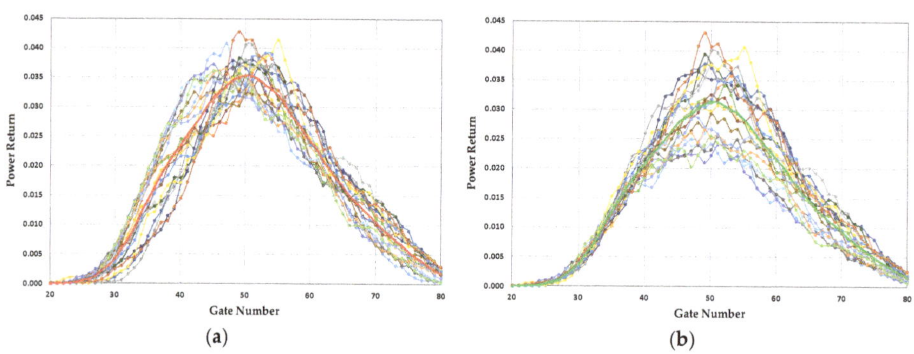

Figure 8. Simulated waveforms on a 20 Hz sample for sinusoidal wave train (wave height H = 4 m, T = 16 s) superimposed to a JONSWAP sea (significant wave height Hs = 4 m; peak period Tp = 8 s): (**a**) without R-effect (red curve represents the average value); (**b**) with R-effect (green curve represents the average value).

The 20 single WFs are reported for both R and non-R-effect simulations. While of course each single waveform is different from the others, it appears that the dispersion of R-effect waveforms is much higher than the dispersion of non-R shapes; also, the comparison of single R and non-R WFs (Figure 9) yields considerable differences.

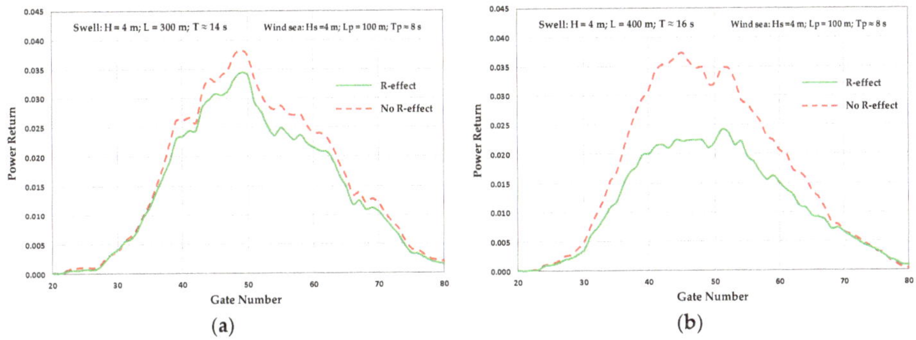

Figure 9. Simulated single waveforms with (solid green line), and without (dashed red line), R-effect for swell superimposed to a wind sea: (**a**) JONSWAP significant wave height Hs = 4 m, peak wave period Tp = 8 s; swell wave height H = 4 m, period T = 14 s; (**b**) JONSWAP significant wave height Hs = 4 m, peak wave period Tp = 8 s; swell wave height H = 4 m, period T = 16 s.

When average 20 Hz WFs are compared, a consistent difference is evident, the R-WF being less steep (Figure 10) in the leading edge.

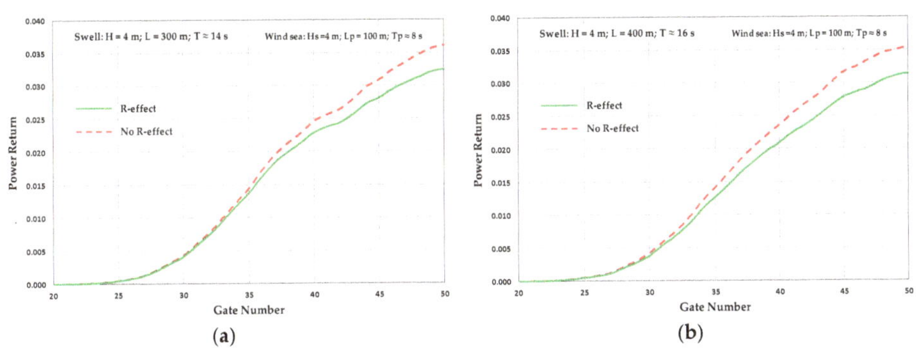

Figure 10. Comparison between averaged 20 Hz WF with (green line), and without (dashed red line), R-effect for swell superimposed to a wind sea: (**a**) JONSWAP significant wave height Hs = 4 m, peak wave period Tp = 8 s; swell wave height H = 4 m, period T = 14 s; (**b**) JONSWAP significant wave height Hs = 4 m, peak wave period Tp = 8 s; swell wave height H = 4 m, period T = 16 s.

The leading edge is actually of particular importance in the extraction algorithms for SWH, such as the ALES (Adaptive Leading Edge Subwaveform) retracker [29,30]. Even if this does not imply an error in the estimation of SWH, since all the algorithms are amply tested and calibrated with experimental data, it could still make sense to verify whether taking the effect into consideration could affect the results.

3.3. 20 Hz Data—Sea Surface Height

An important geophysical parameter obtained through satellite altimeter measurement is SSH; in particular, the 20 Hz standard deviation of SSH, also known—perhaps inappropriately—as "SSH error", is often used as an index to evaluate the quality of this parameter. In order to investigate the possible mechanisms which govern this phenomenon, the simulation has been carried out as above, with a realistic JONSWAP wind sea state, providing 20 results per second. For each of the 20 measurements, the average value $\overline{\eta}$ of the sea surface height over each resolution strip was calculated, both by taking into account, and by neglecting, the R-effect. A typical result is reported in Figure 11.

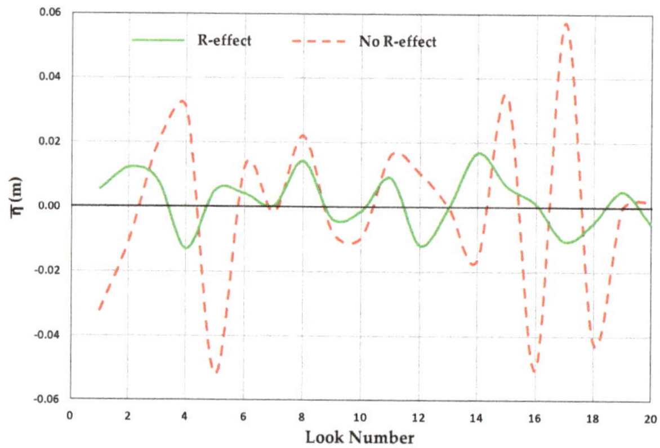

Figure 11. Sea surface height average values $\overline{\eta}$ by taking in account (solid green line), and by neglecting (dashed red line), the R-effect.

The SSH errors, i.e., the standard deviations $\sigma_{\overline{\eta}}$ of the $\overline{\eta}$ over a second (20 looks) can thus be calculated. Figure 12 yields the results for Hs = 4 m and various values of the peak spectral wavelength Lp, both for main wave direction perpendicular (across-track), and aligned (along-track), to the satellite flight direction.

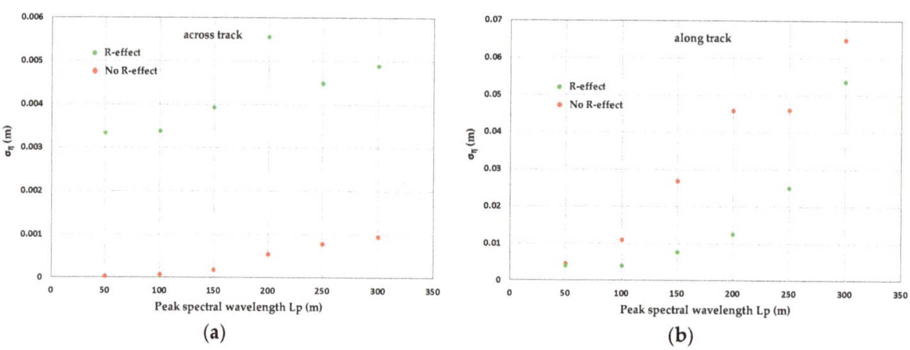

Figure 12. Standard deviation values $\sigma_{\overline{\eta}}$ as a function of peak spectral wavelength Lp by considering (green diamonds), and by neglecting (red circles), R-effect: (**a**) main wave direction across satellite track; (**b**) main wave direction along satellite track.

The results are interesting, because they provide an entirely different picture according to whether R effect is taken into account or not. Along-track R results provide—for a 200 m peak wavelength—a $\sigma_{\overline{\eta}}$ of about 5 mm, while non-R give a negligible value. The reverse is true for across-track results. This leads to the possibility that SSH oscillations, or at least parts of them, might be caused by an aliasing effect. Available experimental information by Fenoglio et al. [8,9], Huang et al. [31], and a simple model by Reale et al. [10] yield very scattered data with an average value of about 1 cm, and a dispersion of the same order of magnitude. Unfortunately, no information is provided in the current literature about the influence of the sea direction with respect to the satellite line of flight, and this is certainly an aspect which should be investigated.

4. Discussion and Conclusions

A number of numerical simulations were performed to evaluate the influence that the R-effect, i.e., the Doppler shift deriving from sea wave orbital movement, can have on the response of delay Doppler altimeters (DDA).

A schematic two-wavelength sea state was first considered; in this condition, the effect is occasionally present and visible in some single waveforms. Another numerical experiment was then carried out, by considering 20 Hz DDA simulated measurements on a more realistic sea state, based on a JONSWAP spectral wind sea superimposed to a long swell. The analysis of the results has shown a minor, but consistent, difference between R and non-R simulations of the 20 Hz averaged waveforms, as well as remarkable differences among some single waveforms; in particular, a complex shape waveform was occasionally found, bearing a similarity with experimental data by other researchers.

A further set of simulations, again based on a JONSWAP wind sea, have provided interesting results on the 20 Hz oscillation $\sigma_{\overline{\eta}}$ of the SSH. For a sea state orthogonal to the satellite track, the SSH simulated values when the R-effect is considered show a $\sigma_{\overline{\eta}}$ which could account, at least in part, for the results reported in the literature.

All the results seem to prove that the vertical orbital velocity component does actually influence DDA response; as per the practical importance, however, the conclusions must be more articulate. As per the evaluation of SWH, current algorithms, at least for wind sea states, should not be affected by any error caused by neglecting the R-effect, considering that they have generally been well tested against sea truth. Further detailed investigations should instead be carried on DDA-based monitoring of SWH in the situation where the sea state includes a long swell, and therefore the Gaussian distribution of the water heights does not hold.

Another aspect where the consideration of the R-effect could be meaningful is the analysis of SSH and of its oscillations $\sigma_{\overline{\eta}}$: on the one hand, the accuracy requirements of SSH monitoring are higher than those of SWH, on the other hand the behavior of $\sigma_{\overline{\eta}}$ is a current, active research field. The effect could be exploited to extract further information on the sea state.

A further line of investigation should lead to an assessment of the possible limits of the resolution attainable in the future with satellite DDA.

The consequences of this work might prove to be of relevance, especially taking into account the new generation of sea monitoring satellites, such as Sentinel-3, which are greatly increasing the availability of DDA data over the oceans.

Author Contributions: Conceptualization: F.R. and E.P.C.; methodology: F.R., E.P.C., and F.D.; software: F.R., E.P.C., and A.D.L.; validation: A.D.L. and E.P.C.; formal analysis: E.P.C. and F.R.; writing—original draft preparation: F.R. and E.P.C.; writing—review and editing: A.D.L. and F.D.; supervision: F.D. and F.R. All authors have read and agreed to the published version of the manuscript.

Funding: This research was carried out with University of Salerno Institutional Funds, as well as with the support of CUGRI (University Consortium for Research on Major Hazards).

Acknowledgments: The authors are grateful to T. Moreau and P. Rieu for help and advice, as well as for authorizing the reproduction of the results reported in Figure 1. Thanks are also due to J. Benveniste and P. Cipollini for useful discussion and encouragement over the years, and to the anonymous reviewers for suggesting significant improvements of the approach. Work was partially carried out within ESA EO-Project 1172 "Remote Sensing of Wave Transformation".

Conflicts of Interest: The authors declare no conflict of interest.

References

1. Rodriguez, E. Altimetry for non-Gaussian oceans: Height biases and estimation of parameters. *J. Geophys. Res.* **1988**, *93*, 14107–14120. [CrossRef]
2. Chelton, D.B.; Ries, J.C.; Haines, B.J.; Fu, L.-L.; Callahan, P.S. Satellite Altimetry. In *Satellite Altimetry and Earth Sciences: A Handbook of Techniques and Applications*, 1st ed.; Fu, L.-L., Cazenave, A., Eds.; Academic Press: San Diego, CA, USA, 2001; pp. 1–131.

3. Martin-Puig, C.; Ruffini, G.; Marquez, J.; Cotton, P.D.; Srokosz, M.A.; Challenor, P.; Raney, K.; Benveniste, J. Theoretical Model of SAR Altimeter over Water Surfaces. In Proceedings of the IGARSS 2008—2008 IEEE International Geoscience and Remote Sensing Symposium, Boston, MA, USA, 7–11 July 2008; pp. 242–245. [CrossRef]
4. Abdalla, S.; Dinardo, S.; Benveniste, J.; Janssen, P.A.E.M. Assessment of CryoSat-2 SAR mode wind and wave data. *Adv. Space Res.* **2018**, *62*, 1421–1433. [CrossRef]
5. Gommenginger, C.P.; Srokosz, M.A. Sea state bias—20 years on. In Proceedings of the Symposium on 15 Years of Progress in Radar Altimetry, Venice, Italy, 13–18 March 2006; Danesy, D., Ed.; ESA-SP 614. ESA Publications Division: Noordwijk, The Netherlands, 2006.
6. Reul, N.; Chapron, B. A model of sea-foam thickness distribution for passive microwave remote sensing applications. *J. Geophys. Res.* **2003**, *108*. [CrossRef]
7. Reale, F.; Dentale, F.; Carratelli, E.P. Numerical Simulation of Whitecaps and Foam Effects on Satellite Altimeter Response. *Remote Sens.* **2014**, *6*, 3681–3692. [CrossRef]
8. Fenoglio-Marc, L.; Dinardo, S.; Scharro, R.; Lucas, B.; Roland, A.; Dutour Sikiric, M.; Benveniste, J.; Becker, M. Validation of CryoSat-2 in SAR Mode data in the German Bight—Open Ocean. In Proceedings of the OSTST 2014 Meeting "New Frontiers of Altimetry", Lake Constance, Germany, 28–31 October 2014.
9. Fenoglio-Marc, L.; Dinardo, S.; Scharoo, R.; Roand, A.; Dutour Sikiric, M.; Lucas, B.; Becker, M.; Benveniste, J.; Weiss, R. The German Bight: A validation of CryoSat-2 altimeter data in SAR mode. *Adv. Space Res.* **2015**, *55*, 2641–2656. [CrossRef]
10. Reale, F.; Dentale, F.; Pugliese Carratelli, E.; Fenoglio-Marc, L. Influence of Sea State on Sea Surface Height Oscillation from Doppler Altimeter Measurements in the North Sea. *Remote Sens.* **2018**, *10*, 1100. [CrossRef]
11. Moreau, T.; Amarouche, L.; Thibaut, P.; Boy, F.; Picot, N. Investigation of swell impact on SAR-mode measurements. In Proceedings of the ESA Living Planet Symposium 2013, Edinburgh, Scotland, 9–13 September 2013.
12. Moreau, T.; Tran, N.; Aublanc, J.; Tison, C.; Le Gac, S.; Boy, F. Impact of long ocean waves on wave height retrieval from SAR altimetry data. *Adv. Space Res.* **2018**, *62*, 1434–1444. [CrossRef]
13. Moreau, T.; Labroue, S.; Thibaut, P.; Amarouche, L.; Boy, F.; Picot, N. Sensitivity of SAR mode Altimeter to swells: Attempt to explain sub-mesoscale structures (0.1-1km) seen from SAR. In Proceedings of the CryoSat Third User Workshop, Dresden, Germany, 12–14 March 2013.
14. Reale, F.; Dentale, F.; Fenoglio-Marc, L.; Pugliese Carratelli, E. On the Effect of Wave Vertical Orbital Velocity on Doppler Radar Altimetry. In Proceedings of the 2016 EUMETSAT Meteorological Satellite Conference, Darmstadt, Germany, 26–30 September 2016.
15. Reale, F.; Dentale, F.; Fenoglio-Marc, L.; Pugliese Carratelli, E.; Buchhaupt, C. Wave vertical orbital velocity effects on Doppler Altimeter waveform and SSH measurement. In Proceedings of the 2016 Ocean Surface Topography Science Team (OSTST) Meeting, La Rochelle, France, 1–4 November 2016.
16. Reale, F.; Dentale, F.; Di Leo, A.; Pugliese Carratelli, E. Wave Orbital Velocity Effect on Doppler Radar Altimetry for off-Nadir Beams. In Proceedings of the EUMETSAT Meteorological Satellite Conference 2017, Rome, Italy, 2–6 October 2017.
17. Boisot, O.; Amarouche, L.; Lalaurie, J.-C.; Guérin, C.-A. Dynamical Properties of Sea Surface Microwave Backscatter at Low-Incidence: Correlation Time and Doppler Shift. *IEEE Trans. Geosci. Remote Sens.* **2016**, *54*, 7385–7395. [CrossRef]
18. Buchhaupt, C.; Fenoglio, L.; Kusche, J. An Investigation of the Impact of Vertical Water Particle Motions on Fully-Focused SAR Altimetry. In Proceedings of the 2019 Ocean Surface Topography Science Team Meeting, Chicago, IL, USA, 21–25 October 2019.
19. Egido, A.; Ray, R. On the Effect of Surface Motion in SAR Altimeter Observations of the Open Ocean. In Proceedings of the 2019 Ocean Surface Topography Science Team Meeting, Chicago, IL, USA, 21–25 October 2019.
20. Buchhaupt, C. Model Improvement for SAR Altimetry. Ph.D. Thesis, Technischen Universität Darmstadt, Darmstadt, Germany, 27 August 2019.
21. Halimi, A.; Mailhes, C.; Tourneret, J.-Y.; Boy, F.; Picot, N.; Thibaut, P. An analytical model for Doppler altimetry and its estimation algorithm. In Proceedings of the 2012 Ocean Surface Topography Science Team Meeting, Venice, Italy, 22–29 September 2012.

22. Halimi, A.; Mailhes, C.; Tourneret, J.-Y.; Thibaut, P.; Boy, F. A Semi-Analytical Model for Delay/Doppler Altimetry and Its Estimation Algorithm. *IEEE Trans. Geosci. Remote Sens.* **2014**, *52*, 4248–4258. [CrossRef]
23. Ray, C.; Martin-Puig, C.; Clarizia, M.P.; Ruffini, G.; Dinardo, S.; Gommenginger, C.; Benveniste, J. SAR Altimeter Backscattered Waveform Model. *IEEE Trans. Geosci. Remote Sens.* **2015**, *53*, 911–919. [CrossRef]
24. Boy, F.; Desjonquères, J.; Picot, N.; Moreau, T.; Raynal, M. CryoSat-2 SAR-Mode Over Oceans: Processing Methods, Global Assessment, and Benefits. *IEEE Trans. Geosci. Remote Sens.* **2017**, *55*, 148–158. [CrossRef]
25. Hasselmann, K.; Raney, R.K.; Plant, W.J.; Alpers, W.; Shuchman, R.A.; Lyzenga, D.R.; Rufenach, C.L.; Tucker, M.J. Theory of synthetic aperture radar ocean imaging: A MARSEN view. *J. Geophys. Res.* **1985**, *90*, 4659–4686. [CrossRef]
26. Pugliese Carratelli, E.; Dentale, F.; Reale, F. Reconstruction of SAR wave image effects through pseudo random simulation. In Proceedings of the Envisat Symposium, Montreux, Switzerland, 23–27 April 2007. European Space Agency, (Special Publication) ESA SP, (SP-636).
27. Benassai, G.; Migliaccio, M.; Montuori, A. Sea wave numerical simulations with COSMO-SkyMed SAR data. *J. Coastal Res.* **2013**, *65*, 660–665. [CrossRef]
28. Migliaccio, M.; Montuori, A.; Nunziata, F. X-band Azimuth cut-off for wind speed retrieval by means of COSMO-SkyMed SAR data. In Proceedings of the 2012 IEEE/OES Baltic International Symposium (BALTIC), Klaipeda, Lithuania, 8–10 May 2012.
29. Schlembach, F.; Passaro, M.; Quartly, G.D.; Kurekin, A.; Nencioli, F.; Dodet, G.; Piollé, J.-F.; Ardhuin, F.; Bidlot, J.; Schwatke, C.; et al. Round Robin Assessment of Radar Altimeter Low Resolution Mode and Delay-Doppler Retracking Algorithms for Significant Wave Height. *Remote Sens.* **2020**, *12*, 1254. [CrossRef]
30. Passaro, M.; Cipollini, P.; Vignudelli, S.; Quartly, G.D.; Snaith, H.M. ALES: A multi-mission adaptive subwaveform retracker for coastal and open ocean altimetry. *Remote Sens. Environ.* **2014**, *145*, 173–189. [CrossRef]
31. Huang, Z.; Wang, H.; Luo, Z.; Shum, C.K.; Tseng, K.-H.; Zhong, B. Improving Jason-2 Sea Surface Heights within 10 km Offshore by Retracking Decontaminated Waveforms. *Remote Sens.* **2017**, *9*, 1077. [CrossRef]

MDPI
St. Alban-Anlage 66
4052 Basel
Switzerland
Tel. +41 61 683 77 34
Fax +41 61 302 89 18
www.mdpi.com

Journal of Marine Science and Engineering Editorial Office
E-mail: jmse@mdpi.com
www.mdpi.com/journal/jmse

www.ingramcontent.com/pod-product-compliance
Lightning Source LLC
Chambersburg PA
CBHW040811250926
51850CB00007BA/1140
* 9 7 8 3 0 3 9 3 6 9 7 2 0 *